Energiepolitik und Klimaschutz
Energy Policy and Climate Protection

Herausgegeben von
L. Mez, Berlin, Deutschland
A. Brunnengräber, Berlin, Deutschland

Weltweite Verteilungskämpfe um knappe Energieressourcen und der Klimawandel mit seinen Auswirkungen führen zu globalen, nationalen, regionalen und auch lokalen Herausforderungen, die Gegenstand dieser Publikationsreihe sind. Die Beiträge der Reihe sollen Chancen und Hemmnisse einer präventiv orientierten Energie- und Klimapolitik vor dem Hintergrund komplexer energiepolitischer und wirtschaftlicher Interessenlagen und Machtverhältnisse ausloten. Themenschwerpunkte sind die Analyse der europäischen und internationalen Liberalisierung der Energiesektoren und -branchen, die internationale Politik zum Schutz des Klimas, Anpassungsmaßnahmen an den Klimawandel in den Entwicklungs-, Schwellen- und Industrieländern, die Produktion von biogenen Treibstoffen zur Substitution fossiler Energieträger oder die Probleme der Atomenergie und deren nuklearen Hinterlassenschaften.

Die Reihe bietet empirisch angeleiteten, quantitativen und international vergleichenden Arbeiten, Untersuchungen von grenzüberschreitenden Transformations- und Mehrebenenprozessen oder von nationalen „best practice"-Beispielen ebenso ein Forum wie theoriegeleiteten, qualitativen Untersuchungen, die sich mit den grundlegenden Fragen des gesellschaftlichen Wandels in der Energiepolitik und beim Klimaschutz beschäftigen.

Herausgegeben von
PD Dr. Lutz Mez
Freie Universität Berlin

PD Dr. Achim Brunnengräber
Freie Universität Berlin

Marco Wedel

The European Integration of RES-E Promotion

The Case of Germany and Poland

Geleitwort Mit einem Geleitwort von PD Dr. Lutz Mez

 Springer VS

Marco Wedel
Berlin, Germany

D 188 – Doctoral Dissertation at Otto-Suhr-Institut of Freie Universität Berlin

Energiepolitik und Klimaschutz Energy Policy and Climate Protection
ISBN 978-3-658-11929-4   ISBN 978-3-658-11930-0 (eBook)
DOI 10.1007/978-3-658-11930-0

Library of Congress Control Number: 2015954466

Springer VS
© Springer Fachmedien Wiesbaden 2016

Printed on acid-free paper

Springer VS is a brand of Springer Fachmedien Wiesbaden
Springer Fachmedien Wiesbaden is part of Springer Science+Business Media
(www.springer.com)

For Ria Siegers and Gertrud Wedel

Preface

Although energy policy belongs to the competencies of the EU since the Lisbon Treaty, the European Commission has tried for more than twenty years to realize an open energy market in Europe according to the developments in the United States. The internal market directives for electricity and gas aimed at the liberalisation and deregulation of the European energy industry, and to establish competition in the energy sector. Previously, the energy sector in the EU was traditionally organized as a stable monopoly or oligopoly. In addition, due to the national energy systems, considerable external costs through energy-related environmental impacts of pollutants such as SO_2 and NO_x, as well as climate change, increased due to the emission of greenhouse gases. External costs can also occur in the form of expenses for the security of energy supplies and the development of technical innovations. The EU's policies aimed at reducing market failure, internalising external costs and promoting the development of renewable energies. The directive 2009/28/EC to promote the use of energy from renewable sources can be described as the most prominent EU action to date.

With this directive, the directives 2001/77/EC and 2003/30/EC were changed and lifted. The directive sets a binding target of 20% for the share of renewable energy sources in energy consumption and a minimum proportion of biofuel in the transport sector by 10%. Because the Member countries must implement binding national targets according to the overall EU target of 20% by 2020 for the first time, this directive can be described as a litmus test for a common policy in the energy sector. Additionally, the directive establishes binding targets for each Member State. The EU Member countries had to draw up and notify national action plans, which were part of the directive.

This book analyses the implementation of directive 2009/28/EC in seven chapters. In the introduction, the author outlines his motivation and the relevance of the chosen topic. His research questions are: 1) How do Member States and the European Institutions interact when it comes to the promotion of renewable energy electricity? 2) How will Member States be encouraged to implement measures and policies for the promotion of renewable energy electricity? 3) What are the constraints and dependencies of the European Integration of RES-(E) promotion? On the basis of five working hypotheses and the triad model, Marco Wedel describes the impact of the EU RES-E policy on the policy process in Poland and the German policy of RES-E. The second chapter "Theory" begins with the national heuristic policy cycle model and presents different political approaches that have criticized this model. Then, supra- and international policy analysis approaches, in particular those which have been developed for the EU, are presented. In Chapter 3, the author depicts the development of a European energy policy since the publication of the Green Paper of 2006, in which the basic points of a sustainable, competitive and secure energy system have been formulated. A subchapter is dedicated to the creation of Directive 2009/28/EC, with which the EU wants to achieve a share of renewable of energy of 20% of final energy consumption in 2020. The target values for Germany and Poland are 18 or 15% - so each is less than the average value of all EU member countries. In addition, the legal framework of the EU for the RES-E policy, as well as the legal basis for the promotion of renewable energy sources - in particular articles 192 and 194 of the Lisbon Treaty - are represented. The author describes in detail the administrative barriers in Poland, as well as the integration of the electricity networks compared to the EU. Future development is represented by the year 2020 with the green-X model. Marco Wedel analyses the implementation of the RES directive with the triad model. Until January 2014 the majority of EU Member States had not yet implemented the directive in its entirety. In chapters 4 and 5, the author examines the implementation of the directive in Germany and in Poland. In this, the national action plans as well as the various laws and law amendments for the promotion of RES-E are analysed. In the 6th chapter, "Inter-National Interdependencies between Germany and Poland", the author compares the RES directive in terms of national RES-E support policies and the development of infrastructure for the exchange of electricity between the two countries. The comparison of target values for 2020 in the national action plans shows that a larger EE capacity by a factor of ten installed in Germany is to make by a factor of seven higher RES-E production than in Poland. In recent years, the already weak Polish power grid has called to resolve loop flows that occur when peaks occur in power production from wind turbines or PV systems in North Germany. Price effects on the European energy exchange EEX are connected with these peaks. The author notes that the Polish electricity sector is influenced by the German Energiewende.

In the conclusion Marco Wedel summarizes the results of the empirical chapters. While Germany has "solidly implemented" the RES directive, the author comes to the conclusion that for Poland the country was "very reticent" in the implementation of the supranational policy consequences of the directive. The author stresses another result: "it can be concluded that the Polish energy economy, representing conventional power plant facilities, mainly based on relatively coal-fired units, is heavily exposed to the impact on increasing German RES-E deployment in both economic and physical (e.g. technical) terms" (p. 241f). This asymmetrical interdependence is an important restriction for Poland's national policy preferences. The dominant position of the German electricity industry on the electricity exchange and in the EI transport infrastructure has economic and physical consequences for the Polish power market. A common EU RES-E policy cannot be determined by comparing the development of the two countries. In summary, the author comes to a clear result and reproducible conclusions.

Lutz Mez   August, 2015

Acknowledgments

This book was written as a doctoral thesis, which I successfully defended at the Freie Universität Berlin on 28 April 2015. Since it wouldn't have been completed without the encouragement, support and assistance of a great many people, my list of acknowledgments is rather long.

First and foremost I would like to thank my supervisors, PD Dr. Lutz Mez and Dr. Behrooz Abdolvand, for their guidance, advice and encouragement over the last four years, throughout the preparation and the defence.

Furthermore, I would like to thank the additional members of the Disputation Committee for their time and consideration of the thesis: Prof. Dr. Eberhard Sandschneider, Prof. Dr. Carina Sprungk, and PD Dr. Achim Brunnengräber.

I would like to especially thank Gunnar Bennewitz whose comments and observations with respect to the subject matter have been truly inspiring and supportive.

It would have been unattainable to comprehend and analyse the specific nature of the incorporated research objectives addressed in the respective chapters without the expertise of PD Dr. Claudio Franzius (on European polity conflicts), Dr. Andrzej Ancygier (on Polish renewable energy policy), and Philip Berthold (on European electricity transmission systems and electricity markets). Thank you very much.

Not least, I am grateful to Prof. Dr. Josef Kallrath, for his valuable comments on my final script, and Prof. Dr. Maarten P. Vink, for his kind response to a research inquiry.

I should not forget to express my appreciation to Daphne Stelter, Henriette Krause, and Ines Stavrinakis. The completion of my doctoral examinations would have been impossible without their administrative support.

I hold my deepest gratitude for my parents, Burkard and Andrea, my sister, Dana, and my granddad, Andreas. Without their love and support this work wouldn't have seen the light of day.

Finally and equally important, I would like to thank Ljubina Krnjajic and all of my friends who have sustained and empowered me in the accomplishment of this intense challenge.

Marco Wedel   Berlin, July 2015

Table of Content

Abbreviations

ACER	Agency for the Cooperation of Energy Regulators
AGEB	Arbeitsgemeinschaft Energiebilanzen
BEMIP	Baltic Energy Market Interconnection Plan
BGBL	Bundesgesetzblatt
BMU	Bundesministerium für Umwelt, Naturschutz, Bau und Reaktorsicherheit
BMWi	Bundesministerium für Wirtschaft und Energie
BPB	Bundeszentrale für politische Bildung
BVerfGE	Bundesverfassungsgericht
CIA	Central Intelligence Agency
CPI	Current Policy Initiative
CT	Constitutional Court
Dena	Deutsche Energie Agentur
DIP	Dokumentations- und Informationssystem des Deutschen Bundestages
EC	European Commission
ECJ	European Court of Justice
EEG	Energy Economics Group
EEX	European Energy Exchange
ENTSO-E	Network of Transmission System Operators for Electricity
EP	European Parliament
EPEX	European Power Exchange
EWEA	European Wind Energy Association
FCC	Federal Constitutional Court
IfIL	Institute for International Law
IPNNW	International Physicians for the Prevention of Nuclear War
Mtoe	Million Tons of Oil Equivalent
NREAP	National Renewable Energy Action Plan
NSI	North-South Electricity Interconnection
NSOG	Northern Seas Offshore Grid
OECD	Organisation for Economic Co-operation and Development
OJoEC	Official Journal of the European Communities
OJoEU	Official Journal of the European Union
PKEE	Polish Electricity Association
PPE	Polish Power Exchange
PPI	Planned Policy Initiative
PWEA	Polish Wind Energy Association
RES	Renewable Energy Source
RES-(E)	Renewable Energy Source Electricity
TEC	Treaty establishing the European Community
TEU	Treaty on European Union
TFEU	Treaty on the Functioning of the European Union
TGPE	Towarzystwo Gospodarcze Polskie Elektrownie
UN	United Nations

Table of Figures

Table of Tables

1. Introduction

In a 2007 Communication to the European Council and the European Parliament, the Commission of what were then the European Communities called for *'An Energy Policy for Europe'* (EC, 2007a: 3). To achieve this goal, the document sketched an Action Plan meant to trigger a *'new industrial revolution'* to *'set pace for a new global industrial revolution'* (EC, 2007a: 5 and 20). With respect to the three main objectives of sustainable, competitive and secure energy identified in a 2006 Green Paper (EC, 2006: 17-18), the Action Plan proposed, inter alia, the creation of an internal European electricity and gas market, and the utilization of local, low emission energy, such as renewable energy (EC, 2007a: 5 and 12). The Commission called for a long-term binding target for renewable energies in the overall energy mix of up to 20% by 2020, suggested reviewing the Emission Trading System (EU ETS), and proposed an ambitious programme of energy efficiency measures (EC, 2007a: 11-13).

In consequence, the EU agreed upon an energy and climate policy package up to 2020 (EC, 2007a: 3). Central to the policy framework are three headline targets to be reached by 2020: (1) The reduction of GHG emissions by 20% relative to emissions in 1990, (2) a 20% share of renewable energy sources in the energy consumed in the EU, and (3) a 20% saving in energy consumption. All of these targets are known as the 20-20-20 targets. In addition, the package calls for a 10% renewable energy target for the transport sector and the decarbonisation of transport fuels by 6% (EC, 2013a: 3). Alongside the Climate and Energy Package, the European Union adopted the so-called third legislative package for an internal EU gas and electricity market (EC, 2007a: 3; EC, 2011b: 3).

Keystone of the 2nd headline target of the energy and climate policy package is Directive 2009/28/EC of the European Parliament and of the Council, on the promotion of the use of energy from renewable sources. It establishes a European framework for renewable energy promotion by setting mandatory national renewable energy targets (EC, 2007a: 3; EC, 2013b: 2). Tracing back Directive 2009/28/EC as one manifest result of the original intention of creating *'An Energy Policy for Europe'* makes an interesting subject matter for analysis (EC, 2007a: 3). Indeed, since roughly four years have passed since the respective European institutions adopted this legislation, empirical insights are at hand to conduct a viable scientific analysis (EP, 2008). The overall scientific interest arising out of these circumstances can best be subsumed in the first guiding question of this thesis: *Can one state the successful achievement of a common European energy policy based on the lessons of the implementation of Directive 2009/28/EC?*

Almost every scientific investigation of European matters proves difficult from the start. This is especially true for political and social science. Whilst this may cause irritation in face of public perception that European institutions exist unquestionably, it is still unclear in principle what the European Union is. There are many attempts to describe the European Union: multilevel governance, regulatory state, federal state to be, confederation, multilevel constitutionalism and so on (Pernice, 2001: 5). But reflections upon the Union defy the classic categories used in definitions of how the three dimensions reflected in the logic of politics – polity, politics, policy – interrelate in the European Union, as well as between the Union and its Member States (Franzius and Preuß, 2012: 18-22; Meyer, 2010: 81; Leuffen, Rittberger and Schimmelfennig, 2013: 2). Consequently, stable system parameters (such as those within a nation state) for any *de facto* policy implementation and theoretical reflection thereof cannot easily be presumed. The Union itself challenges the dichotomous categories of two basic types of contemporary polities: modern nation states, on the one hand, and international organizations, on the other (Leuffen, Rittberger and Schimmelfennig, 2013: 2 and 32). Since the process of European Integration is unprecedented and its development ongoing, any ontological descriptions based on 19th and 20th century theories fail to comprehend the process of integration, let alone the status quo of the European Union. Whilst challenging, it is this unique and relatively unknown, yet heavily debated, status that makes the EU so intriguing for scientific research.

The EU's peculiar status – which will be elaborated in greater detail in the on-going discussion – makes the question concerning a common European energy policy, exemplified through Directive 2009/28/EC, relevant for detailed analysis. It is not so much the emergence of European policies, which can be treated in principle by established approaches to policy analysis – for they are fairly similar to nation-bound policy cycles – rather the *effects* of these policies within the European arena that call for analysis and investigation. This is because a generally binding path-dependency of adopted legislation in the supranational arena cannot be expected a priori within the EU, owing to the dichotomy of supranational and national policy arenas. The factual adoption of a European policy agenda, and legislation in particular, is not backed by the same executive authority as is national legislation, derived within stable system parameters enshrined in national polities. Thus, in a broader European context, the question is not only whether a policy does or does not work, but also whether it can be expected to be wholly incorporated – that is, transposed by the Member States – or not. Both aspects are of course heavily intertwined. Taking the perspective of the European legislator, the goal of this thesis is to research constraints and dependencies of the European Integration of the promotion of renewable energy sources (RES), and in particular, renewable energy source electricity (RES-E). To gain insightful results, it is necessary not only to take a look at the implementation effectiveness of Directive 2009/28/EC in the Member States, but much more generally to analyse the potentially conflicting polity implications of the entities involved, as well as their respective policy preferences and processes in light of (renewable) energy matters. Combined, these give rise to multi-level constraints and interdependencies for the European Integration of RES-(E) promotion.

Whilst the implementation of Directive 2009/28/EC through Member State policies has led to a significant growth in renewable energies in the period up to 2012, a renewable energy progress report reveals a less optimistic outlook for 2020. Owing to the nature of national renewable energy policies implemented up to 2013, an on-going economic and financial crisis in the European Union, prevailing administrative and infrastructure barriers, as well as policy and support schemes disruptions, Member States need to take further measures to ensure the achievement of their promotion targets (EC, 2013b: 3; EC, 2013c: 3). Furthermore, the European Commission (2011: 2) acknowledges that there is *'inadequate direction as to what should follow the 2020 agenda'*. Therefore, whilst analysing the progress of the promotion of renewable energies in the context of the 2020 policy framework, the European Commission also released a 2013 Green Paper reflecting upon a new framework for climate and energy policies up to 2030 (EC, 2013a: 3; EC 2013c: 3). In addition, a communication on a general energy roadmap up to 2050 – to *'explore routes towards decarbonisation'* and to indicate the level of change needed, both structural and social, whilst keeping a competitive and secure energy sector – has been commonly released by the Commission, the Parliament, the Council, the Economic and Social Committee and the Committee of the Regions (EC, 2011a). Whilst the latter is a rather vague collection of ideas, the Green Paper is, if not a proposal for a European climate and energy policy beyond 2020, a insightful survey of the challenges of a common European climate and energy policy framework (EC, 2013a). The paper states 22 overall questions to evaluate the existing 20-20-20-policy framework, in order to draw lessons for the future design of the Union's energy and climate change policy (EC, 2013a: 13). The following questions – as can be found in the fourth section of the Green Paper – are deemed particularly interesting (2013a: 13):

- Which lessons from the 2020 framework and the present state of the EU energy system are most important when designing policies for 2030?

- Which targets for 2030 would be most effective in driving the objectives of climate and energy policy? At what level should they apply (EU, Member States, or sectoral), and to what extent should they be legally binding?

- How can targets reflect better the economic viability and the changing degree of maturity of technologies in the 2030 framework?

- Are changes necessary to other policy instruments, and how do they interact with one another, including between the EU and national levels?

- How should specific measures at the EU and national levels best be defined to optimise the cost-efficiency of meeting climate and energy objectives?

- How can fragmentation of the internal energy market best be avoided, particularly in relation to the need to encourage and mobilise investment?

- How should the new framework ensure an equitable distribution of effort among Member States? What concrete steps can be taken to reflect their different abilities to implement climate and energy measures?

Based on the collection of questions stated above, one can identify two major concerns: (i) *How do Member States and the supranational EU levels interact on the matter of climate and energy policy?* (At what level should objectives apply? To what extent should they be legally binding? How do necessary policy instruments interact between the EU and national levels? How do we ensure an equitable distribution of effort among Member States?) (ii) *How will the Member States be encouraged to implement climate and energy measures?* (How do targets better reflect economic viability? How should measures be defined to optimise cost-efficiency? How can investment be encouraged and mobilised? How can fragmentation of an EU internal energy market be avoided? How do we reflect the different Member States' abilities to implement climate and energy measures?)

In light of this paper and the scientific interest described above, the overall questions stated here can be restated as the guiding questions of this thesis accordingly:

(i) *How do Member States and the European Institutions interact on the matter of the promotion of renewable energy electricity?*

and

(ii) *How will the Member States be encouraged to implement measures and policies for the promotion of renewable energy electricity?*

This leads to

(iii) *What are the constraints and dependencies of the European Integration of RES-(E) promotion?*

Whilst the latter question is considered a first necessary step to narrow down the overall subject, the following working hypothesis shall serve as a starting point for the scientific analysis. Please be advised that the description of proposed explanations will be subject to the on-going discussion:

The implementation of polices for the promotion of renewable energy electricity (RES-E) in the Member States of the European Union is (i) fundamentally instigated by supranational renewable energy measures and policies leading to a general but not sufficient vertical Europeanization; (ii) predominantly influenced by national preferences determined by geopolitical and energy-specific economic interests; and (iii) primarily exposed to incremental European inter-state competition leading to horizontal adaptational pressure.

This hypothesis translates as follows in terms of the empirical analysis of this study:

If (i) the European Union maintains an overall adaptational pressure by deciding on RES-E measures and policies for its Member States; (ii) pioneering Member States with sufficient power in the international system with regard to RES-E promotion (such as Germany) succeed in implementing an energy policy with a high share in electricity from renewable energy sources; and (iii) neighbouring Member States (such as Poland) are exposed to a pioneering Member State's RES-E policy and/or RES-E market design – **then**, non-pioneering Member States' RES-E measures and policies will adapt to the vertical (EU level) and horizontal (inter-state level) pressures. (These multi-level policy dependencies will encourage the harmonization of a European promotion of renewable energies and describe a reasoned path to a common promotion of renewable energies in all 28 Member States of the European Union.)

The hypothesis is based on empirical observations that lead to three subordinate hypotheses. These will be regarded as explanatory factors in the overall theory design and shall briefly be outlined here.
As mentioned before, the implementation of the renewable energy Directive 2009/28/EC led to a significant growth in renewable energies in the period up to 2012. Whilst 20 Member States achieved or exceeded the planned 2010 renewable energy shares – bindingly established in the Directive (Part B of Annex I) – and 12 Member States exceeded their planned targets for renewable energy electricity in 2010, 15 countries missed their targets (EC, 2013c: 3-4). Altogether, and independent of the Member States' indicative targets since 2004, the share of renewable energy in final energy consumption grew in all Member States. With the exception of Luxembourg, Romania and Slovenia, the same is true for the period 2010-2011 (Eurostat, 2013a: 1-2). The European Commission concludes that *'spurred on by the adoption of the 2009 Renewable Energy Directive and the legally binding renewable energy targets, renewable energy grew strongly'* (EC, 2013b: 12).
In light of the overall hypothesis, this leads to a first explanatory hypothesis (subordinate hypothesis (i)):

(i) The supranational RES-(E) measures and policies of the European Union, when expressed in legislation, excite change in Member States' RES-(E) policies only when there is a degree of misfit between EU and domestic RES-(E) policies that causes adaptational pressure, leading to the national promotion of RES-(E) in accordance with EU level implications.

Nevertheless, the European Commission concludes in its 'Renewable energy progress report' (EC, 2013b) that there is reason for concern with regard to the future progress of the promotion of renewables energies.

The transposition of the Directive 2009/28/EC has been *'slower than desirable'* and current Member State policies are insufficient to trigger the required renewable energy deployment (EC, 2013b: 13). Overall: more action is needed to ensure the Europeanization of energy (EC, 2013b: 9). It will therefore examine Member States' implementation of Directive 2009/28/EC and will launch infringement cases if implementation is incomplete (EC, 2013b: 7-10).

By February 2013, the Commission had opened 86 infringement procedures for non-communication of national transposition measures in the context of energy. Front-runner is Poland, with seven pending infringement procedures, followed by Cyprus with six pending procedures. Denmark and Sweden are the only Member States with no pending procedures for failing to transpose supranational energy related directives (EC, 2013d). With regard to the implementation of the renewable energy directive on 21 March 2013, the Commission referred Poland and Cyprus to the Court of Justice for failing to transpose the Directive 2009/28/EC (EC, 2013e). For the same reason, Belgium and Estonia were referred to the Court on 30 May 2013 (EC, 2013f). Whilst the mere fact of an infringement procedure is not conclusive in terms of the quality of the violation, it nevertheless shines a light on the matter of unresolved conflicts between the supranational and national levels of the European Union.

The case of a *'slower than desirable'* transposition of the renewable energy directive and infringement procedures for non-transposition in light of the legally binding nature of the directive and legally binding renewable energy targets calls for a second hypothetical explanation (ii), rivalling subordinate hypothesis (i):

> (ii) The RES-E policy of a Member State is an exclusive result of the domestic policy process that reflects energy-specific economic interests of the dominant domestic group and is determined by geopolitical and economic interest.

If there is a national policy outcome for which there exists a corresponding, yet different, European policy, two general deviations are possible, a positive and a negative. In light of the scope of the thesis, two empirical examples are chosen to evaluate the above-suggested suppositions. The first, Germany (a), is an example of an evolved national policy assumed to be beyond European transposition necessities. The second, Poland (b), is an example of an evolved national policy assumed to fall short of meeting European implications. The analysis will show whether these assumptions are accurate or – more importantly – *why* they were.

a) With an energy dependence of 68,4%, Germany's energy mix is dominated by fossil fuels, with coal (hard coal: 19,1%, lignite: 25%) as a major source for electricity generation (BMWi, 2014b, based on AGEB, 2013; BMWi, 2014c, based on AGEB, 2013 and BDEW, undated). In May 2011, in light of the Fukushima Daiichi nuclear disaster, the German government decided on an complete phase-out of nuclear energy technology by 2022. In consequence, on 30th June 2011, the German parliament adopted, based on a draft proposal by the Conservative and Liberal parliamentary group, the 13th law on the amendment of the Atomic Energy Act (DIP, 2011e: 13404; DIP, 2011f: 1-8; BGBL, 2011b: 1704-1705). When this law entered into force on 31st July 2011, the seven oldest German nuclear power plants lost their operating license and the three newest facilities must end electricity production by 31st December 2022 (BGBL, 2011b: 1704; BMU, 2011: 1). The entire phase out of nuclear energy, states Chancellor Angela Merkel in a 2011 government declaration, is to start the *'era of renewable energies'*, making them the foundation of the future German energy supply (Bundesregierung, 2011b: 4). According to this new energy strategy, whilst in 2011 roughly 80% of power production was generated on the basis of fossil fuels, by 2050 80% of electricity production is to come form renewable energy sources (compared to 1990 levels). In addition, greenhouse-gas emissions are to be cut by 80% in comparison to 1990 levels (BMWi, 2012: 4-5). The new German energy policy, or 'Energiewende' as it is called, requires a substantial transformation of the German energy system.

b) Poland is the largest coal producer in the European Union. It accounts for 80% of its energy production (Paiz.gov.pl, 2013: 4; Cwiek-Karpowicz, Gawlikowska-Fyk and Westphal, 2013: 3; EC, 2011d: 125). In 2012, Poland had to import 94,7% of its demand for oil and 73,8% of its gas demand (Eurostat, 2014). However, with the rather large occurrence of indigenous coal resources – out of which the country exports large quantities to other EU Member States – the Polish score with regard to energy dependence (30,7%) is well below the EU average (53,3%) (Eurostat, 2014; Cwiek-Karpowicz, Gawlikowska-Fyk and Westphal, 2013: 3; EC, 2007c: 2). Combined heat and power plants using traditional fuels produced 82% of the Polish electricity in 2012 (hard coal: 54,5% and lignite 25,2%). Wind power plants accounted for 6,3%, and hydro power plants for 5,7% (Paiz.gov.pl, 2013: 5). Overall, the Polish energy sector is shaped by a high demand of energy, and an inadequate transmission infrastructure as well as fuel and energy generation (Ministry of Economy, 2009a: 4) One of Poland's official energy objectives is to diversify its electricity generation by introducing nuclear energy (Ministry of Economy, 2009a: 15). In 2013, Polish Prime Minister Donald Tusk declared that the country *'would fulfil the 15% renewable energy target but "nothing more"'* (Ancygier and Szulecki, 2013a: 1).

On 21 March 2013, the European Commission notified the public in a press release that it refers Poland to the European Court of Justice (ECJ) for failing to transpose EU rules (EC, 2013e). The order sought from the ECJ is meant to *'declare that, by failing to adopt the laws, regulations and administrative provisions necessary to ensure compliance with Directive 2009/28/EC of the European Parliament and of the Council of 23 April 2009 on the promotion of the use of energy from renewable sources and amending and subsequently repealing Directive 2001/77/EC and 2003/30/EC, and in any event by not notifying the Commission of such provisions, the Republic of Poland has failed to fulfil its ob-*

ligation under Article 27 of that directive' (OJoEU, 2013b: 8). The Commission furthermore proposes to impose a penalty payment of EUR 133.228,80 from the day on which the judgment is delivered and to obtain an order for the Republic of Poland to pay the costs of the proceedings (OJoEU, 2013b: 8). The penalty suggested, the Commission argues, is based on the duration and the gravity of the infringement. A formal notice of non-transposition of Directive 2009/28/EC was sent to Poland as early as January 2011, followed by a Reasoned Opinion in March 2012 (EC, 2013e). In response, the country adopted a temporary bill (labelled "Small Tri-pack") to ensure full implementation of Directive 2009/28/ EC (Ancygier and Szulecki, 2013a: 2; Rybski and Stoczkiewicz, 2013: 1). A first legal analysis of the respective legislation by Rybski and Stoczkiewicz (2013: 15) points to an on-going noncompliance with the renewable energy directive.

The (iii) aspect of the working hypothesis introduced above refers to horizontal aspects of European inter-state interconnectedness. This aspect will be incorporated in the overall theory design by subordinate hypothesis (iii), which reads as follows:

> (iii) European inter-state connectedness, competition and economic pressure will foster the promotion of RES-E in Europe beyond pioneering Member States **if** a pioneering state with sufficient power in the international system establishes a dominant market position, reflected in its RES-E market design.

As before, the qualitative empirical examples are Poland and Germany. Overall, according to the indicative trajectories of the National Renewable Energy Action Plans of both Germany and Poland, by 2020 Germany will have installed ten times the capacity of Poland, whilst producing seven times as much electricity from renewable energies (Zepeda, 2014: 3; NREAP Germany, 2010; NREAP Poland, 2010). The German RES-E deployment, in particular wind and solar feed-in, heavily affects the Polish power system. In interconnected grids – as is the case between Germany and Poland (Krajnik (PL) – Vierraden (PL) and Mikulowa (PL) – Hagenwerder (DE) (50Hertz, 2014: 1) – electricity flows independent of national borders from the generator to the consumer on paths of lowest electrical resistance (PSE and 50Hertz, 2014: 6). In a zonal market approach, as used in Europe, the difference between a physical flow and a schedule (commercial transactions scheduled on a given border) is called an unplanned flow (PSE and 50Hertz, 2014: 6). In the case of Germany and Poland, unplanned flows can reach significant volumes, with schedules going from Poland to Germany, whilst physical flows go in the opposite direction (PSE and 50Hertz, 2014: 7).

Between January 2010 and December 2012, measured physical flows on the border between the two countries in question had the opposite direction to the commercial schedules. During this period there was a permanent and high level of unplanned flows from Germany to Poland in between 500 and 1.500 MW (CEPS+MAVIR+PSE+SEPS, 2013: 13).

Whilst the main concern of Transmission System Operators (TSO) in the context of unplanned flows is the maintenance system security, another equally important aspect regards the economic implications of growing RES feed-in for national and inter-national energy markets. On the basis of these briefly-sketched realities, together with aspects of the distinct nature of the power markets, – which are characterised by unique price patterns as a result of the seasonality of underlying demand, the non-storability of electricity, the need for a balanced grid at all times, the challenge of intermittent renewable energy sources, and the need for base-load capacities – the assumptions reflected in subordinate hypothesis (iii) can be subject to further analysis (Erni, 2012: 13).

To identify the constraints and dependencies of the European Integration of RES-(E) promotion, the explanatory factors implicitly describe three different influencing factors that follow a top-down, that is, vertical logic, in turn leading to a horizontal logic. The first subordinate hypothesis (i) claims supranational renewable energy measures and policies produce compliance at the level of Member States (Radaelli, 2003: 19). Subordinate hypothesis (ii) counterpoints the first by claiming the European Union lacks the executive authority to expect compliance at the Member States level. Instead, it is national preferences that dominate the Member States' policy outcomes. Finally, subordinate hypothesis (iii) integrates the third aspect of inter-state interdependencies into the theory framework. The underlying assumption is that successful promotion of renewable energy electricity in country A will produce adaptational economic and regulatory pressure in country B, and ultimately lead to the alignment of both systems.

The three different possible origins for renewable energy electricity promotion in the European Union, as reflected in the abovementioned explanatory factors (hypotheses), are: (i) a supranational extrinsic logic for the national promotion of renewable energy, (ii) a national intrinsic logic in line or at odds with the supranational implications, and (iii) a logic of inter-state interdependencies leading to a horizontal momentum triggered by the market, the diffusion of ideas, and a discourse about policies and practices (Radaelli, 2003: 17). The analysis will show to what degree the respective subordinate hypotheses can be falsified or verified. In case any of the rivalling subordinate hypotheses – (i) and (ii) – can be verified in total, the working hypothesis must be regarded falsified.

The brief outline of these three explanatory factors defines the very nature of this study. It is an empirical analysis, embedded in an overall theoretical framework, based on the supranational entities of the European Union, the Member State Germany, and the Member State Poland. They are chosen because, whilst the overall research question is concerned with the European Union and not the Member State, three empirical examples are needed, including the supranational European level and at least two neighbouring Member States, to represent the vertical and horizontal realities of European Integration. Since any progress in the promotion of renewable energy electricity can only be measured at the Member States level, the focal point needs to be the Member State. Of course, a sovereign nation is the source for the emergence and implementation of national policies. But the same is true for supranational legislation being executed by the adoption of national transposition laws. However, all three origins of constraints and dependencies for RES-E promotion in the Member States (supranational extrinsic, national intrinsic, inter-state interdependencies) will be considered, making it an analysis of European mechanisms, rather than a mere Member State analysis or comparative case study. Each of the three identified entities, representing the model used in this analysis, indicates a possible influencing factor for the promotion of renewable energy electricity in the European Union. The following figure exemplifies their interrelations.

Fig. 1 Triad-model of different impact origins on EU Member States for RES-E promotion and theoretical frameworks

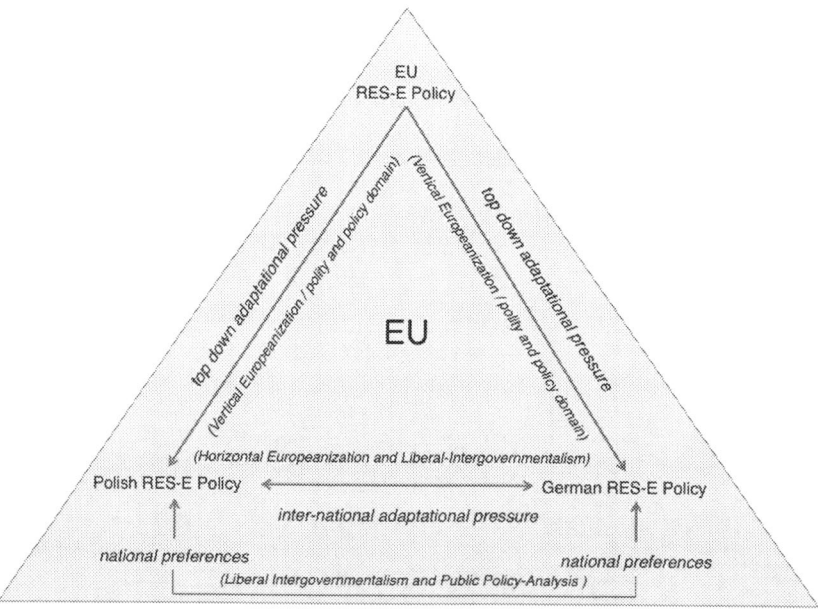

Regarding the overall theoretical framework, Fig. 1 illustrates the necessity for a multi-causal explanation. The terminology used so far hints at the theoretical frameworks that will be applied in this study. Since one cannot explain integration in the European Union by limiting it to a mono-causal explanation, multiple theories and factors are needed to create a coherent approach (Moravcsik and Schimmelfennig, 2009: 68). Moravcsik and Schimmelfennig (2009: 68) argue that a minimum of three theories, organized in a multistage model and with respect to context, are needed to explain integration. Regarding the triadic logic of impact origins for RES-S promotion, three theoretical overall frameworks will inform the empirical analysis: Europeanization, Liberal Intergovernmentalism, and Public Policy-Analysis.

In prinicple, Europeanization in its various forms raises the question on the impact of the European Union on its Member States (Radaelli, 2003: 27). It provides the theoretical framework for the analysis of supranational extrinsic adaptational pressure. However, Europeanization research designs to date are very broad in scope. There is a risk of prejudging the significance of supra-national variables, which requires systematic analysis of rival hypothesis and complementary explanations (Radaelli, 2003: 27). Liberal Intergovernmental-ism, based on the idea that EU integration can be understood best as a series of rational choices made by nation states, becomes the alternative theoretical framework for an intrinsic national logic (Moravcsik and Schimmelfennig, 2009: 69). Additionally, the understanding of the toolbox of (domestic) Public Policy-Analysis will inform the metatheoretical language applied in this thesis. Finally, both Europeanization and Liberal Intergovernmentalism theoretically inform the empirical analysis of the supposed inter-national interdependencies. This mul-ti-causal analysis makes sure to identify the multi-level origins for constraints and dependencies of the promotion of renewable energies in the European Un-ion (EU, Member State, inter-national). As will become apparent in the outline of this thesis, the theory design specifically involves the analysis of policy and polity constraints. Taking into account polity implications of the different entities that compose the European Union is deemed to be of significant importance in understanding their complex interplay.

The outline and ultimately the chapters of this study will follow the vertical-to-horizontal-logic as reflected in the explanatory factors (subordinate hypoth-eses). New input to the canon of studies with regard to the subject matter is ex-pected to arise theoretically and empirically. The dichotomous nature of most theory frameworks in the context of European Integration (Supranational vs. Member State, Member State vs. Member State) will be substituted by a triadic logic, since – arguably – the smallest possible reduction to represent European Integration can be achieved by taking into account the vertical and horizontal realities of the European Union, incorporating at least two neighbouring Mem-ber States. The study draws on primary sources where possible, existing quali-tative and quantitative analysis where needed, and takes into account interdis-ciplinary reflections. Linking acquired knowledge to complete the underlying triadic logic and closing the gap by accounting for missing analysis will ideally result in a stimulating and constructive contribution for future debates.

2. Theory

Natural science is concerned with correlating empirical observations so as to 'objectively' reflect 'reality', providing explanations and predictions of phenomena. Reflecting on how scientific practice is to be conducted is the philosophy of science, the methodological fundament of all scientific disciplines (Schnell, Hill and Esser: 2005: 49). Scholars, however, often talk past each other because of terminological confusion (Barkin, 2010: 13). The first step of this study therefore shall be a description of how the various terms of its methodological approach are used, including the understanding and acceptance of its frameworks, theories and models.

The governing discipline here is political science. This refers to the analysis of the policy processes in the context of authoritative execution *and* the unfolding of contexts, structures, constraints and dynamics in multi-level political realities (Schlager, 2007: 293). As has been stressed in the introduction, there are multiple disciplines and multiple levels of analysis involved. Analysing a political process leads to theories and models, which are grounded in a theoretical framework (Ostrom, 2007: 21; Schlager, 2007: 293). Of course – and for the sake of social science's integrity it should clearly be acknowledged – it is impossible to construct scientific principles similar to those employed in natural science. This makes all social science biased analysis, based on an extremely complex set of elements that interact over time (Sabatier, 2007: 3), necessarily heuristic in nature. This should not, however, justify, as Schlager (1997: 14, quoted in Sabatier, 2007: 323) argues, an *'ocean of descriptive'* works surrounding *'mountain islands of theoretical structure [...] occasionally attached together by foothills of shared methods and concepts'.* Sabatier (2007: 8) argues that a promising theoretical framework for policy analysis must meet four criteria. (i) It must meet the criteria of scientific theory – that is, concepts and propositions must be clear and internally consistent, it must identify causal drivers, it must give rise to falsifiable hypotheses and be fairly broad in scope. (ii) Each framework must be considered in the light of recent conceptual development and testing. (iii) It must be a positive theory and should contain some explicitly normative elements. (iv) It must take into account a comprehensive set of factors that have been traditionally regarded in public policy-making.

For Ostrom (2007: 25) frameworks provide *'a metatheoretical language that can be used to compare theories'.* Whilst frameworks bound inquiry, and focus attention on critical features of the social and physical landscape by specifying classes of variables and their interrelations, they cannot in and of themselves provide explanations and predictions (Schlager, 2007: 293). It is theories that *'place values on some of the variables identified as important in a framework, and make predictions about likely outcomes'* (Schlager, 2007: 296). Theories are compatible with different frameworks (Ostrom, 2007: 26). Indeed, as argued by Moravcsik and Schimmelfennig (2009: 68), a minimum of three theories, organized in a multistage model and paying respect to the issue and its circumstances are needed to explain a matter such as European Integration, wherein the promotion of RES-E mechanisms is to be located. Whilst Popper (1976: 31-46 quoted in Schnell, Hill and Esser: 2005: 53) uses the terms *hypothesis* and *theory* synonymously, the starting point of this study will be – as is common in social science (Schnell, Hill and Esser: 2005: 53) – a working hypothesis. Proposed explanations, based on precise assumptions about a limited set of parameters, reflected in the working hypothesis, will be developed into a model (Ostrom, 2007: 26). This allows the scholar to test specific parts of theories by *'fixing a limited number of variables at specific setting and exploring the outcome produced'* (Schlager, 2007: 294). Sabatier (2007: 322-323) sums up this hierarchy as follows: *'A conceptual framework identifies a set of variables and relationships that should be examined to explain a set of phenomena. [...] A theory provides a denser and more logically coherent set of relationships, including direction and hypothesis that self-consciously seek to explain a set of phenomena. [...] A model is a representation of a specific situation.'*

After a general introduction to the logic of politics, frameworks and theories applied in this research, and a concise methodological layout with respect to the subject matter, will be introduced.

2.1 Nation-Bound Policy Analysis Frameworks

One cannot understand why, for example, a policy implementation, is not successful, if one does not understand the fundamental nature of the political system. This becomes especially true in the context of European Integration. In order to focus on any scientific analysis in the context of the political, it is necessary to understand the theoretical framework that defines what is political. Today, it is widely acknowledged in political science that the political process takes place in three dimensions: polity, politics and policy. All three dimensions describe what is commonly referred to as politics.

Whilst the logic of politics is a constructed result of scientific reflection, the concept itself is of empirical relevance and can help to gain insightful observations of political developments. Not only does it help to understand the complexity of the political process, it enables the researcher to describe adequately interrelations in the political sphere (Meyer, 2010: 81). Especially in the context of this thesis, it is imperative to understand the differentiation between polity – the constitutionally enshrined principles defining the organization of a given political system, as well as the operation of a political culture around that system – and policy – the dimension referring to any kind of topic, agenda, programme, instrument, decision, etc., that has been developed in a specific political field based on the attempt to solve a politically acknowledged problem (Meyer, 2010).

A good framework defining stages to approximate the policy dimension is the *stages heuristic approach,* better known as the policy cycle. It is considered to be the most influential framework up until the mid-1980s. Although harshly criticised as a non-theory, it is the first approach that divides the complex policy process into discrete stages (Sabatier, 2007: 6-7). Since its terminology and – in parts – its layout have implicitly been incorporated in almost all policy theory, it is considered a good starting point for a brief theoretical discussion. The framework itself has been subject to extensive development and finds its greatest 21st century advocates in Howlett, Ramesh and Perl (2009). According to scholars of this approach, the following stages are commonly identified: problem perception, agenda setting, policy formulation, decision making, implementation, and policy evaluation (Howlett, Ramesh and Perl, 2009: 10-14; Jänicke, Kunig and Stitzel, 2003: 53-66). Please refer to the cited sources for further information.

As mentioned before, the policy cycle has been subject to harsh criticism, because it is not considered a causal theory that could give rise to hypotheses. Its stages are descriptively inaccurate and oversimplify the process of multiple interacting cycles (Sabatier, 2007: 7). Notwithstanding these legitimate objections, this study is based on the understanding that the policy cycle produces the guiding logic for alternative theoretical framework proposals, such as the *advocacy-coalition framework* (Sabatier and Jenkins-Smith, 1988), *the multiple-stream framework* (Kingdon, 1984) or the *network approach* (Windhoff-Heritier, 1987). The advocacy-coalition framework aims to identify political sub-systems and their core beliefs in order primarily to analyse the 'decision-making' process, whilst the multiple-streams framework refers to 'decision-making' (Cohen, March and Olsen, 1972: 2) and 'agenda-setting' (Brunner, 2008: 51) in a national context. The network approach creates stable patterns which take shape around 'policy problems'. All of these domains are stages of the policy-cycle (Sabatier and Weible, 2007: 198; Soritrov and Memmler, 2011: 4; Brunner, 2008: 501; Adam and Kriesi 1999: 146). For detailed information on the respective frameworks, please refer to the cited sources.

With respect to the above-mentioned approaches, it becomes apparent that the frameworks draw heavily on sociological explanations of the policy process. Their explanations, however, differ substantially (Schlager, 2007: 302). The key concept of 'advocacy-coalitions' is defined by (i) a macro-level assumption that policy-making occurs among specialists; (ii) a model of the individual as heavily influenced by social psychology; and (iii) a meso-level conviction that the best way of dealing with social interactions in a subsystem is to aggregate them into 'advocacy-coalitions' (Sabatier and Weible, 2007: 191). The 'multiple-streams' framework focuses on 'policy entrepreneurs' and 'policy communities' investing *'resources of various hopes'* that return a *'favourable political outcome'* (Kingdon, 1984: 151; Brunner, 2008: 501). The 'network-approach' emphasizes the predominance of informal and decentralised relations (Windhoff-Heritier, 1987: 43).

The emphasis on sociological explanations means that the nation-bound policy analysis approaches can be incorporated in the overall framework of the applied theory design. Independent of whether one talks of 'advocacy-coalitions', 'policy communities' or 'policy-networks', it is human beings and collective actions that shape the policy process against the backdrop of a given polity (Schlager, 2007: 302; Böcher, 2011: 4). It is vital to take into account the assumptions about actors and their relationships for the applied analysis.

However, considering the frameworks introduced here as exclusive would neglect a particular set of variables that should be examined to explain the observed phenomena. All policy analysis as described so far is *'deeply implicated in facilitating the stability of the political institutions of the Western nation state'* (Hajer, 2003: 175). But in the context of the European Union, the policy process cannot be explained within the boundaries of sovereign polities (Hajer, 2003: 175). The European Union's policy process implies a new spatiality to which policy analysts and policy theory adjust only slowly (Meny et al., 1996 quoted in Hajer, 2003: 179). Following Howlett, Ramesh and Perl (2009: 75) introduce the 'stages-heuristic': *'policy-making is very much a domestic concern involving national governments and their citizens'*. For Zahariadis (2007: 65) 'multiple streams' is *'[...] a lens, perspective or framework [...] that explains how policies are made by national governments [...]'*. Sabatier and Weible (2007: 199) argue that the 'advocacy-coalition framework' has stable system parameters (e.g. constitution). Adam and Kriesi (2007: 146) conclude that the analytical value of the 'network approach' *'lies in the fact that it conceptualizes policy making as a process involving a diversity of actors who are mutually interdependent. Taking into account the role of state and societal actors in policy making, it synthesizes state- and society-centred approaches'*.

It is not surprising that the advocates of these frameworks claim the emphasis of these approaches to be on the unfolding of the policy process, rather than on executive authority or a preoccupation with power and influence (Schlager, 2007: 293; Cohen, March and Olsen, 1972: 2). Nevertheless, what has long been considered stable in policy analysis – clear rules and norms according to which policy measures are to be implemented, in short, established constitutional frameworks (polity) – becomes a topic for analysis itself (Hajer, 2003: 175-177). It becomes necessary to define what stable system parameters are.

The 'state' seems to be the focal point for describing stable system parameters. The cited framework implications for nation-bound policy analysis frameworks are evidence thereof (Hajer, 2003: 175; Howlett, Ramesh and Perl, 2009: 75; Zahariadis, 2007: 65; Weible, 2007: 199; Adam and Kriesi, 2007: 146). Indeed, there is no generally accepted answer as to what the 'state' is (Maier, 2001: 27-28). Against this backdrop it does not seem surprising, as Daniel Gaus (2001: 2) argues, that some political analysts do not even consider the concept of 'state' important for the analysis of politics. Even if taken into account, the problem remains *'that political analysis is mostly driven by a diffuse view on the ontological nature of the object 'state''* (Gaus, 2001: 24).

A good starting point to approximate the concept of 'state' is Jellinek, who defines a triad of indicators that define what a state is: 'territoriality', 'people' and 'executive authority' (Maier, 2001: 29). Whilst 'territoriality' (as a geographically defined scope of executive authority) and 'people' (sum of people defined by the same citizenship) can be regarded as self-explanatory in this very context, 'executive authority' cannot. It is Weber (1919, reprinted 1994) who introduced the concept of 'executive authority' as a major force to explain both politics and the state. Weber primarily refers to patterns of empirical social action when he states that *a compulsory political organization with continuous operations will be called "state" insofar as its administrative staff successfully upholds the claim to the monopoly of the legitimate use of physical force in the enforcement of its order'* (Weber, 1978: 54, quoted in Gaus, 2011: 19). Whilst by following this definition, power becomes the basic means to define the modern state sociologically, a vital link is missing: on which ground is the execution of power based? It is the enforcement of the 'law', to which the modern Verfassungsstaat (constitution-based state) is the exclusive origin (Staatsgewalt) (Weber, 1994: 35; Maier, 2001: 36-37). And it is the Staatsgewalt, that upholds, bound by law, the exclusive, undivided and self-originating force upon its citizens within a territoriality (Maier, 2001: 35). The triad of indicators that define what a state is needs, if the executive authority is bound by law, an addition. This addition must be law itself (Maier, 2001: 29). The unification of law and politics, that is, law and the enforcement of its order through executive authorities within the nation state, have become self-evidential fundamentals in the teachings of political science (Franzius, 2010: 6). Yet the relation of law and politics and the term 'law' especially need further clarification. It is Grimm (2001: 13-22) who defines three development phases that clarify the relation between law and politics. Whilst the following excursus might appear a little off target, it establishes the explanation of a very fundamental problem of European Integration, which many scholars of political science, both in theory and empiricism, misleadingly disregard in their analysis.

In pre-modern societies (first phase), the idea that men could create law did not exist. Instead law was understood to originate *qua* transcendental or sacred endowment, and was therefore irrevocable. Neither its content nor its prevalence could be influenced or regulated by politics. Consequently, law could not have been subject to bargaining or decision-making and the areas of political influence were limited to enforcing and preserving the sacred order (Grimm, 2001: 15-16; Franzius, 2010: 6-7).

The second phase that defines the relation of law and politics is described as *'politicising the law'* (Grimm, 2001: 16). During religious conflicts in the 16th century societies began to fight over the question of what kind of order god intended for mankind. What started as a fight between groups of different confessions soon developed into a civil war. To find a solution that could pacify the involved parties and restore peace, a new order could not have been based in just another sacred order, to whose validity the defeated parties would never agree, but an independent and secular order. Bearer of that order would be, in accordance with the feudal pyramid of this time, the reigning aristocrats (Grimm, 2001: 16-17). In consequence, it became possible for men to create law, and it is authority, not genuine truth, that creates and regulates this law *(auctoritas non veritas facit legem* – Hobbes, quoted in Franzius, 2010: 5) This reaction to pre-modern realties lead to a system based on politics and positive law, manifest at first in the reigning aristocrats, and later differentiated in administrative, financial and military sub-systems. These were understood, at the time, as a 'state'. But the fact that the monopoly to create and enforce the law was in the hands of the monarch did not guarantee the societies' demand for justice. The relation of law and justice was turned around, for instead of sacred law being superior to political execution (first phase), it was political power that created law (second phase). Hence, the law, being dependent on the subjective practice of the reigning aristocrat, could not be the unalienable source of truth and justice (Grimm, 2001: 16-19, Franzius, 2010: 7). As Hobbes (1929 : 204) states: *'The Soveraign of a Common-wealth, be it an Assembly, or one Man, is not Subject to the Civill Lawes. For having power to make, and repeale Lawes, he may when he pleaseth, free himselfe from that subjection, by repealing those Lawes that trouble him, and making of new; and consequently he was free before. For he is free, that can be free when he will: Nor is it possible for any person to be bound to himselfe; because he that can bind, can release; and therefore he that is bound to himself onely, is not bound'* (see also Franzius, 2011: 8).

Whilst by politicising the law, law could be created by authority, its enforcement could not be protected from despotism. To restrict and define how political powers were legitimised, how they could interact, and to what degree they could execute their power, an unalienable source was necessary, which, because it could not be based in transcendental truth, would become a law superordinate to the law made by political powers (Grimm, 2001: 20, Meier, 2010:). This third phase Grimm (2001: 19) calls *'juridification of politics'*, and it was achieved by establishing a constitution that would normatively bind the process of positive law-making. It would normatively bind the process of creating norms by defining framework conditions that differentiate between principles for the process of political decision-making and decision-making itself (Grimm, 2001: 20-21). It is Franzius (2010: 8) who vividly illustrates the matter by referring to Spinoza and his Ulysses parable. Spinoza (1883 : Chapter VII) writes: *'And to this end the first point to be noted is, that it is in no way repugnant to experience, for laws to be so firmly fixed, that not the king himself can abolish them [...] . Which we can make plain by the example of Ulysses. For his comrades were executing his own order, when they would not untie him, when he was bound to the mast and captivated by the Sirens' song, although he gave them manifold orders to do so, and that with threats. And it is ascribed to his forethought, that he afterwards thanked his comrades for obeying him according to his first intention. And, after that example of Ulysses, kings often instruct judges, to administer justice without respect of persons, not even of the king himself, if by some singular accident he order anything contrary to the established law. For kings are not gods, but men, who are often led captive by the Sirens' song.'*

The importance of understanding the triadic logic of politics, and in particular of the polity aspect – as has been pointed out in the introductory to this chapter – now becomes apparent. Polity characterises the type and organization of a given political system as well as the operation of a political culture around that system. Type and organization are most likely described in a constitution (Meyer, 2010). It is the constitution that defines fundamental rules (primary law) applicable to a society within a state (Stone, 2000; Vorländer, 2009: 9f, quoted in Hönninge, Kneip and Lorenz, 2011: 8). The 'Western nation state' (Hajer, 2003: 175) is the ideal-type representation of the relation of politics and the law defined above (Franzius, 2013: 206), and as such – characterized by territoriality, people and executive authority bound by a constitution – defines, in a normative sense, stable system parameters for policy analysis. It is this very school of thought, represented in the political and cultural traditions of the Anglo-Saxon and European academies, in particular the interpretation of law as the only legitimate result of collective political decision-making and its execution *within* a nation state, that results in constraints on the application of nation-bound policy analysis in a European context. The European Union based on the Lisbon treaty is not a nation state.

2.1.1 Summary

A good framework defining stages to approximate the phenomenon of the policy process is the stages-heuristic approach, better known as the policy cycle. It is considered to be the most influential framework up until the mid-1980s. According to scholars of this approach, the following stages are commonly identified: problem perception, agenda setting, policy formulation, decision making, implementation, and policy evaluation (Howlett, Ramesh and Perl, 2009: 10-14; Jänicke, Kunig and Stitzel, 2003: 53-66).
The policy cycle approach has been subject to harsh criticism, because it is not considered a causal theory that could give rise to hypotheses. It nevertheless produces the guiding logic for alternative theoretical framework proposals, such as the advocacy-coalition framework (Sabatier and Jenkins-Smith, 1988), the *multiple-stream framework* (Kingdon, 1984) or the *network approach* (Windhoff-Heritier, 1987).
Studying these approaches, it becomes apparent that their frameworks draw heavily on sociological explanations of the policy process. The emphasis on sociological explanations allows the nation-bound policy analysis approaches to be incorporated in the overall framework of the applied theory design. However, considering these frameworks as exclusive metatheoretical sources would neglect a particular set of variables that should be examined to explain the observed phenomena. All such policy analysis is *'deeply implicated in facilitating the stability of the political institutions of the Western nation state'* (Hajer, 2003: 175). But in the context of the European Union, the policy process cannot be explained within the boundaries of sovereign polities (Hajer, 2003: 175). The European Union's policy process implies a new spatiality to which policy analysts and policy theory adjust only slowly (Meny et al., 1996 quoted in Hajer, 2003:

179). Howlett, Ramesh and Perl (2009: 75) introduce the 'stages-heuristic': *'pol-icy-making is very much a domestic concern involving national governments and their citizens'*. For Zahariadis (2007: 65) 'multiple streams' is *'[...] a lens, perspective or framework [...] that explains how policies are made by national governments [...]'*. Sabatier and Weible (2007: 199) argue that the 'advocacy-coalition framework' has stable system parameters (e.g. constitution). Adam and Kriesi (2007: 146) conclude that the analytical value of the 'network approach', *'lies in the fact that it conceptualizes policy making as a process involving a di-versity of actors who are mutually interdependent. Taking into account the role of state and societal actors in policy making, it synthesizes state- and society-centred approaches'*.

The 'Western nation state' (Hajer, 2003: 175) is the ideal-type representation of the defined relation of politics and the law as part of the overall logic of politics (Franzius, 2013: 206) and as such – characterized by territoriality, people and executive authority bound by a constitution – it defines, in a normative sense, stable system parameters for policy analysis. It is this very school of thought, represented in the political and cultural traditions of the Anglo-Saxon and Euro-pean academies, in particular the interpretation of law as the only legitimate re-sult of collective political decision-making and its execution within a nation state, that results in constraints on the application of nation-bound policy analysis in a European context. The European Union based on the Lisbon treaty is not a nation state.

2.2 Trans-, Supra-, and International Bound Policy Analysis Frameworks

Stable system parameters based on the rule of law with an executive authority that upholds the monopoly of the legitimate use of physical force can be found in nation states. As a basis for policy analysis, nation states amount to an ordered system that could give rise to independent variables (Weber, 1978: 54 in Gaus, 2011: 19). But reflections upon the European Union defy the classic categories of the aforementioned definitions as to how the three dimensions reflected in the logic of politics – polity, politics, policy – interrelate in the European Union and between the European Union and its Member States (Franzius and Preuß, 2012: 18-22; Meyer, 2010: 81; Leuffen, Rittberger and Schimmelfennig, 2013: 2). There are many attempts to describe the European Union: multilevel governance, regu-latory state, federal state to be, confederation, multilevel constitutionalism, and so on (Pernice, 2001: 5). Since the process of European integration is unprecedent-ed and its development on-going, any ontological descriptions derived from 19th and 20th century thinking fail to comprehend the process of integration, let alone the status quo of the European Union. Regardless, or in spite, of these difficulties, theoreticians approached the matter searching for theoretical solutions.

It is theories of European Integration that try to grasp and reflect upon the unsettled condition of the European Union. There are two general sources of theoretical development dealing with European Integration. One source is clas-sical, or nation-bound, policy analysis, as reflected in Political Science. The other source is rooted in the field of International Relations (Leuffen, Rittberger and Schimmelfennig, 2013: 29). But since the European Union challenges the dichotomous categories of two basic types of contemporary polities – modern nation states, on the one hand, and international organizations, on the other – the explanatory foci of the original framework approaches must be broadened (Leuffen, Rittberger and Schimmelfennig, 2013: 2 and 32).

Leuffen, Rittberger and Schimmelfennig (2013: 2-7) developed a cluster of five typical characteristics to explain why the European Union cannot be classified as either a nation state or an international organization, three of which are quite clearly influenced by Jellinek's state concept (Maier, 2001: 29):

- Membership: States are composed of individuals that mostly acquire citizenship by birth. Members of international organizations are states. The European Union is defined by Member States, who are also free to leave the Union. Yet, Art. 20 of the Treaty on the Functioning of the European Union (TFEU) establishes that every person holding the nationality of a Member State shall be a citizen of the Union (TFEU, 2008: Art. 20).

- Territoriality/Delimitation: A state is a territorially ordered system. International organizations are functionally delimited. The EU has a territory and a physical border. People enter the EU in principal like they enter the USA but they cannot enter the World Trade Organization or the North Atlantic Treaty Organization.

- Executive Authority: States uphold the executive authority and the monopoly of the legitimate use of physical force in the enforcement of their order (Weber, 1978: 54 in Gaus, 2011: 19). International Organizations do not hold these attributes of sovereignty. Whilst the EU system is highly integrated, its separation of powers is more similar to a state, and around 75 per cent of all EU laws result from ordinary legislative procedure; the execution of Union policies is left to national administrations (Knill and Lenschow, 2005: 581). The EU is not sovereign and cannot order the use of force.

- Governance: As opposed to states, bureaucracies of international organizations are small, weak, without independent income or power of taxation. The administrative capacity of the EU is strong, in contrast with international organizations, but weak in comparison to European States. The administration of Switzerland amounts to 33,000 staff and serves 7.7 million citizens. The European Commission has about the same number of staff but serves about 580 million citizens.

- Legitimacy: The modern state reflects the idea of a distinct community. International organizations do not have strong identity and cultural fundaments. The EU reflects the idea of a distinct community, but has a very volatile identity and cultural fundament.

In conclusion, Leuffen, Rittberger and Schimmelfennig (2013: 5) argue that the EU is in between anarchy and hierarchy. Whilst rule-making and adjudication are hierarchical, the EU is not sovereign and cannot order the use of force. Hierarchy in the shadow of anarchy could therefore characterize it. *'Whether and why the EU will remain more like an international organization or become more like a state is also the traditional core question and debate of integration theories'* (Leuffen, Rittberger and Schimmelfennig, 2013: 7).

Intergovernmentalists argue that states are and will remain the dominant actors in European Integration. Member States remain the *'Masters of the Treaties'*, having conferred the exercise of some of their powers to the EU, whose primary treaty law merely remains a derivative legal order (BVerfGE, 1993: 89 (155); Tatham, 2013: 83; Leuffen, Rittberger and Schimmelfennig, 2013: 32 and 40). With reference to Grimm (2001: 13-22) and the three development phases of the relation of law and politics, intergovernmentalists would likely fall into the second phase. Tapering the words of Hobbes (1929: 204) in an intergovernmentalist fashion: The masters of the treaty are not subject to the (development of) European law or European authority, for they may, if they please, free themselves from this subjection, by repealing their membership. It is not possible for any state to be bound to itself, for who can bind, can release.

Supranationalists, on the other hand, emphasize the potential of European Integration to develop into a new kind of European polity, ultimately leading to a European federal state (Leuffen, Rittberger and Schimmelfennig, 2013: 62). With reference to Grimm (2001: 13-22), they would likely fall into the third category. Citing the words of Spinoza (1998: Chapter VII), supranationalists could claim that it is not repugnant to experience for laws and the authority of the EU to be so firmly fixed, and that the Member States themselves cannot abolish them. After the example of Ulysses, the EU should administer justice without respect to Member States, if by some singular accident they are led captive by the Sirens' song.

In the following section, two frameworks will be introduced. *Liberal Intergovernmentalism* will be the guiding framework for this study. And although the Supranationalist approach has always been the main competitor to intergovernmentalist approaches, the complementary framework for this study will be derived from the *Europeanization* approaches.

2.2.1 Liberal Intergovernmentalism

Liberal Intergovernmentalism represents *'the most prominent, up-to-date, complete, and theoretically elaborate intergovernmentalist integration theory'* (Leuffen, Rittberger and Schimmelfennig, 2013: 42). It will hence serve as an Intergovernmentalist synthesis, or 'baseline theory' (Moravcsik and Schimmelfennig, 2009: 67) to inform one aspect of the overall theory framework of this study. The theory of Liberal Intergovernmentalism was developed by Moravcsik in the last decade of the 20th century. The cornerstones of this theory presented here are based on a detailed analysis published under the title of 'The choice for Europe' by Moravcsik (1998). His central argument is: '[...] European integration can best be explained as a series of rational choices made by national leaders. These choices responded to constraints and opportunities stemming from the economic interest of powerful domestic constituents, the relative power of each state in the international system, and the role of international institutions in bolstering the credibility of interstate commitments' (Moravcsik, 1998: 18). The school of thought applied by Moravcsik is derived from an overall rationalist framework of international cooperation, in which the EU is seen as an international regime for policy coordination, as a result of international negotiations being disaggregated into three stages: national preference formation, interstate bargaining, and institutional choice (Moravcsik, 1998: 20; Leuffen, Rittberger and Schimmelfennig, 2013: 46).

The primary political instrument by which national citizens and societies try to influence and ultimately alter the international agenda in their favour are nation states. The nation state in Liberal-Intergovernmentalism is considered to be a unitary and rational actor. This assumption is not to be mistaken for the idea that states are unitary in their internal politics. The unitary-actor assumption, which maintains the idea that each nation state acts in international negotiations 'as if' with a single voice, subsumes the idea that 'once particular objectives arise out of [...] domestic competition, states strategize as unitary actors vis-à-vis other states in an effort to realize them' (Moravcsik, 1998: 22). The same holds true for the assumption that nation states act rationally. The idea is that states make internal decisions 'as if' efficiently conducting a weighted choice based on a stable set of underlying principles. Both assumptions, however, should not be overemphasized (Moravcsik, 1998: 23). Fundamental goals of states are neither uniform nor fixed. State preferences vary over time and in response to changes in economic, ideological, or geopolitical environments (Moravcsik, 1998: 22-23).
The theory does not claim a logic of general economic dominance over politics, rather a political economic perspective focusing on the distributional as well as the efficiency consequences of policy decisions (Moravcsik and Schimmelfennig, 2009: 69; Moravcsik, 1998: 36). In total, Moravcsik (1998: 24) defines national preferences '[...] as an ordered and weighted set of values placed on future substantive outcomes [...], that might result from international political influence. Preferences reflect the objectives of those domestic groups which influence the state apparatus; they are assumed to be stable within each position advanced on each issue by each country in each negotiation [...]'.

There are two fundamental determiners of national preferences: geopolitical and economic interests. Geopolitical explanations predict that governments tend to favour integration, if they are expected to generate positive geopolitical externalities to perceived threats to national sovereignty or territorial integrity.

Economic integration under a geopolitical approach is supposed to be triggered by indirect security implications, for expected positive geopolitical externalities foster governments' willingness for integration (Moravcsik, 1998: 24-35). Direct economic implications are reflected in political and economic interests. The underlying theories explain international cooperation justified by the conviction that policy coordination across countries leads to mutually beneficial patterns of economic policy externalities. In other words, governments integrate in order to secure and enrich commercial advantages for economic groups, whilst taking into account the consequences of political economic decisions (Leuffen, Rittberger and Schimmelfennig, 2013: 47; Moravcsik, 1998: 35-38). Based on empirical findings, Moravcsik (1998: 472) comes to the conclusion that the dominant motivation influencing national preferences are political and economic interests, commercial interests in particular. However, geopolitical interests are also an important aspect of European integration. A theory of political and economic preferences only would have predicted a pan-European free trade area, rather than a more differentiated Lisbon Treaty–based European Union (Moravcsik and Schimmelfennig, 2009: 70; Moravcsik, 1998: 6 and 472-474).

Whilst national preferences alone cannot sufficiently explain (European) integration, Liberal Intergovernmentalism introduces a bargaining theory to describe the relevant negotiation processes concerning the distribution of gains from cooperation (Leuffen, Rittberger and Schimmelfennig, 2013: 48). In order to overcome suboptimal outcomes of coordination and cooperation, governments must negotiate how mutual gains are distributed among states. These negotiations, e.g. in case of the European Community – as analysed by Moravcsik – are characterized by *'hard interstate bargaining, in which credible threats to veto proposals, to withhold financial side-payments, and to form alternative alliances excluding recalcitrant governments carried the day. The outcomes reflected the relative power of states – more precisely patterns of asymmetrical interdependence. Those who gained the most economically from integration compromised the most on the margin to realize it, whereas those who gained the least or for whom the costs of adaptation were highest imposed conditions'* (Moravcsik, 1998: 3; see also Leuffen, Rittberger and Schimmelfennig, 2013: 49). Again, the focus is on states as the dominant actor in the process. According to Liberal Intergovernmentalism, governments do not need the help of supranational (European) organizations to agree upon a feasible solution for the distribution of mutual gains, because supranational entrepreneurs do not hold superior capacities that are vital to reaching an inter-state agreement (Leuffen, Rittberger and Schimmelfennig, 2013: 49; Moravcsik and Schimmelfennig, 2009: 71).

Finally, national preference formation and interstate bargaining will lead to institutional choice (Moravcsik, 1998: 20). The design of international organizations is explained by a regime-theoretical approach (Moravcsik and Schimmelfennig, 2009: 72). The necessity of supranational delegation arises out of governments' concerns about each other's future compliance with the cooperation agreement on how mutual gains are distributed. The establishment of those institutions is meant to prevent future transaction cost for further international negotiations regarding the agreed upon deal, and to bolster the credibility of interstate commitments (Moravcsik, 1998: 472). Pooling of competencies arises, *if '[...] governments [...] find it too costly or technically impossible to specify all future contingencies involved in legislating or enforcing [...]'* agreed-upon broader goals (Moravcsik, 1998: 73). Pooling or delegating, being the a result of a cost-benefit analysis, reflects a certain amount of national willingness to commit to being outvoted or overruled in individual supranational decisions (Moravcsik, 1998: 75). In summary, Liberal Intergovernmentalism as a theory of European Integration assumes intergovernmental decision-making under anarchy and argues that European Integration is not a preordained movement towards a federal Europe, but a series of pragmatic bargains among national governments based on concrete national interests, relative power, and carefully calculated transfer of sovereignty (Moravcsik, 1998: 472; Moravcsik and Schimmelfennig, 2009: 73).

Critics say Liberal Intergovernmentalism is limited to treaty-amending decisions and not appropriate to describe every-day decision-making (Moravcsik and Schimmelfennig, 2009: 73) or, more importantly, legally binding development of secondary law by EU institutions beyond the carefully calculated intentions of international coordination. This research focuses on bottom-up perspectives to account for institution-building processes and an emerging European polity (Börzel and Risse, 2000: 1). What can be considered advantageous in order to derive stable system parameters for policy analysis is disadvantageous in terms of taking into account critical features of Lisbon Treaty–based EU realities (Schlager, 2007: 293). The same is true of the main competitor of Intergovernmentalism, namely Supranationalism. *'Supranationalism explains European integration as a progressive, self-reinforcing process of institutionalization. Integration may initially correspond to intergovernmentalist expectations of a formal agreement negotiated among interdependent states, and shaped by their preferences and power. Once supranational organizations and rules are in place, however, integration produces unintended and unanticipated consequences, and escapes the exclusive control of the states'* (Leuffen, Rittberger and Schimmelfennig, 2013: 75). Supranationalists' expectations differ substantially from Intergovernmentlists' predictions: states do not remain the dominant actors, rather supranational entrepreneurs have the capacity to strengthen supranational rules and expand their powers. Nevertheless, the theory initially takes a bottom up perspective to explain – in its original formulation – the formation of a European federal polity with state-like stable system parameters (Leuffen, Rittberger and Schimmelfennig, 2013: 72 and 76).

However, in order to fully capture the Lisbon Treaty–based European Union it is important to add a top-down perspective to the theory framework. Therefore, the Supranationalist account will be disregarded explicitly, but taken into account implicitly by giving merit to the Europeanization approach. This approach focuses on the policy and politics dimension of European integration by asking how the European Union affects, based on primary and secondary European law, the policies and politics of EU Member States.

2.2.2 Europeanization

Europeanization is a relatively new concept (Ancygier, 2013: 27; Graziano and Vink, 2013: 32-36). Whilst it certainly draws on basic aspects of European Integration theories, notably Supranationalism, both study approaches have not often been clearly linked (Graziano and Vink, 2013: 39). Europeanization, however, is not equal to integration theories. Whilst the latter are predominantly concerned with why and how countries pool sovereignty (Moravcsik, 1998: 73-75), Europeanization focuses on the understanding of what happens once EU institutions are in place and produce their effects (Radaelli, 2003: 8-9). In short, it is concerned with the impact of policy outcomes and institutions at the supranational level on domestic Member States' policies, politics and (to a certain extend) polities. (Schimmelfennig and Sedelmeier, 2005: 5; Börzel and Risse, 2000: 1; Heritier, 2001: 3). In analytical terms, Radaelli (2003: 9) argues that EU policy formation – as a result of preferences, bargaining and interaction between national and supranational actors – should be kept distinct from impact assessment. Whilst bottom-up and top-down dimensions are connected in 'real life' and feed back into each other (Börzel, 2002: 193; Radaelli, 2003: 9), one should distinguish analytically between the process leading to the formation of supranational policies and the reaction they cause in the policies and politics arena of Member States (Radaelli, 2003: 9).

This does not preclude the linkage of Europeanization frameworks to other frameworks in empirical research. But it is a matter of scope, rather than a matter of definition (Radaelli, 2003: 9-10). Opposed to Radaelli (2003: 9), Heritier (2005: 200) explicitly includes Member States' strategies to influence the formation of EU policy measures within the framework of Europeanization. This aspect will be disregarded in this study, for it finds representation the abovementioned frameworks. Europeanization in this study will hence be understood as a process in which states adopt EU rules (Schimmelfennig and Sedelmeier, 2005: 7). This definition makes domestic impact of EU policies the main analytical target (Graziano and Vink, 2013: 34).
There is widespread consensus that Europeanization covers differentiated phenomena (Börzel and Risse, 2000: 4). European policymaking has different impacts on different Member States and consequently produces different domestic reactions (Heritier, 2001: 9; Heritier and Knill, 2001: 286), predicting which is the function of the analytical explanation mechanisms applied (Knill and Lehmkuhl, 2002: 256). But: *'If everything is Europeanized to a certain extent, what is not Europeanized?'* (Radaelli, 2003: 7).

The *'single most identified proposition'* (Ladrech, 2010: 31) or *'most interesting (and well investigated)'* contribution in that respect is the 'goodness of fit' approach and its 'mediating factors' aspect (Graziano and Vink, 2013: 40). The approach restricts itself to a top-down logic and identifies two conditions for expecting domestic change. First, there must be an inconvenient level of misfit between EU-level policies, politics (and polities) with regard to the domestic-level counterparts. Second, the adaptational pressure (misfit of supranational vs. national level policies, politics, etc.) has to cause reverberation in the national arena, involving actors, institutions, etc. (Börzel and Risse, 2000: 1). Owing to an on-going debate between rationalism and constructivism, there are at least two conceivable logics of action that national rule adaption follows (Schimmelfennig and Sedelmeier, 2005: 9). (i) The rationalist perspective follows a 'logic of consequences', where the misfit between EU level and domestic level provides local societal and political actors with new opportunities for, or constraints on, pursuing their interests. Domestic change is thus dependent on national actors' capacities to exploit the opportunities or avoid the constraints (Börzel and Risse, 2000: 2; Schimmelfennig and Sedelmeier, 2005: 9). With respect to Liberal Intergovernmentalism and following the 'logic of consequences', it would be challenging to argue that domestic impact becomes the dependent variable of Europeanization (a term which in that context does not refer to the theory but to the process the theory reflects). This is because if EU-level policies enable or disable national actors with capacities to adopt or neglect a policy choice, then ultimately EU-level policies and politics proposals are dependent on national actors benevolence, and not on EU-level pressure for national transposition.

(ii) The constructivist's 'logic of appropriateness' argues that European policies and norms exert adaptational pressure on domestic processes. Following this logic, a misfit leads to a process of arguing about the legitimacy or appropriateness of EU rule-transfer and rule-adoption (Börzel and Risse, 2000: 2; Schimmelfennig and Sedelmeier, 2005: 9). Whilst the degree of rule-transfer and rule-adoption remains a topic for discourse between supranational entrepreneurs and national actors, there is a general consensus that 'adaptational pressures are generated by the fact that the emerging European polity encompasses structures of authoritative decision making which might clash with national structures of policy making [...]' but '[...] EU member states have no exit option given that EU law constitutes the law of the land' (Börzel and Risse, 2009: 5).

Following this logic, and given that the European polity does not leave Member States an exit option, for EU law constitutes the law of the land, Europeanization might lead to a policy misfit or cause an institutional misfit (Börzel and Risse, 2009: 5-6). This could cause four possible Member State reactions (Radaelli, 2003: 13): inertia, absorption, transformation or retrenchment. Heritier (2001: 9) argues that it ultimately depends on the reform capacity whether or not a domestic policy is transposed in the sense of EU requirements or not. Of course, there is no need for domestic reform, if EU demands and national practices are concurrent. It is furthermore reasonable to argue that the lower the compatibility, the higher the adaptational pressure (Börzel and Risse, 2009: 5). However, Heritier claims (quoted in Ladrech, 2010: 32) that an initial policy fit might lead to strengthening a domestic constellation of actors that oppose the European objectives, and promote a domestic approach instead. The four possible domestic reactions to policy or institutional misfit, as aggregated by Radaelli (2003: 13), could nevertheless be a vague indicator of the approximate potential reform capacities at the domestic level:

- Inertia (no change) is a situation where EU demands are too dissimilar to domestic practice. Resistance and non-transposition of EU directives would be an example. However, inertia should produce crisis and abrupt change (Radaelli, 2003: 13; Ladrech, 2010: 36).

- Absorption (low degree of change) refers to non-fundamental change or accommodation of policy requirements without real modification (Radaelli, 2003: 13, Ladrech, 2010: 36).

- Transformation (substantial change) occurs when the fundamental logic of political behaviour changes (Radaelli, 2003: 13; Ladrech, 2010: 36)

- Retrenchment (resistance to change) implies that national policy becomes less in accordance with European demands than it was before (Radaelli, 2003: 13, Ladrech, 2010: 36).

There are other analysts who categorize domestic responses in a threefold manner, in which the degree of pressure roughly equals the degree of change (Ladrech, 2010: 36):

- Absorption (low): Member States respond to a low misfit situation and incorporate policies or ideas into their domestic structures without substantially modifying existing processes. Domestic core beliefs and institutional logics are kept intact (Börzel and Risse, 2009: 14; Ladrech, 2010: 36-37)

- Accommodation (modest): More pronounced adaptational pressure leads to a defined, but modest, domestic change. There is a medium amount of change suggesting an empowerment of certain actors or advocacy-coalitions (Ladrech, 2010: 37; Börzel and Risse, 2009: 14).

- Transformation (high): Member States replace existing policies, processes or institutions with new ones, or alter existing ones with substantially different core features (Börzel and Risse, 2009: 14; Ladrech, 2010: 37).

What brings about domestic change is – as mentioned above – highly dependent on differentiated state structures (Ladrech, 2010: 34; Heritier, 2001: 9; Heritier and Knill, 2001: 286). But identifying domestic mediating factors helps to explain why domestic change might or might not be stimulated. With respect to the goodness-of-fit argument, there are five mediating factors, as Ladrech (2010: 34) summarizes:

- Multiple veto points: The degree to which power is dispersed in a domestic institutional setting can empower actors with diverse interests to avoid constraints emanating from Europeanization pressure (Börzel and Risse, 2009: 8, Börzel and Risse, 2000: 7; Ladrech, 2010: 34).

- Formal institutions: Existing domestic institutions can enable domestic actors to exploit EU opportunities by providing material and ideational resources, and thus promote domestic adaptation (Börzel and Risse, 2009: 9, Börzel and Risse, 2000: 7; Ladrech, 2010: 34).

- Political and organizational cultures: A confrontational or consensual political culture can hinder or encourage the process of domestic adaption (Ladrech, 2010: 34).

- Empowerment of actors: The redistribution of power, for example, to change agents or norm entrepreneurs, can mobilize policy-makers at the domestic level to initiate change (Ladrech, 2010: 34; Börzel and Risse, 2009: 11).

- Learning: A reassessment of domestic policy-agendas can lead to changed goals and preferences, which will then be in line with EU policy proposals or legislation (Ladrech, 2010: 34).

In short: *'the mechanisms of change, or how the EU causes change, are varied, but depend on some degree of misfit between the EU and domestic institutions and policies. Whether adaption/change occurs in response to the pressure that occurs between the two levels is dependent on the presence or absence of mediating factors, factors that vary according to each member state, thus making any sweeping generalizations about convergence or homogenizations of member states' domestic structure unfounded'* (Ladrech, 2010: 35).
A top-down EU impact on Member States is also dependent on the means by which the EU tries to achieve its goals. In terms of EU policies, the impact can be direct (hard EU policy) or indirect (soft EU policy). With respect to the former, the majority of EU policy falls under what could be labelled positive integration. The EU derives a policy template from inter-governmental negotiation (involving the Commission and the European Parliament) from which the Member States are obliged to download and implement the resulting legislation (Ladrech, 2010: 30). Whilst positive integration aims at removing barriers like market deficiencies or problems that arise from national market process and the development of a Single Market, negative integration aims at removing constraints, such as barriers to trade (Ladrech, 2010: 30 and 172-175). With respect to the latter, soft EU policy consists of non-binding forms of regulation (recommendations, declarations, resolutions) that do not fall under the authority of the Commissions whose role is closer to a facilitator and promoter of ideas and networks (Ladrech, 2010: 30 and 182).
Because the focus of attention for academic scholars has been to take the debate beyond its long-standing concerns with the bottom-up logic of integration theories (Blumer and Radaelli, 2013: 376), all of the above-mentioned

approaches have referred to a top-down logic. But Europeanization cannot be limited to a vertical, Brussels-induced top-down adaptation logic. It should be expanded theoretically to include a horizontal logic as well (Graziano and Vink, 2013: 47). Whilst little theoretical literature is available to specify the explanatory mechanisms of a horizontal approach, Radaelli (2003: 17) sums it up as follows: *'Horizontal Europeanization is a process of change triggered by the market and the choice of the consumer or by diffusion of ideas and discourses about the notion of good policy and best practice. More precisely [...] the horizontal mechanisms involve different forms of framing.'* He illustrates a vague idea of the process by claiming that, for example, regulatory competition is a *'mechanism starting with vertical prerequisites but that has horizontal consequences'* (Radaelli, 2003: 17). This results from the fact that in a more integrated Europe, civil servants, lobbyists, entrepreneurs and, most importantly, market participants and investors increasingly have cross-border contacts, enterprises and undertakings, enhancing the exchange of information and expertise (Graziano and Vink, 2013: 46-47). This logic of the interconnectedness of inter-state actors places adaptational pressure beyond a merely top-down or bottom-up logic. Ladrech (2010: 20) concludes that the status of Member State of the EU reflects a level of participation both vertically (with EU institutions) and horizontally (with other Member States) in which *'boundaries are permeable depending upon the specific linkage that is in question'*.

So far, the concept of Europeanization, its dimension and mechanisms have been introduced, but there remains the need to clarify how its outcomes can be measured empirically (Radaelli, 2003: 14). Radaelli (2003: 27-28) warns that research designs are still too rigid because they do not control for rival alternative hypothesis. Applying a rigid top-down approach can hence produce serious fallacies when the aim is to find out the domestic effects of independent variables defined at EU level. Ladrech (2010: 40-41) recommends counterfactual reasoning by questioning if a particular domestic outcome would have occurred were it not for EU-level initiatives. This can be achieved by considering alternative reasons or factors for empirical evidence. In an attempt to account for an EU effect, a combination of three approaches allows for a theoretically informed starting point (Ladrech, 2010: 41). This argument is indeed compatible with the argument by Moravcsik and Schimmelfennig (2009: 68) that a minimum of three theories, organized in a multistage model and with respect to context are needed to explain a matter such as European Integration. It is hence time, after a brief summary of the introduced trans-, supra- and international bound policy analysis frameworks, to elaborate further the frameworks, theories and models applied for the empirical research of this study.

2.2.3 Summary

Reflections upon the European Union defy definitions as to how the three dimensions reflected in the logic of politics – polity, politics, policy – interrelate

in the European Union itself and between the European Union and its Member States (Franzius and Preuß, 2012: 18-22; Meyer, 2010: 81; Leuffen, Rittberger and Schimmelfennig, 2013: 2).
There are two general sources of theoretical development dealing with European Integration. One source feeding into European Integration theory development is classical, or nationbound, policy analysis, as reflected in Political Science. The other source is rooted in the field of International Relations (Leuffen, Rittberger and Schimmelfennig, 2013: 29). Since the European Union challenges the dichotomous categories of two basic types of contemporary polities – modern nation states, on the one hand, and international organizations, on the other – the explanatory foci of the original frameworks must be broadened (Leuffen, Rittberger and Schimmelfennig, 2013: 2 and 32).

Intergovernmentalists argue that states are and will remain the dominant actors in European Integration. Member States remain the 'Masters of the Treaties', having conferred the exercise of some of their powers to the EU, whose primary treaty law merely remains a derivative legal order (BVerfGE, 1993: 89 (155); Tatham, 2013: 83; Leuffen, Rittberger and Schimmelfennig, 2013: 32 and 40). With reference to Grimm (2001: 13-22) and the three development phases of the relation of law and politics, Intergovernmentalists would likely fall into the second phase. Tapering the words of Hobbes (1929: 204) in an Intergovernmentalist fashion: The masters of the treaty are not subject to the (development of) European law or European authority, for they may, if they please, free themselves from this subjection, by repealing their membership. It is not possible for any state to be bound to itself, for who can bind, can release.
Supranationalists and theoreticians in favour of Europeanization (top-down), on the other hand, emphasize the potential of European Integration to develop into a new kind of European polity (Leuffen, Rittberger and Schimmelfennig, 2013: 62). With reference to Grimm (2001: 13-22), they would likely fall into the third category. Citing the words of Spinoza (1998: Chapter VII), Supranationalists could claim that it is not repugnant to experience for laws and the authority of the EU to be so firmly fixed, and that the Member States themselves cannot abolish them. After the example of Ulysses, the EU should administer justice without respect of Member States, if by some singular accident they are led captive by the Sirens' song.
Liberal Intergovernmentalism as a theory of European Integration assumes intergovernmental decision-making under anarchy and argues that European Integration is not a preordained movement towards a federal type Europe, but a series of pragmatic bargains among national governments based on concrete national interests, relative power, and a carefully calculated transfer of sovereignty (Moravcsik, 1998: 472; Moravcsik and Schimmelfennig, 2009: 73). Supranationalists' expectations differ substantially from Intergovernmentlists' predictions: states do not remain the dominant actors, rather supranational entrepreneurs have the capacitiy to strengthen supranational rules and expand their powers. Nonetheless, the theory initially takes a bottom-up perspective to explain – in its original formulation – the formation of a European federal polity

with state-like stable system parameters (Leuffen, Rittberger and Schimmelfen-nig, 2013: 72 and 76).

Whilst Europeanization certainly draws on basic aspects of European Integration theories, notably Supranationalism, these study approaches have not often been clearly linked (Graziano and Vink, 2013: 39). Europeanization, however, is not equal to integration theories. Whilst the latter is predominantly concerned with why and how countries pool sovereignty (Moravcsik, 1998: 73-75), Europeanization focuses on the understanding of what happens once EU institutions are in place and produce their effects (Radaelli, 2003: 8-9). *'The Europeanization framework begins from the observation that the policy output of the EU has an effect on member states, but more specifically this effect may generate pressure for a domestic response arising from the degree of misfit between the logic of the EU policies and/or policy-making process and the corresponding domestic policy and/or institution'* (Ladrech, 2010: 213).

Because the focus of attention for academic scholars has been to take the debate beyond its long-standing concerns with the bottom-up logic of integration theories (Blumer and Radaelli, 2013: 376), their approaches have primarily referred to a top-down logic. But Europeanization cannot be limited to a vertical, Brussels-induced top-down adaptation logic. It should be expanded theoretically to include a horizontal logic as well (Graziano and Vink, 2013: 47).

2.3 Applied Framework, Theory and Model

As has been stressed in the introduction, this study is a foremost empirical observation in the context of the promotion of RES-E in the European Union. The empirical analysis is based on three objects and their interrelation: the European Union and the impact of its polity implications and policy outcomes on Member States' policies (Schimmelfennig and Sedelmeier, 2005: 5; Börzel and Risse, 2000: 1; Heritier, 2001: 3); the *Member State Germany*, its polity implications with respect to European Integration, its reaction to EU adaptational pressure, and its national preferences (Moravcsik, 1998: 24-35); the *Member State Poland*, its polity implications with respect to European Integration, its reaction to EU adaptational pressure, and its national preferences; and the horizontal consequences in between the Member State Germany and Poland with respect to adaptational pressure (Radaelli, 2003: 17). These three *objects* are chose because, whilst the overall research question is concerned with the European Union and not the Member States, three empirical examples are needed, including the supranational European level and at least two neighbouring Member States, to represent the vertical and horizontal realities of European Integration. Since any progress in the promotion of renewable energy electricity can only be measured at the Member States level, the focal point needs to be the Member State. It is a matter of course that a sovereign nation is the source of the emergence and implementation of national policies. But the same is true for supranational legislation being executed by the adoption of national transposition laws. The dependent variable in this analysis are the

Member State RES-E policies of Germany and Poland, with an ultimate focus on the vertical and horizontal impacts on the RES-E policy of the latter. It has been well stressed that a mono-causal approach cannot explain the impact on the dependent variable *Member State RES-E*. There hence cannot be a single independent variable (e.g. defined at EU level) (Radaelli, 2003: 27-28). Instead, there will be a set of three decisive variables, occurring at the EU level, the national level and the inter-state level. Fig. 2 illustrates their interrelations. Since all three influencing factors for the promotion of RES-E in Member States (supranational extrinsic, national intrinsic, inter-state interdependencies) will be considered, the analytical focus aims at identifying general insights about constraints and dependencies of European RES-E promotion mechanisms.

Fig. 2 Triad-model of different impact origins on EU Member States for RES-E promotion

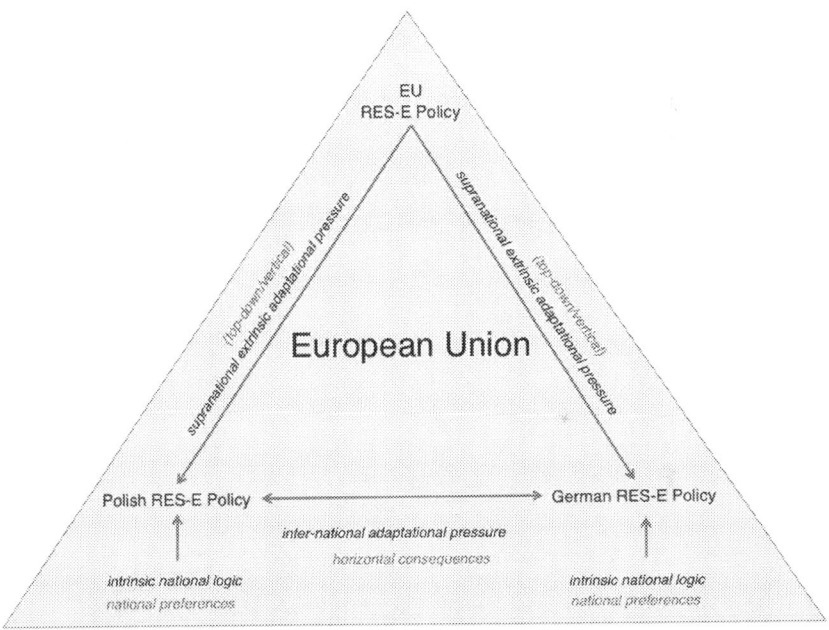

2.3.1 Applied Framework

In order to meet the multi-level realities of the subject matter, the applied framework design will be informed by various theoretical approaches. To avoid mono-causal explanations and serious analytical fallacies (Ladrech, 2010: 40-41), theoretical approaches that might seem exclusive in fact need to be combined. There is a need for a trans-, supra- and international bound policy analysis framework in order to reflect the implications of RES-E promotion in the European Union. For this study, Liberal Intergovernmentalism and Europeanization have been chosen. Both draw on sociological explanations in order to explain change. For Moravcsik (1998: 18 and 22) it is rational choices made by national leaders in response to powerful domestic constituents and domestic competition that shape the state as a unitary-actor. For Börzel and Risse (2000: 5) Europeanization is characterized by a discourse between supranational entrepreneurs and national actors. A good means to define human interaction beyond the vague idea of *'discourse between actors'* (Börzel and Risse, 2005: 5) or *'response to powerful actors'* (Moravcsik, 1998: 18) is to incorporate findings of nation bound policy analysis frameworks. The 'metatheoretical language' (Ostrom, 2007: 25) used to describe critical features of the actor-dominated multi-level environment of the European Union will hence be informed by features of the advocacy-coalition framework (Sabatier and Weible, 2007: 191), the multiple-stream framework (Cohan, March and Olsen, 1972: 2) and the network-approach (Windhoff-Heritier, 1987: 43), as described above.

It has been made clear that nation bound policy analysis is *'deeply implicated in facilitating the stability of the political institutions of the Western nation state'* (Hajer, 2003: 175). But in the context of the European Union, the policy process cannot be explained within the boundaries of sovereign polities (Hajer, 2003: 175). *'Whether and why the EU will remain more like an international organization or become more like a state is [...] the traditional core question and debate of integration theories'* (Leuffen, Rittberger and Schimmelfennig, 2013: 7). Intergovernmentalists argue that states remain the *'Masters of the Treaties'* (BVerfGE, 1993: 89 (155), Tatham, 2013: 83) and decisive actors in European Integration. Scholars of Europeanization – which will be given priority over Supranationalism in the applied framework design – argue that the emerging European polity is superior to Member States because EU law constitutes the law of the land. Both arguments have been linked to the three development phases that clarify the relation of law and politics in terms of stable system parameters, as described by Grimm (2001: 13-22).

Moravcsik and Schimmelfennig (2009: 68) and Ladrech (2010: 41) argue that a minimum of three theories, organized in a multistage model and with respect to context, are needed to explain a matter such as European Integration. Counterfactual reasoning – that is, considering alternative reasons or factors for empirical evidence (Ladrech, 2010: 40-41) – will be accounted for by incorporating both Liberal Intergovernmentalism and Europeanization into the applied theory framework. The understanding and acceptance of the applied frameworks is based on the descriptions above. However, since this is not a theoretical study, a detailed outline of each framework can be found in the sources indicated above. The overall framework of this study will be informed by sociological explanations of nation bound policy analysis, namely, the advocacy-coalition framework (Sabatier and Weible, 2007: 191), the multiple-stream framework (Cohan, March and Olsen, 1972: 2) and the network-approach (Windhoff-Heritier, 1987: 43), as well as the basic findings of Liberal Intergovernmentalism as defined by Moravcsik (1998) and Europeanization, incorporating the 'goodnes of fit' approach (Börzel and Risse, 2000) and horizontal Europeanization (Radaelli, 2003).

2.3.2 Applied Theory

A theory provides a denser and more logically coherent set of relationships, in-cluding direction and hypotheses, that self-consciously seek to explain a set of phenomena' (Sabatier, 2007: 322-323). The overall framework includes theo-ries derived from Liberal Intergovernmentalism and Europeanization. The so-ciological explanations incorporated from nation bound policy analysis do not give rise to falsifiable theory-hypotheses in the applied framework, but inform the metatheoretical language.

A simple theory hypothesis reflecting a top-down Europeanization framework would be:

The EU causes a change of national policies in accordance with EU level implications if there is some degree of misfit between EU and domestic policies, for EU law constitutes the law of the land (Ladrech, 2010: 35; Börzel and Risse, 2009: 5).

In light of the subject matter, the theoretical hypothesis could be translated as follows:

(i) The supranational RES-(E) measures and policies of the European Union, when expressed in legislation, excite change in Member States' RES-(E) policies only when there is a degree of misfit between EU and domestic RES-(E) policies that causes adaptational pressure, leading to the national promotion of RES-(E) in accordance with EU level implications.

A simple theoretical hypothesis reflecting a Liberal-Intergovernmentalist frame-work would be:

The policy of a nation state is an exclusive result of the domestic policy process that reflects the issue-specific interests of the dominant domes-tic group (national preferences) and is determined by geopolitical and economic interests (Leuffen, Rittberger and Schimmelfennig, 2013: 42; Moravcsik, 1998: 24-35).

In light of the subject matter, the theoretical hypothesis could be translated as follows:

(ii) The RES-E policy of a Member State is an exclusive result of the domes-tic policy process that reflects energy-specific economic interests of the dominant domestic group and is determined by geopolitical and economic interest.

A simple theoretical hypothesis reflecting both Liberal Intergovernmentalism and Europeanization in terms of horizontal integration would be:

The relative power of states in the European Union, reflected in asymmetri-cal interdependence (Moravcsik, 1998:3), and national preferences trig-gered by the market and the dominant domestic group (Radaelli, 203: 17) will lead to the establishment of the dominant market position, reflected in a particular market design (as a result of policy implications).

In light of the subject matter, the theoretical hypothesis could be translated as follows:

(iii) European inter-state connectedness, competition and economic pres-sure will foster the promotion of RES-E in Europe beyond pioneering Member States if a pioneering state with sufficient power in the international system establishes a dominant market position, reflected in its RES-E market design.

The applied theoretical hypothesis cannot be reduced to a mono-causal design with a single independent variable (Radaelli, 2003: 27-28). None of the theory-hypotheses above can be applied exclusively. I therefore suggest a multi-level policy dependency theory, which takes into account all of the aforementioned subordinate-hypotheses and framework constraints. In this theory design, the dependent variable will be explained by three decisive variables (which by the very logic of the design cannot be labelled independent). Taking into account the sociological explanations as well as the bottom up, top down and horizontal logic of the described frameworks, the overall theory-hypothesis reads as follows:

The implementation of polices for the promotion of renewable energy electricity (RES-E) in the Member States of the European Union is (i) fundamentally instigated by supranational renewable energy measures and policies leading to a general but not sufficient vertical Europeanization; (ii) predominantly influenced by national preferences determined by geopolitical and energy-specific economic interests; and (iii) dominantly exposed to incremental European inter-state competition leading to horizontal adaptational pressure.

Guided by the overall research question – *What are the constraints and dependencies of the European Integration of RES-(E) promotion?* – the working hypothesis for the empirical analysis translates as follows:

If (i) the European Union maintains an overall adaptational pressure by deciding on RES-E measures and policies for its Member States; (ii) pioneering Member States with sufficient power in the international system with regard to RES-E promotion (such as Germany) succeed in implementing an energy policy with a high share in electricity from renewable energy sources; and (iii) neighbouring Member States (such as Poland) are exposed to a pioneering Member State's RES-E policy and/or RES-E market design – **then,** non-pioneering Member States' RES-E measures and policies will adapt to the vertical (EU level) and horizontal (inter-state level) pressures. (These multi-level policy dependencies will encourage the harmonization of a European promotion of renewable energies and describe a reasoned path to a common promotion of renewable energies in all 28 Member States of the European Union.)

The working hypothesis implicates three explanations, which, if occurring simultaneously, will lead to an on-going RES-E promotion in Europe. In order to grasp each explanation analytically, there is need for defined explanatory factors. In this theory design, there are three subordinate hypotheses that serve as explanatory factors to investigate the proposed theory-hypothesis. Following suggestions by Ladrech (2010: 40-42), the explanatory factors proposed consider alternative reasons or factors for empirical evidence in order to allow for counterfactual reasoning. As described above, these three subordinate hypotheses reflect (i) top-down Europeanization; (ii) Liberal-Intergovernmentlism; and (iii) horizontal integration as theoretically covered by both Liberal Intergovernmentalism and Europeanization.

2.3.3 Applied Model

Since the mechanism behind the proposed multi-level policy dependency theory should, if verified, apply in the entire European Union and the interaction of its institutions with EU Member States, as well as the interactions between Member States, the empirical analysis should involve all 28 Member States. This, however, lies well beyond the scope of this study. Instead, a qualitative analysis will be conducted including three objects, as mentioned in the introduction to this chapter: the *European Union* and the impact of its polity implications and of its policy outcomes on Member State policies (Schimmelfennig and Sedelmeier, 2005: 5; Börzel and Risse, 2000: 1; Heritier, 2001: 3); the *Member State Germany,* its polity implications with respect to European Integration, its reaction to EU adaptational pressure, and its national preferences (Moravcsik, 1998: 24-35); the *Member State Poland,* its polity implications with respect to European Integration, its reaction to EU adaptational pressure, and its national preferences; and the horizontal consequences between the *Member State Germany and Poland* (Radaelli, 2003: 17). The ultimate focus lies on Poland. The model as visualized in Fig. 3 illustrates the relation of decisive and dependent variables. The arrows indicate the direction of adaptational pressure. Much has been said about the origins of multi-level adaptational pressures, which in the applied model are represented by the decisive variable. The dependent variables defined as *'change'* of Polish RES-E policies and measures, and of German RES-E policies and measures, have also been identified. Little has been said about the indicators that will be used to validate the proposed model empirically. An analysis scheme as can be seen in Fig. 4 demonstrates the approach derived for this thesis.

With respect to the top-down and bottom-up implications of the proposed framework and the theories therein, a chapte r will be designated to the EU-level, followed by Germany, Poland and Germany-Polish interdependencies. Whilst the order of the European chapter is slightly changed for the benefit of introducing the overall subject of European renewable energy policies, the basic logic of each chapter is identical, with the exception of the chapter on German–Polish interdependencies. Key to understanding the interplay of the involved entities is identifying the respective norms and obligations that are enshrined in their interacting polities. Polities, as understood in this thesis, are the constitutional or quasi-constitutional primary sources that define the freedom of action of the politics and policy dimension, and consequently the constraints of the political institutions and their legal and executive power. Fundamental constrains and dependencies are enshrined here. Consequently, all involved polities, the primary law of the European Union, as well as the constitution of Germany and Poland, will be analysed in light of European Integration and with respect to energy and renewable energy implications. Not only will legal texts serve as indicators, but also the case law judgements of the respective constitutional courts.

Fig. 3 Applied model of adaptational pressure for RES-E policies and measures on Poland

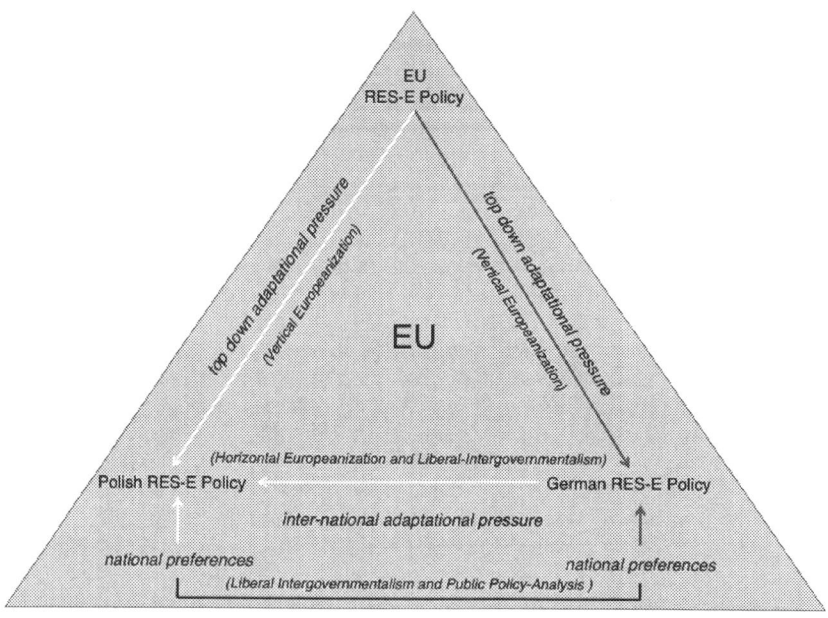

The EU chapter, followed by a European polity analysis, will describe the EU-level supranational policy for the promotion of renewable energies, as reflected in the Directive 2009/28/EC and as part of the overall Climate and Energy Package of the European Union (EC, 2008), as well as the relevant aspects of the third legislative package for an internal EU gas and electricity market (EC, 2011b: 3). In a next step, an overall implementation effectiveness evaluation of Directive 2009/28/EC will be displayed. Predominantly, it will be based on a 'renewable energy progress and biofuels sustainability' report produced for the European Commission by ECOFYS, Fraunhofer, Becker Bütter Held

(BBH), Energy Economics Group and Windrock International (Ecofys, 2012).[1]
The chapter will be concluded with an analysis.

Fig. 4 Analysis Scheme

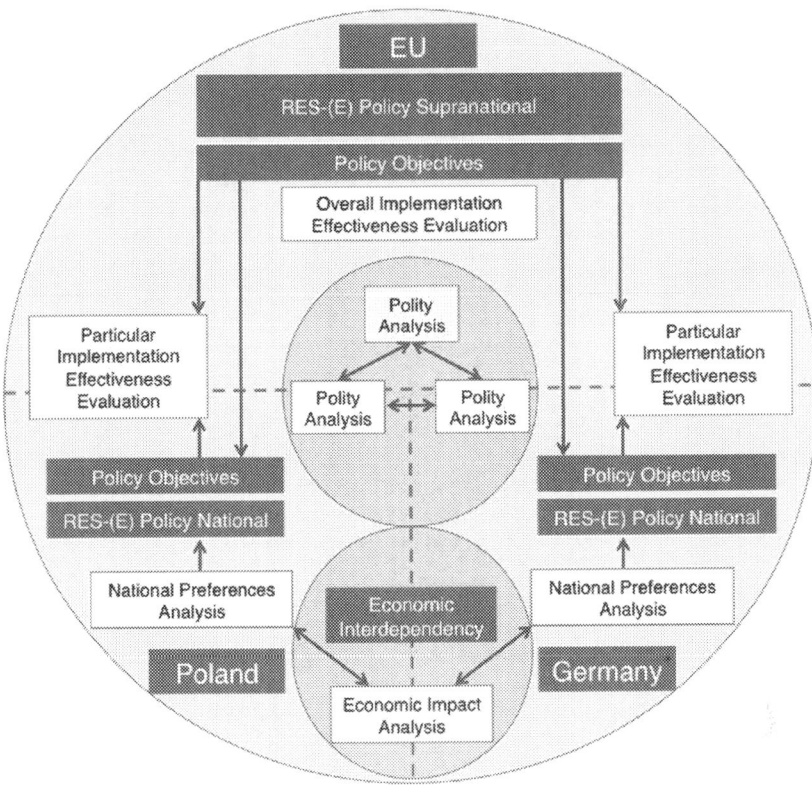

[1] Whilst this might provoke criticism for a lack of counterfactual reasoning – given that there
is only one major source – it is deemed adequate for the following reasons: First of all, empirical information about the implementation effectiveness serves as one aspect in the over all
thesis design. It lies beyond the scope of this thesis to conduct a new comprehensive implementation analysis for all European Member States to counterpoint the study in question.
Secondly, whilst additional information on certain aspects reflected in the evaluation report
might be available from different sources (e.g. statistic agencies from the Member States in
question) they do not apply the same evaluation methodology, which makes a data synopsis
illegitimate. Thirdly, the research in question has been carried out for the Commission and
informs its renewable energy progress report (EC, 2013b) and the 'commission staff working
document' (EC, 2013c). It therefore not only reflects the basis for official evaluation statements
in the political process, but also guarantees a common analysis basis for inter-state comparability. Furthermore, it provides the elements of evidence for additoinal decision in the context
of infringement procedures for non-communication of national transposition measures and/or
non-transposition of supranational provisions.

The chapters on Germany and Poland start with an introduction to each national constitutional context in light of European Integration, taking into account energy-specific implications. In a second step, based on the same sources used for the evaluation of overall European implementation effectiveness, the particular national implementation of Directive 2009/28/EC will be evaluated. In a further step, the national renewable energy policies of Germany and Poland will be analysed in the context of their trajectories as defined in Directive 2009/28/EC (including national renewable energy action plans, national energy strategies and national energy legislation etc.). Since policy goals on the supranational level are clear and explicit, it is possible to examine national policy affects on the achievement of supranational goals as defined on the EU level (Howlett, Ramesh and Perl, 2009: 179). In addition, unique national preferences as reflected in national policy strategies and legislation will be taken into account. Consequently it will be possible to identify whether the Member State's RES-E policy has been shaped top-down (testing subordinate hypothesis (i)) or – if it is not concurrent with EU-level implications – whether or to what extinct it has been shaped by national preferences (testing subordinate hypothesis (ii)). Both chapters will be concluded by an analysis.

The chapter on German–Polish interdependencies will start with a reiteration of supranational framework policies and conditions for inter-state interconnectedness. In a next step, German and Polish RES-E deployment, described in the respective chapters on national developments, will be compared. Since the third aspect of the working hypothesis and of corresponding subordinate hypothesis (iii) emphasizes – as suggested by Liberal Intergovernmentalism – the importance of energy markets and RES-E market design, the analytical focus will change from political to economic factors. To understand economic indicators, such as electricity prices, it will be necessary to touch upon physical and technical characteristics of the energy market. The chapter will be concluded by an analysis.

The analysis scheme and the indicators used will expose the information needed to conduct a final analysis with respect to the working hypothesis, which will be evaluated in a final conclusion.

3. European Union
3.1 Developing a European Energy Policy

In 2006, the Commission of what were then European Communities published a Green Paper on a European strategy for sustainable, competitive and secure energy (EC, 2006). Based on seven findings, the paper claims that Europe has entered a new energy era: *'There is an urgent need for investment. [...] Our import dependency is rising. [...] Reserves are concentrated in a few countries. [...] Global demand for energy is increasing. [...] Oil and gas prices are rising. [...] Our climate is getting warmer. [...] Europe has not yet developed fully competitive internal markets'* (EC, 2006: 3). In consequence, the Green Paper identifies six key areas, specified by follow-up questions in order to identify whether there is agreement on the need for a new energy strategy (EC, 2006: 4-5):

1. Is there agreement on the fundamental importance of a single energy market? (Competitiveness and the internal energy market)

2. What should the EU do to promote the climate-friendly diversification of energy supplies? (Diversification of the energy mix)

3. Which measures need to be taken at Community level? (Solidarity)

4. How can a common energy strategy best address climate change, balancing the objectives of environmental protection, competitiveness and security of supply? (Sustainable development)

5. What action should be taken and at what level to ensure Europe remains a world leader in energy technologies? (Innovation and technology)

6. Should there be a common external policy on energy? (External policy)

These six core questions addressing three main objectives (sustainability, competitiveness and security of supply) are met with a number of concrete proposals (EC, 2006: 17-19). These proposals include measures to (1) complete the internal gas and electricity market, (2) ensure an internal energy market guarantees security of supply and solidarity between the Member States, (3) foster a debate on different energy sources, (4) deal with the objectives so that they are compatible with the Lisbon objectives,[2] (5) develop a strategic energy technology plan, and (6) establish a common external energy policy (EC, 2006: 17-19).

---

[2] On 23rd and 24th of March 2000, the European Council *'set itself a new strategic goal for the next decade: to become the most competitive and dynamic knowledge-based economy in the world, capable of sustainable economic growth with more and better jobs and greater social cohesion'* (Council of the European Union, 2000).

The Green Paper was followed by a 2007 communication of the Commission of the European Communities proposing a European Energy Policy in response to submissions received during the consultation period on the Green Paper (EC, 2007a: 5). Therein the Commission proposes *an EU objective in international negotiations of 30% reduction in greenhouse gas emissions by developed countries by 2020 compared to 1990 [...]'* as well as *'an EU commitment now to achieve, in any event, at least a 20% reduction of greenhouse gases by 2020 compared to 1990'* (EC, 2007a: 5). In addition, the document sketches an action plan meant to trigger a *'new industrial revolution',* which should *'set pace for a new global industrial revolution'* (EC, 2007a: 5 and 20).

Key to the action plan, with respect to the three main objectives identified in the Green Paper (EC, 2006: 17-18), is the creation of an internal market and a long-term target for the promotion of renewable energies (EC, 2007a: 6 and 12). Requirements for an internal energy market include, for example, unbundling (where vertically integrated companies remain owners of network assets but without responsibility for their operations, and where network companies are wholly separate from supply and generation companies (EC, 2007a: 7)), harmonised energy regulators, and discrimination-free transparent markets (EC, 2007a: 7-10). The Commission furthermore proposes a binding target for renewable energies in the overall energy mix of up to 20% by 2020, suggests reviewing the emission trading system (EU ETS) to reduce carbon emissions, and calls for an ambitious programme of energy efficiency measures (EC, 2007a: 11-13).

Simultaneously, in 2007, the Commission published the communication *'Limiting Global Climate Change to 2 degrees Celsius. The way ahead for 2020 and beyond.'* Here, it explicitly addressed the Spring 2007 European Council, aiming to decide on an integrated and comprehensive approach for a European energy and climate change policy (EC, 2007b: 2). In response, the Presidency of the Council of the European Union (2007) concluded in its May 2007 Cover Note that the Council should adopt a comprehensive energy Action Plan, as envisioned in the Commissions communication *'An Energy Policy for Europe'* (EC, 2007a), fully respecting *'Member States' choice of energy mix and sovereignty over primary energy sources'* (Council of the European Union, 2007: 11), yet underpinned by solidarity of Member States in terms of security of supply, ensuring competitive economies with affordable energy and the promotion of environmental sustainability, and combating climate change (Council of the European Union, 2007: 11 and 13). The council stresses the need to increase energy efficiency in the EU up to 20% compared to projections for 2020 (Council of the European Union, 2007: 20). It furthermore reaffirms the long-term commitment to renewable energy development by endorsing a binding target of a 20% share of renewable energies in the EU's overall energy consumption by 2020, with a 10% binding target of biofuels share in the EU's transport consumption by 2020. The Council calls for an overall framework for renewable energies by establishing a new comprehensive directive on the use of all renewable energy resources (Council of the European Union, 2007: 22). Furthermore, the Council underlines the central role of the EU Emission Trading Scheme in providing a market-based and cost-effective means to reduce GHG emissions (Council of the European Union, 2007: 22).

Based on the same 2007 Commission communication, the European Parliament adopted, in February 2007, a resolution on climate change. The Parliament (EP, 2007a: Point 18) *'stresses that energy policy is a crucial element of the EU global strategy on climate change and that diversification of the renewable energy resources and a switch to the most energy-efficient technologies has great potential for emission reduction while ensuring less energy dependency on external sources'*. Whilst it considers the huge energy efficiency potential with regard to emission reduction, the Parliament calls on the Commission and Member States to consider reduction targets above the 20% proposed by the Commission (EP, 2007a: Point 20). It notes the proposed binding target for 20% renewable energy in the energy mix by 2020 to be a good starting point, but deems a 25% target more appropriate (EP, 2007a: Point 27). It furthermore notes the absence of binding sectoral renewable targets (EP, 2007a: Point 28) and repeats its proposal to revise the Emission Trading Scheme (EP, 2007a: Point 15).
In response to the European Council's agreement (Council of the European Union, 2007) and the European Parliament's resolution (EP, 2007a), the European Commission came up with a *'20 20 by 2020 - Europe's climate change opportunity'* strategy (EC, 2008) to translate the EU's political agenda into action (EC, 2008: 3). By *'respecting the principles set out by the European Council'* (EC, 2008: 4) the Commission defines key principles for specific legislative proposals (EC, 2008: 4-5):

- Targets must be met to show the EU's seriousness of intent by adopting proposals with mechanisms for monitoring and compliance;

- Effort required must be fair by accounting for different Member States' starting points and circumstances;

- Costs must be limited to guarantee competitiveness, employment and social cohesion;

- The EU must be prepared to take measures beyond 2020 to make deeper cuts in greenhouse gases;

- The EU must promote international agreement to cut greenhouse gas emis-

Based on six legislative proposals by the Commission, Members of the European Parliament, before the first-reading vote on 17th December 2008, reached an informal agreement with the French Presidency of the Council of the European Union to give backing to the Climate Change Package. It falls under the co-decision procedure or, ever since the entry into force of the Lisbon Treaty, the ordinary legislative procedure (see Text Box:1), making the European Parliament and the Council equal co-legislators (EP, 2008: 1).

Text Box 1: Ordinary Legislative Procedure pursuant to Article 294 TFEU (OJoEU, 2008: 173-175)

Article 294
(ex Article 251 TEC)

1. Where reference is made in the Treaties to the ordinary legislative procedure for the adoption of an act, the following procedure shall apply.

2. The Commission shall submit a proposal to the European Parliament and the Council.

First reading

3. The European Parliament shall adopt its position at first reading and communicate it to the Council.

4. If the Council approves the European Parliament's position, the act concerned shall be adopted in the wording which corresponds to the position of the European Parliament.

5. If the Council does not approve the European Parliament's position, it shall adopt its position at first reading and communicate it to the European Parliament.

6. The Council shall inform the European Parliament fully of the reasons which led it to adopt its position at first reading. The Commission shall inform the European Parliament fully of its position.

Second reading

7. If, within three months of such communication, the European Parliament:

(a) approves the Council's position at first reading or has not taken a decision, the act concerned shall be deemed to have been adopted in the wording which corresponds to the position of the Council;

(b) rejects, by a majority of its component members, the Council's position at first reading, the proposed act shall be deemed not to have been adopted;

(c) proposes, by a majority of its component members, amendments to the Council's position at first reading, the text thus amended shall be forwarded to the Council and to the Commission, which shall deliver an opinion on those amendments.

8. If, within three months of receiving the European Parliament's amendments, the Council, acting by a qualified majority:

(a) approves all those amendments, the act in question shall be deemed to have been adopted;

(b) does not approve all the amendments, the President of the Council, in agreement with the President of the European Parliament, shall within six weeks convene a meeting of the Conciliation Committee.

9. The Council shall act unanimously on the amendments on which the Commission has delivered a negative opinion.

Conciliation

10. The Conciliation Committee, which shall be composed of the members of the Council or their representatives and an equal number of members representing the European Parliament, shall have the task of reaching agreement on a joint text, by a quali

fied majority of the members of the Council or their representatives and by a majority of the members representing the European Parliament within six weeks of its being convened, on the basis of the positions of the European Parliament and the Council at second reading.

11. The Commission shall take part in the Conciliation Committee's proceedings and shall take all necessary initiatives with a view to reconciling the positions of the European Parliament and the Council.

12. If, within six weeks of its being convened, the Conciliation Committee does not approve the joint text, the proposed act shall be deemed not to have been adopted.

Third reading

13. If, within that period, the Conciliation Committee approves a joint text, the European Parliament, acting by a majority of the votes cast, and the Council, acting by a qualified majority, shall each have a period of six weeks from that approval in which to adopt the act in question in accordance with the joint text. If they fail to do so, the proposed act shall be deemed not to have been adopted.

14. The periods of three months and six weeks referred to in this Article shall be extended by a maximum of one month and two weeks respectively at the initiative of the European Parliament or the Council.

Special provisions

15. Where, in the cases provided for in the Treaties, a legislative act is submitted to the ordinary legislative procedure on the initiative of a group of Member States, on a recommendation by the European Central Bank, or at the request of the Court of Justice, paragraph 2, the second sentence of paragraph 6, and paragraph 9 shall not apply.

In such cases, the European Parliament and the Council shall communicate the proposed act to the Commission with their positions at first and second readings. The European Parliament or the Council may request the opinion of the Commission throughout the procedure, which the Commission may also deliver on its own initiative. It may also, if it deems it necessary, take part in the Conciliation Committee in accordance with paragraph 11.

Beforehand, on 11th and 12th December 2008, the Council of the European Union had invited itself to seek agreement with the European Parliament and wrote that it *'applauds the results of the work undertaken with the European Parliament, under the co-decision procedure, which has given rise to broad agreement in principle on the bulk of the four proposals in the energy/climate legislative package'* and *'applauds the full agreement on the legislative proposals concerning light vehicles' CO2 emissions and fuel quality and the Renewables Directive'* (Council of the European Union, 2008: 8). On 17th December 2008, after 11 months' legislative work and roughly four years after the Green Paper on a European strategy for sustainable, competitive and secure energy (EC, 2006) was published by the Commission, the following six legislative resolutions (stated as published in the Official Journal of the European Union) were adopted by the European Parliament (EP, 2008: 1-14):

- Regulation (EC) No 443/2009 of the European Parliament and the Council of 23 April 2009, setting emission performance standards for new passenger cars as part of the Community's integrated approach to reduce CO2 emissions from light-duty vehicles (OJoEU, 2009a: 1-15).

- Directive 2009/28/EC of the European Parliament and the Council of 23 April 2009, on the promotion of the use of energy from renewable sources and amending and subsequently repealing Directives 2001/77/EC and 2003/30/EC (OJoEU, 2009b: 16-62).

- Directive 2009/29/EC of the European Parliament and the Council of 23 April 2009, amending Directive 2003/87/EC so as to improve and extend the greenhouse gas emission allowance trading scheme of the Community (OJoEU, 2009c: 63-87).

- Directive 2009/30/EC of the European Parliament and the Council of 23 April 2009, amending Directive 98/70/EC as regards the specification of petrol, diesel and gas-oil, introducing a mechanism to monitor and reduce greenhouse gas emission, amending Council Directive 1999/32/EC as regards the specification of fuel used by inland waterway vessels, and repealing Directive 93/12/EEC (OJoEU, 2009d: 88-113).

- Directive 2009/31/EC of the European Parliament and of the Council of 23 April 2009, on the geological storage of carbon dioxide, and amending Council Directive 85/337/EEC, European Parliament and Council Directives 2000/60/EC, 2001/80/EC, 2004/35/EC, 2006/12/EC, 2008/1/EC, and Regulation (EC) No 1013/2006 (OJoEU, 2009e: 114-135).

- Decision No 406/2009/EC of the European Parliament and of the Council of 23 April 2009, on the effort of Member States to reduce their greenhouse gas emissions to meet the Community's greenhouse gas emission reduction commitments up to 2020 (OJoEU, 2009f: 136-148).

The Council adopted the climate and energy legislative package on 6th April 2009. Each above-mentioned legislative act was signed by the President of the European Parliament and the President of the Council on 23rd April 2009 and published in the Union's Official Journal on the 5th June 2009 (see also Tab. 1) (OJoEU, 2009a-f: 11-145).

The energy and climate policies of the European Union described above are subsumed within the 2020 strategy, indicating that the policies are limited to achieving targets due in the year 2020. In March 2013, the European Commission released a new Green Paper to reflect upon a new framework for climate and energy policies up to 2030 (EC, 2013a: 3). In addition, a communication on a general energy roadmap up to 2050, to *explore routes towards decarbonisation'* and to indicate the level of change needed, both structural and social, whilst keeping a competitive and secure energy sector, was released by the Commission, the Parliament, the Council, the Economic and Social Committee and the Committee of the Regions (EC, 2011a).

Tab. 1 Legislative Observatory – Climate Change Package [3]

Legislative Act	Regulation 433/2009	Directive 2009/28	Directive 2009/29	Directive 2009/30	Directive 2009/31	Decision 406/2009
Legislative proposal published	19.12.07	23.01.08	23.01.08	31.01.07	23.01.08	23.01.08
Committee referral announced in Parliament,1st reading/ single reading	17.01.08	19.02.08	19.02.08	13.03.07	19.02.08	19.02.08
Debate in Council	25.02.08	28.02.08	28.02.08	28.06.07	28.02.08	28.02.08
Debate in Council	03.03.08	03.03.08	03.03.08	30.10.07	03.03.08	03.03.08
Referral to associated committees announced in Parliament	12.04.08	10.04.08	10.04.08	-	10.04.08	10.04.08
Debate in Council	05.06.08	05.06.08	05.06.08	-	05.06.08	05.06.08
Debate in Council		06.06.08	06.06.08	-	06.06.08	06.06.08
Vote in committee,1st reading/single reading	25.09.08	11.09.08	07.10.08	27.11.07	07.10.08	07.10.08
Debate in Council	20.10.08	-	-	-	-	-
Committe report tabled for plenary, 1st reading/single reading	28.10.09	26.09.08	-	06.12.07	-	
Amended budget adopted by Council	-	09.10.08				09.10.08
Debate in Council	-	-	09.10.08	20.12.07	09.10.08	-
Committee report tabled for plenary, 1st reading/single reading	-	-	15.10.08	-	-	15.10.08
Debate in Council	-	20.10.08	20.10.08	-	-	20.10.08
Debate in Council	-	04.12.08	04.12.08	-	-	04.12.08
Debate in Council	-	08.12.08	08.12.08	-	-	08.12.08
Debate in Parliament	16.12.08	16.12.08	16.12.08	16.12.08	16.12.08	16.12.08
Decision by Parliament, 1st reading/single reading	17.12.08	17.12.08	17.12.08	17.12.08	17.12.08	17.12.08
Results of vote in Parliament	17.12.08	17.12.08	17.12.08	17.12.08	17.12.08	17.12.08
Act adopted by Council after Parliament's 1st reading	06.04.09	06.04.09	06.04.09	06.04.09	06.04.09	06.04.09
End of procedure in Parliament	22.04.09	22.04.09	22.04.09	22.04.09	22.04.09	15.04.09
Final act signed	23.04.09	23.04.09	23.04.09	23.04.09	23.04.09	23.04.09
Final act published in Official Journal	05.06.09	05.06.09	05.06.09	05.06.09	05.06.09	05.06.09

---

[3] For further Information, see Legislative Observatory of the European Parliament (http://www.europarl.europa.eu/oeil/home/home.do), procedure file: 2007/0297 (COD) for Regulation (EC) No 443/2009; 2008/0016 (COD) for Directive 2009/28/EC; 2008/0013 (COD) for Directive 2009/29/EC; 2007/0019 (COD) for Directive 2009/30/EC; 2008/0015 (COD) for Directive 2009/31/EC; 2008/0014 for Decision No 406/2009/EC.

With respect to an internal gas and electricity market and based on, *inter alia,* the Commission's Green Paper (EC, 2006) and its communication *'An Energy Policy for Europe'* (EC, 2007a), the European Parliament (2007b: Point 48) in a July 2007 resolution wrote that it *'agrees with the Commission that there is no alternative to the liberalisation process and calls on Member States to ensure full and effective transposition of existing liberalisation directives'* and *'endorses the Commission's proposal to address the malfunctioning of the market by applying both competition-based and regulatory remedies'.* It all together set out a critical and detailed resolution addressing 58 points regarding the prospects of the internal gas and electricity market, taking into account matters of transmission unbundling, cooperation between national regulators, regulated tariffs, social impact and consumer protection, interconnections, long-term contracts, electricity grid and gas networks, strategic stocks, transparency, and the implementation of community legislation (EP, 2007b).

With regard to the same matter, the Presidency of the Council of the European Union (2007) concluded in its May 2007 Cover Note that a competitive, interconnected and single internal energy market would be a major benefit for the EU, and invited the Commission to come forward with relevant proposals building on existing legislation where possible (Council of the European Union, 2007: 16-17).

Following the co-decision procedure, the Commission published on 19th September 2007 five legislative proposals (stated as published in the Official Journal of the European Union):

- Regulation (EC) No 713/2009 of the European Parliament and of the Council of 13th July 2009, establishing an Agency for the Cooperation of Energy Regulators (OJoEU, 2009g: 1-14).

- Regulation (EC) No 714/2009 of the European Parliament and of the Council of 13th July 2009, on conditions for access to the network for cross-border exchanges in electricity, and repealing Regulation (EC) No 1228/2003 (OJoEU, 2009h: 15-35).

- Regulation (EC) No 715/2009 of the European Parliament and of the Council of 13th July 2009, on conditions for access to the natural gas transmission networks and repealing Regulation (EC) No 1775/2005 (OJoEU, 2009i: 36-54).

- Directive 2009/72/EC of the European Parliament and of the Council of 13th July 2009, concerning common rules for the internal market in electricity, and repealing Directive 2003/54/EC (OJoEU, 2009j: 55-93).

- Directive 2009/73/EC of the European Parliament and of the Council of 13th July 2009, concerning common rules for the internal market in natural gas, and repealing Directive 2003/55/EC (OJoEU, 2009k: 94-136).

In a first parliamentary reading on 18th June and 9th July 2008, the European Parliament adopted legislative resolutions amending the above-mentioned proposals of the Commission. However, on 9th January 2009, the Council adopted a common position rejecting some of the European Parliament amendments. After a compromise was reached between the Members of the European Parliament and the Presidency of the Council, and after a second parliamentary reading in April 2009, the Council approved the Third Energy Package on 25th June 2009 (EP, 2009: 1). Each above-mentioned legislative act was signed by the President of the European Parliament and the President of the Council on 13th July 2009 and published in the Union's Official Journal on the 14th August 2009 (see Tab. 2) (OJoEU, 2009i-k: 14-132).

The deadline for the implementation of the third legislative internal energy package is the year 2014. A communication of the Commission *'Making the internal energy market work'* reveals, however, that the EU is not on track to meet this deadline (EC, 2012a: 2). The European Parliament concludes that, despite the fact of common rules *'for an internal market in energy, the market remains fragmented due to insufficient interconnections between national energy networks and to the suboptimal utilisation of existing energy infrastructure. However, Union-wide integrated networks [...] are vital for ensuring a competitive and properly functioning integrated market [...] and [the] integration of distributed renewable energy sources [...]'* (EP, 2013: 11). As a result, the President of the European Parliament and of the Council signed Regulation (EU) No 347/2013 of the European Parliament and the Council of 17th April 2013 on guidelines for trans-European energy infrastructure, repealing Decision No 1364/2006/EC, and amending Regulations (EC) No 713/2009, (EC) No 714/2009 and (EC) No 715/2009 (see Tab.2) (OJoEU, 2013a: 39).

Tab. 2 Legislative Observatory (EP, 2014) – Third Energy Package [4]

Legislative Act	Regulation 713/2009	Regulation 714/2009	Regulation 715/2009	Directive 2009/72	Directive 2009/73	Regulation 347/2013
Legislative proposal published	19.09.07	19.09.07	19.09.07	19.09.07	19.09.07	19.10.11
Committee referral announced in Parliament,1st reading/single reading	11.10.07	11.10.07	11.10.07	11.10.07	11.10.07	15.11.11
Debate in Council	28.02.08	28.02.08	28.02.08	28.02.08	28.02.08	24.11.11
Debate in Council	03.03.08	03.03.08	03.03.08	30.10.07	03.03.08	-
Vote in committee, 1st reading/single reading	28.05.08	28.05.08	28.05.08	06.05.08	05.06.08	18.12.12
Committee report tabled for plenary, 1st reading/single reading	05.06.08	05.06.08	-	19.05.08	-	08.02.13

Legislative Act	Regulation 713/2009	Regulation 714/2009	Regulation 715/2009	Directive 2009/72	Directive 2009/73	Regulation 347/2013
Debate in Council	06.06.08	06.06.08	06.06.08	06.06.08	06.06.08	-
Committee report tabled for plenary, 1st reading/single reading	-	-	13.06.08	-	13.06.08	-
Debate in Parliament	17.06.08	17.06.08	08.07.08	17.06.08	08.07.08	11.03.13
Decision by Parliament, 1st reading/single reading	18.06.08	18.06.08	09.07.08	18.06.08	09.07.08	12.03.13
Act adopted by Council after Parliament's 1st reading	-	-	-	-	-	21.03.13
Results of vote in Parliament	18.06.08	-	-	18.06.08	-	-
Council position published	09.01.09	09.01.09	09.01.09	09.01.09	09.01.09	-
Committee referral announced in Parliament, 2nd reading	15.01.09	15.01.09	15.01.09	15.01.09	15.01.09	-
Vote in committee, 2nd reading	31.03.09	31.03.09	31.03.09	31.03.09	31.03.09	-
Committee recommendation tabled for plenary, 2nd reading	03.04.09	02.04.09	03.04.09	02.04.09	03.04.09	-
Debate in Parliament	21.04.09	21.04.09	21.04.09	21.04.09	21.04.09	-
Decision by Parliament, 2nd reading	22.04.09	22.04.09	22.04.09	22.04.09	22.04.09	-
Results of vote in Parliament	-	22.04.09	22.04.09	-	22.04.09	-
Act approved by Council, 2nd reading	25.06.09	25.06.09	25.06.09	25.06.09	25.06.09	-
End of procedure in Parliament	09.07.09	09.07.09	09.07.09	09.07.09	09.07.09	17.04.13
Final act signed	13.07.09	13.07.09	13.07.09	13.07.09	13.07.09	17.04.13
Final act published in Official Journal	14.08.09	14.08.09	14.08.09	14.08.09	14.08.09	25.04.13

[4] For further Information, see Legislative Observatory (EP, 2014) of the European Parliament procedure file: 2007/0197(COD) for Regulation (EC) No 713/2009; 2007/0198 (COD) for Regulation (EC) No 714/2009; 2007/0199 (COD) for Regulation (EC) No 715/2009; 2007/0195 (COD) for Directive 2009/72/EC; 2007/0196 (COD) for Directive 2009/73/EC; 2011/0300(COD) for Regulation (EC) No 347/2009.

3.1.1 Summary

The overall EU-level energy policies and measures with relevance to, *inter alia,* RES-E promotion in the period up to 2020 are clustered in what are called the Climate Change Package and the Third Energy Package (EP, 2008: 1; EP, 2009: 1). They are comprised of eleven legislative acts, out of which six belong to the Climate Change Package, and five belong to the Third Energy Package (see above).

The six pieces of the Climate Change Package legislation aim at: (1) setting emission performance standards for new passenger cars registered in the Community (Regulation (EC) No 443/2009; OJoEU, 2009a: 1); (2) establishing a framework for the promotion of energy from renewable sources by setting mandatory national targets for the overall share of energy from renewables in the gross final energy consumption, and for the share of renewable sources in transport, with an overall Europe-wide target of 20% in gross final energy consumption and 10% in transport (Directive 2009/28/EC; OJoEU, 2009b: 17 and 27); (3) improving the greenhouse gas emission allowance trading scheme EU-ETS (Directive 2009/29/EC; OJoEU, 2009c: 63); (4) establishing minimum specifications for petrol and diesel fuels for use in road and non-road applications (excluding waterway vessels when at sea) and a reduction target of life cycle GHG emissions (Directive 2009/30/EC; OJoEU, 2009d: 88 and 93); (5) establishing a framework for safe geological storage of carbon dioxide (Directive 2009/31/EC; OJoEU, 2009e: 119); and (6) contributing to Member States' greenhouse gas emission reduction commitments up to 2020 (20% for the EU as a whole) (Decision No 406/2009/EC; OJoEU, 2009f: 140).

The five pieces of legislation of the Third Energy Package aim at: (1) establishing an Agency for the Cooperation of Energy Regulators (Regulation (EC) No 713/2009; OJoEU, 2009g: 4); (2) setting fair rules for cross-border exchanges in electricity and facilitating the emergence of a wholesale electricity market (Regulation (EC) No 714/2009; OJoEU, 2009h: 17-18); (3) setting non-discriminatory rules for access conditions to natural gas transmission systems and to LNG facilities, as well as facilitating the emergence of a wholesale gas market (Regulation (EC) No 715/2009; OJoEU, 2009i: 39); (4) establishing common rules for the generation, transmission, distribution and supply of electricity (Directive 2009/72/EC; OJoEU, 2009j: 62); and (5) establishing common rules for the generation, transmission, distribution, supply and storage of natural gas as well as LNG, biogas, gas from biomass and other types of gas (Directive 2009/73/EC; OJoEU, 2009k: 101).

Since the EU is not on track to meet its 2014 deadline for the completion of the internal energy market (EC, 2012a: 2), the European Parliament and the Council agreed in 2013 to lay down guidelines for a timely development of a trans-European energy infrastructure and to address identification of regional energy projects of common interest in the areas of electricity, gas, oil, and carbon dioxide (Regulation (EU) no 347/2013; OJoEU, 2013a: 39 and 65).

In light of this study, Directive 2009/28/EC and – to a certain extent – Directive

2009/72/EC and Regulation No 347/2013 are deemed especially relevant. They will hence be analysed in greater detail in the following sections.

3.2 Directive 2009/28/EC

The Directive 2009/28/EC is the key legislative act at EU level promoting renewable energies in the Union. As its title, *'on the promotion of use of energy from renewable sources and amending and subsequently repealing Directives 2001/77/EC and 2003/30/EC'*, indicates, it is based on two preliminary legal acts (EP, 2008: 11). Whilst the objective of Directive 2001/77/EC was the promotion of electricity produced from renewable energy sources (OJoEC, 2001: 33), Directive 2003/30/EC was designed to promote biofuels or other renewable fuels, to replace diesel and petrol as main fuel for transport (OJoEU, 2003a: 44). In comparison, Directive 2009/28/EC aims at increasing the share of energy from renewable energy sources[5] in the Union's gross final consumption as a sum of gross final electricity consumption, gross final consumption for heating and cooling, and final consumption in transport (OJoEU, 2009b: 29).

Article 5 of the Directive in question sets legally binding targets and measures for each Member State in order to reach, in accordance with the 20-20 by 2020 strategy (EC, 2008), a share of at least 20% renewable energies in the Union's gross final energy consumption by 2020 (OJoEU, 2009b: 28). The national overall targets are defined in Annex I of the Directive (see Tab. 3), whereas the indicative target takes into account the year 2005 as its starting point (OJoEU, 2009b: 18 and 46).

Tab. 3 National overall RES targets as defined in Annex I of Directive 2009/28/EC (OJoEU, 2009b: 45)

Country	Share of RES in gross final energy consumption by 2005	Shares of RES in gross final energy consumption by 2020
Belgium	2,2%	13%
Bulgaria	9,4%	16%
Czech Republic	6,1%	13%
Denmark	17%	30%
Germany	5,8%	18%
Estonia	18%	25%
Ireland	3,1%	16%
Greece	6,9%	18%

---

[5] Renewable energy sources, as defined in Directive 2009/28/EC, are non-fossil fuels, including wind, solar aerothermal, geothermal, hydrothermal and ocean energy, as well as hydropower, biomass, landfill gas, sewage treatment plant gas and biogases (OJoEU, 2009b:27).

Country	Share of RES in gross final energy consumption by 2005	Shares of RES in gross final energy consumption by 2020
Spain	8,7%	20%
France	10,3%	23%
Italy	5,2%	17%
Cyprus	2,9%	13%
Latvia	32,6%	40%
Lithuania	15%	23%
Luxembourg	0,9%	11%
Hungary	4,3%	13%
Malta	0%	10%
Netherlands	2,4%	14%
Austria	23,3%	34%
Poland	7,2%	15%
Portugal	20,5%	31%
Romania	17,8%	24%
Slovenia	16%	25%
Slovak Republic	6,7%	14%
Finland	28,5%	38%
Sweden	39,8%	49%
United Kingdom	1,3%	15%

Whilst the Directive does not prescribe a definite design for measures such as supporting schemes for RES-E promotion, Member States' implementation, as pointed out in Article 3(2), must lead to designs effective to fulfil or exceed the indicative trajectories (OJoEU, 2009b: 28). A non-exhaustive definition of possible instruments, schemes or mechanisms is laid out in letter 'k' of Article 2, referring to *'investment aid, tax exemptions or reductions, tax refunds, renewable energy obligation support schemes including those using green certificates, and direct price support schemes including feed-in tariffs and premium payments'* (OJoEU, 2009b: 27).

According to Article 4, each Member State is obliged, based on a template prepared by the Commission, to adopt a National Renewable Energy Action Plan (NREAP) to communicate Member States' national targets for transport, electricity, heating and cooling (OJoEU, 2009b: 28). The action plan includes an overview of all national policies and measures to promote the use of energy from renewable sources, an assessment of the total contribution expected from each renewable energy technology to meet the indicative interim trajectories (as defined in the NREAPs), and the binding overall targets, as well as an assessment of the contribution expected from energy efficiency and energy saving measures with respect to interim trajectories and binding overall targets (EC, 2009: 4-5).

Complementary to the NREAPs, the Directive Article 4(3) calls on Member States to prepare a forecast document for the Commission six months before the NREAP, and (Article 22) to prepare a report on the progress of the promotion of renewable energies by 31st December 2011, and every two years thereafter (OJoEU, 2009b: 28 and 41). In particular, Article 22(3) demands that Member States outline in their first report whether they (a) intend to establish a single administrative body processing authorisation, certification and licensing for renewable energy application; (b) provide for an automatic approval of planning and permit applications; or (c) indicate geographical locations suitable for exploitation of energy from renewable sources (OJoEU, 2009b: 42).

With respect to RES-E promotion, Directive 2009/28/EC prepares for Member States to cooperate on joint projects (including heating or cooling) (Article 7) and joint projects with one or more third countries, if the electricity is consumed in the Community (OJoEU, 2009b: 30-31). Furthermore, access to and operation of electricity grids (Article 16) must meet the demands of an increasing RES-E production. In consequence, Member States are responsible for taking appropriate measures for the development of transmission and distribution infrastructure, storage facilities and the secure operation of the electricity system (OJoEU, 2009b: 35). Of particular relevance for the promotion of RES-E are the transparent and non-discriminatory criteria (Article 16(2) a, b and c) leading to a guarantee of transmission and distribution of electricity produced from RES, priority access or guaranteed access to the grid-system for electricity produced from RES, and priority given to generating installations using RES when dispatching electricity generating installations, as long as the secure operation of the national electricity system permits it (OJoEU, 2009b: 35).

Administrative procedures concerning national rules for authorisation, certification and licensing procedures for plants producing electricity from RES and transmission and distribution of RES-E are outlined in Article 13 of the Directive (OJoEU, 2009b: 32). All together, the Article calls for a comprehensive, transparent, proportionate, discrimination-free and simple authorisation process, which should be limited in time and facilitate smaller, decentralised projects (OJoEU, 2009b: 33; EC, 2013b: 7).

Details outlined in the Directive concerning a 10% share of RES in the final consumption of energy in transport in each Member State (Article 3(4)), including sustainability criteria for biofuels and bioliquids, can be found starting Article 17 (OJoEU, 2009b: 28 and 36).

3.3 Regulation 714/2009 and Regulation 347/2013

Regulation (EC) No 714/2009 on conditions for access to the network for cross-border exchanges in electricity, and Regulation (EU) No 347/2013 on guidelines for trans-European energy infrastructure, are important pieces of framework legislation in the context of RES-E promotion in the European Union (OJoEU, 2009h: 15-35; OJoEU, 2013a).

The aim of Regulation 714/2009 as laid out in Article 1 is to establish fair rules for cross-border exchange in electricity by facilitating the emergence of a well functioning and transparent wholesale market for electricity, with a high level of energy security, compensation mechanisms for cross-border flows, harmonised transmission charges and principles for the allocation of capacities of interconnections between national transmission systems (OJoEU, 2009h: 17-18). Key of the Regulation (Article 4) is the establishment of a European Network of Transmission System Operators for Electricity (ENTSO-E). The Network, established ahead of the official implementation of the Regulation in December 2008, became operational in July 2009 (ENTSO-E, 2009: 8). Based on Regulation Article 6, together with the Agency for the Cooperation of Energy Regulators as established through Regulation (EC) No 713/2009 and the European Commission, ENTSO-E shall develop network codes covering areas specified as follows from Article 8 (OJoEU, 2009h: 19-21): (a) network security and reliability rules, (b) network connection rules, (c) third-party access rules, (d) data exchange and settlement rules, (e) interoperability rules, (f) operational procedures in an emergency, (g) capacity-allocation and congestion-management rules, (h) rules for trading, (i) transparency rules, (j) balancing rules, (k) rules regarding harmonised transmission tariff structures, and (l) energy efficiency (OJoEU, 2009h: 20-21).

The ENTSO-E shall furthermore adopt, as outlined in Article 8(3), (a) common network operation tools, (b) a non-binding Community-wide ten-year network development plan, (c) recommendations for the technical cooperation with third-country transmission system operators, (d) an annual work programme, (e) an annual report and (f) an annual summer and winter generation adequacy outlook (OJoEU, 2009h: 20). According to Regulation 714/2009, the Community-wide network development plan (Article 8(3) b), shall build on national investment plans and community aspects of network planning. With respect to the latter, electricity network projects of priority and of common interest have been defined in Annex I and II of Decision No 1364/2006/EC, laying down guidelines for trans-European energy networks (OJoEU, 2006a: 8-14). Furthermore, the network development plan shall take into account the needs of different system users and identify investment gaps with respect to cross-border capacities (OJoEU, 2009h: 21).

Responsible for monitoring the ENTSO-E and especially the implementation of network codes is the Agency for the Cooperation of Energy Regulators (OJoEU, 2009h: 21). The Agency was launched in practice in March 2011, two years after its establishment in theory in Article 1 of the Regulation (EC) No 713/2009, establishing an Agency for the Cooperation of Energy Regulators (ACER, 2013: 2; OJoEU, 2009g: 4). Since Regulation (EU) No 1227/2011 came into force, on wholesale energy market integrity and transparency, establishing *'rules prohibiting abusive practices affecting wholesale energy markets which are coherent with the rules applicable in financial markets and with the proper functioning of those wholesale energy markets whilst taking into account their specific characteristics'*, the Agency is also responsible for overseeing the wholesale energy market by detecting and preventing trading based on inside information and market manipulation (OJoEU, 2011: 5 and 9; ACER, 2013: 2). The European Parliament and the Council of the Union conclude that, despite Directive 2009/72/EC and Directive 2009/73/EC of the Third Energy Package, concerning common rules for the internal energy market, the market remains fragmented owing to insufficient interconnections between national networks (OJoEU, 2013a: 40). Furthermore, the ten-year network development plan published by ENTSO-E, including the trans-European energy networks for electricity framework (TEN-E) as defined in Decision No 1364/2006/EC (OJoEU, 2006a: 8-14), fails its intention by lacking 'vision, focus, and flexibility to fill identified infrastructure gaps' (OJoEU, 2013a: 39). Hence Regulation (EU) No 347/2013 of the European Parliament and the Council of 17 April 2013, on guidelines for trans-European energy infrastructure and repealing Decision No 1364/2006/EC, and amending Regulations (EC) No 713/2009, (EC) No 714/2009 and (EC) No 715/2009, were adopted (OJoEU, 2013a: 39).

The scope of this Regulation is to deliver guidelines for a timely development and interoperability of priority corridors and areas of trans-European energy infrastructure (OJoEU, 2013a: 44). The Regulation (a) addresses the identification of projects of common interest, (b) facilitates timely implementation of projects, (c) provides rules and guidance for cross-border allocation of costs, and (d) determines the conditions of eligibility for projects of common interest requiring financial assistance by the Union (OJoEU, 2013a: 45).

With respect to electricity, four priority electricity corridors are defined in Annex I of Regulation 347/2013 (OJoEU, 2013a: 62). These are:

(1) Northern Seas offshore grid (NSOG) concerning Member States: Belgium, Denmark, France, Germany, Ireland, Luxemburg, Netherlands, Sweden, United Kingdom.

(2) North–South electricity interconnections in Western Europe (NSI West Electricity) concerning: Austria, Belgium, France, Germany, Ireland, Italy, Luxembourg, Netherlands, Malta, Portugal, Spain, United Kingdom.

(3) North–South electricity interconnections in Central Eastern and South Eastern Europe (NSI East Electricity) concerning: Austria, Bulgaria, Croatia, Czech Republic, Cyprus, Germany, Greece, Hungary, Italy, Poland, Romania, Slovakia, Slovenia.

(4) Baltic Energy Market Interconnection Plan in electricity (BEMIP Electricity) concerning: Denmark, Estonia, Finland, Germany, Latvia, Lithuania, Poland, Sweden.

In case a project of common interest – identified to implement priority corridors – encounters implementation difficulties, Article 6 of the Regulation provides for appointing a European coordinator to promote the projects by consulting concerned stakeholders, obtaining necessary permits and, if need be, ensuring strategic support by the Member States involved. The coordinator is furthermore obliged to submit a report to the Commission on the progress of the projects in question (OJoEU, 2013a: 45 and 49).

3.3.1 Summary

The Directive 2009/28/EC, on the promotion of renewable energies, and Regulation (EC) No 714/2009 on conditions for access to the network for cross-border exchanges in electricity, both within the Climate Change and Third Energy Packages, together with Regulation (EU) No 347/2013, on guidelines for trans-European energy infrastructure and repealing Decision No 1364/2006/EC and amending Regulations (EC) No 713/2009, (EC) No 714/2009 and (EC) No 715/2009 account, if implemented accordingly, for a legislative package with substantial EU-level impact on Member States.

Directive 2009/28/EC defines legally binding targets and measures for each Member State in order to reach a share of at least 20% renewable energies in the Union's gross final energy consumption by 2020 (OJoEU, 2009b: 28). Each Member State is obliged to adopt a National Renewable Energy Action Plan (NREAP) to communicate its national targets for transport, electricity, heating and cooling (OJoEU, 2009b: 28). In the context of RES-E (Article 16(2) a, b and c), Member States have to implement legislation in order to guarantee the transmission and distribution of electricity produced from RES, guarantee priority access to the grid-system for electricity produced from RES, and guarantee priority for generating installations using RES when dispatching electricity generating installations, as long as the secure operation of the national electricity system permits it (OJoEU, 2009b: 35). Transposition of this Directive was due on 5th December 2010 (OJoEU, 2009b: 44). Concerning the Third Energy Package, ownership unbundling – that is, separating the operation of gas and electricity networks from the business of providing gas or generated power, and creating a wholesale integrated energy market with a high level of network interconnectedness and harmonised supranational network codes – describes the core of the reform (EP, 2009: 3; OJoEU, 2009h: 17-18; OJoEU, 2013a: 62). With respect to the latter two aspects, the legislation in question provides for the establishment of a European Network of Transmission System Operators for Electricity (ENTSO-E), an Agency for the Cooperation of Energy Regulators, and guidelines for a timely development of trans-European energy infrastructures (OJoEU, 2009g: 4; OJoEU, 2009h: 21; OJoEU, 2013a: 44). The Regulations are binding in their entirety and directly applicable in all Member States (OJoEU, 2009g: 14; OJoEU, 2009h: 28; OJoEU, 2013a: 61).

3.4 EU Constitutional Context in light of EU Renewable Energy Policy

In order to understand the likelihood that the EU Climate Change Package and Third Energy Package have an impact on Member States, it is necessary to understand the European judicial framework for the adopted secondary provisions at the supranational level. The set of problems in question should be linked at this point to the overall framework of this study, namely Liberal Inter-governmentalism (Moravcsik, 1998) and Europeanization (Börzel and Risse, 2000; Radaelli, 2003). It furthermore should be linked to Grimm's (2001: 13-22) definitions about the relation of law and politics (see, inter alia, 2.2 Trans-, Su-pra-, and International Bound Policy Analysis Frameworks).

Based on the jurisdiction of the European Court of Justice (ECJ), there is a recognised constitutional principle of the European Union reflected in the sui generis nature of EU law, its supremacy, direct effect, pre-emption, and protection of fundamental rights (Tatham, 2013: 15). The ECJ validates the legal order of the community treaties in a constitutional mode, as opposed to interpreting them by applying a traditional international law methodology (Franzius, 2010: 39; Tatham 2013: 14). In a case law judgement of the ECJ from 1963, the Court argues that *'the Community constitutes a new legal order of international law for the benefit of which the states have limited their sovereign rights, albeit within limited fields, and the subjects of which comprise not only Member States but also their nationals. Independently of the legislation of Member States, community law therefore not only imposes obligations on individuals but is also intended to confer upon them rights which become part of their legal heritage. These rights arise not only where they are expressly granted by the treaty, but also by reason of obligations which the treaty imposes in a clearly defined way upon individuals as well as upon the Member States and upon institutions of the community'* (ECJ, 1963: II B, Franzius, 2010: 39).

Fifteen years later, after the ECJ had underlined its position in various judge-ments (Franzius, 2010: 39-41), the ECJ found rather scathing words to define the interrelations of the Union and Member States' legal order, claiming that *'in accordance with the principle of the precedence of community law, the rela-tionship between provisions of the treaty and directly applicable measures of the institutions on the on hand and the national law of the Member States on the other is such that those provisions and measures not only by their entry into force render automatically inapplicable any conflicting provisions of current national law but – in so far as they are an integral part of, and take precedence in, the legal order applicable in the territory of each of the Member States – also preclude the valid adoption of new national legislative measures to the extent to which they would be incompatible with the community provisions. Any recognition that national legislative measures which encroach upon the field within which the community exercises its legislative power or which are otherwise incompatible with the provisions of community law had any legal effect would amount to a corresponding denial of the effectiveness of obliga-tions undertaken unconditionally and irrevocably by Member States pursuant to the treaty and would thus imperil the very foundation of the Community'* (ECJ, 1978: Summary 3.; Franzius, 2010: 43). Altogether, and according, *inter alia*, to the case law judgments cited above, in little over a decade, the ECJ has laid down basic principles for the interrelation of Community law and Member State law (Tatham, 2013: 17). In consequence, the supremacy of EU law requires national judges to become Union judges. In case of conflicting national provisions, including constitutional law, judges need to apply Union law over domestic norms at any level (Tatham, 2013: 17). This, as will be shown later, does not go without objection by national constitutional courts (Franzius, 2010: 43). Since the 17th Declaration of the Final Act of the Treaty of Lisbon, *'in accordance with well settled case law of the Court of Justice of the European Union, the Treaties and the law adopted by the Union on the basis of the Treaties have primacy over the law of Member States, under the conditions laid down by the said case law'* (OJoEU, 2007a: 256), it is necessary to focus on the nature of the Treaties and their consequences for the promotion of RES-E in the Union.

The principles of the conferral of powers are regulated in Article 3(6), 4, 5 and 13(2) of the Treaty on European Union (OJoEU, 2010: 17-22; Tobler and Beglinger, 2012: 82):

- The Union shall only act where it enjoys powers given to it by the Member States (Art. 5(2) TEU, OJoEU, 2010: 18).

- In accordance with Art. 5, all competencies not conferred upon the Union remain with the Member States (Art. 4; OJoEU, 2010: 18).

- The only field entirely excluded from EU law is national security, which remains the sole responsibility of each Member State (Art. 4(2); OJoEU, 2010: 18).

There are three different types of Union competencies, which are listed in the Treaty on the Functioning of the European Union (TFEU) (Tobler and Beglinger, 2012: 83):

- Exclusive competences[6] of the Union, where Member States shall no longer act, except for the implementation of Union acts or where they are empowered by the Union (Art. 2(1);OJoEU, 2008: 50).

- Shared competences between the Union and the Member States, where Member States exercise competence to the extent the Union has not exercised its competence. Member States have lost their competence if the field is occupied by Union law (pre-emption) (Art. 2(2); OJoEU, 2008: 50; Tobler and Beglinger, 2012: 83).

- Supporting, coordinating or supplementing competencies of the Union, where Union action does not supersede Member States' competencies (Art. 2(5); OJoEU, 2008: 50).

---

[6] Areas of exclusive competence are: customs union, competition rules for functioning of internal market, monetary policy for EMU Member States, conservation of marine biological resources under common fisheries policy, and common commercial policy (Art 3 (1), OJoEU, 2008: 51).

According to Article 4 TFEU, policy areas including the internal market, the environment and energy fall within the scope of shared competencies between the Union and the Member States (OJoEU, 2008: 51).

In order to facilitate the attainments of the primary goals of the Treaties (Art. 288 TFEU), that is, to exercise the Union's competencies, the Union can adopt secondary measures identical to regulations, directives, decisions, recommendations and opinions (Tobler and Beglinger, 2012: 91; OJoEU, 2008: 171).

- Regulations are binding in their entirety and directly applicable in Member States (OJoEU, 2008: 171).

- Directives are binding regarding their goal, but leave Member States the freedom to choose how to achieve these results (OJoEU, 2008: 172).

- Decisions are binding in their entirety but can be limited to a number of addressees (OJoEU, 2008: 172).

- In terms of recommendations and opinions, the TFEU is not exhaustive, only telling us for certain that these are non-binding.recommendations and opinions, including resolutions, communications and policy papers are subsumed under the term 'soft law' (Tobler and Beglinger, 2012: 91; OJoEU, 2008: 172)

The Lisbon Treaty distinguishes between the ordinary legislative procedure and the special legislative procedure (Tobler and Beglinger, 2012: 95). Union legislative acts may, unless the Treaties provide otherwise, only be adopted on the basis of a Commission proposal (OJoEU, 2010: 25). The ordinary legislative procedure demands both the European Parliament and the Council to adopt a Commission proposal (Tobler and Beglinger, 2012: 94; OJoEU, 2008: 172). As described under Article 294 TFEU, the ordinary legislative procedure consists of up to three readings and might involve a Conciliation Committee. The Council has to approve or reject the act with a qualified majority[7] (OJoEU, 2008: 173-175). In light of the subject matter, the ordinary legislative procedure applies, as mentioned before, in the areas of the environment (Art. 192(1) TFEU), energy (Art. 194(2) TFEU) and aspects of the internal market (Art. 46, 50, 62 (2), 75(1) TFEU) (Tobler and Beglinger, 2012: 94). For a detailed listing of the procedure in the context of the Climate Change Package and Third Energy Package, see Tab. 1, Tab. 2, and Text Box 1.

---

[7] As defined in Art. 16(4) TEU, 'a qualified majority shall be defined as at least 55% of the members of the Council, comprising at least fifteen of them and representing Member States comprising at least 65% of the population of the Union' (Official Journal of the European Union, 2010: 24).

3.5 EU Legal Basis for Energy and RES-E promotion

With regard to energy policy competencies, energy and environmental policies are inextricably bound up to each other, which *'raises a number of issues concerning horizontal competency overlaps and the attendant issue of
vertical competency delimitation in terms of leeway allowed Member States to
set their own energy policies'* (German Advisory Council on the Environment,
2011: 175). In the pre–Lisbon Treaty EU there were no provisions for regulatory
authority over the energy sector. The EU shaped its energy policies through
competencies in the fields of the environment (ex Article 174, 175, 176 Treaty
establishing the European Community, TEC), the internal market (ex Article 94,
95, 96, 97 TEC) and trans-European Networks (ex Article 154, 155, 156 TEC)
(German Advisory Council on the Environment, 2011: 175; Czeberkus, 2013:
23; OJoEU, 2006b: 79-81, 116-117 and 123-125).

Text Box 2: Article 192 (ex Article 175 TEC) (OJoEU, 2008: 134-135)

Article 192
(ex Article 175 TEC)

1. The European Parliament and the Council, acting in accordance with the ordinary legislative procedure and after consulting the Economic and Social Committee and
the Committee of the Regions, shall decide what action is to be taken by the Union in
order to achieve the objectives referred to in Article 191.

2. By way of derogation from the decision-making procedure provided for in paragraph 1
and without prejudice to Article 114, the Council acting unanimously in accordance
with a special legislative procedure and after consulting the European Parliament, the Economic and Social Committee and the Committee of the Regions, shall adopt:

(a) provisions primarily of a fiscal nature;
(b) measures affecting:

— town and country planning,
— quantitative management of water resources or affecting, directly or indirectly, the avail-
ability of those resources,
— land use, with the exception of waste management;
—
(c) measures significantly affecting a Member State's choice between different energy
sources and the general structure of its energy supply.
The Council, acting unanimously on a proposal from the Commission and after consulting
the European Parliament, the Economic and Social Committee and the Committee of the
Regions, may make the ordinary legislative procedure applicable to the matters referred to
in the first subparagraph.

3. General action programmes setting out priority objectives to be attained shall
be adopted by the European Parliament and the Council, acting in accordance
with the ordinary legislative procedure and after consulting the Economic and Social Committee and the Committee of the Regions.
The measures necessary for the implementation of these programmes shall be adopted
under the terms of paragraph 1 or 2, as the case may be.

4. Without prejudice to certain measures adopted by the Union, the Member States shall finance and implement the environment policy.

5. Without prejudice to the principle that the polluter should pay, if a measure based on the provisions of paragraph 1 involves costs deemed disproportionate for the public authorities of a Member State, such measure shall lay down appropriate provisions in the form of:
— temporary derogations, and/or
— financial support from the Cohesion Fund set up pursuant to Article 177.

Based on Article 191 TFEU (ex Article 174 TEC), Union policy on the environment shall, *inter alia,* improve the quality of the environment, and promote measures at international level to deal with regional and worldwide environmental problems, in particular climate change (OJoEU, 2008: 132). Environmental policy measures require a qualified majority vote in Council and are subject to the ordinary legislative procedure. However (Article 192(2), ex Article 175 TEC): (a) provisions primarily of fiscal nature, (b) measure affecting town and country planning, quantitative management of water resources, land use (except waste management), and (c) *'measures significantly affecting a Member State's choice between different energy sources and the general structure of its energy supply'* are all subject to a special legislative procedure, where the Council acts unanimously after consulting the Parliament, the Economic and Social Committee and the Committee of the Regions (OJoEU, 2008: 133; German Advisory Council on the Environment, 2011: 175). In other words, EU measures significantly affecting Member States' choices can be adopted, if unanimity is provided for in the Council (Ballesteros, 2010: 11). Therefore, following the German Advisory Council on the Environment, Article 192(2) lays the basis for EU authority to adopt environmental policy measures, *'even in cases where such measures infringe on Member States' freedom of action'* (German Advisory Council on the Environment, 2011: 177).

Article 114 TFEU (ex Article 95 TEC) empowers the EU to adopt measures with the purpose of establishing an internal market, and has been the basis for numerous energy policy measures, in particular with respect to the Third Energy Package (Fouquet, Nysten, Held and Johnston, 2012: 7; German Advisory Council on the Environment, 2011: 177; EP, 2009: 1). The Article in question calls for the ordinary legislative procedure with a qualified majority vote in Council (Fouquet, Nysten, Held and Johnston, 2012: 7; OJoEU, 2008: 94-95). Following Fouquet, Nysten, Held and Johnston (2012: 8-9), whilst worded to sound broad, powers conferred to the Union based on Article 114 TFEU are limited. In the context of measures with budget implications or fiscal provisions, procedural safeguards for the Member States are ensured in the above-mentioned Article 192(2) TFEU or Article 113 [8] TFEU.

---

[8] Article 113 TFEU: 'The Council shall, acting unanimously in accordance with a special legislative procedure[...], adopt provisions for the harmonisation of legislation concerning turnover taxes, excise duties and other forms of indirect taxation to the extent that such harmonisation is necessary to ensure the establishment and the functioning of the internal market and to avoid distortion of competition' (OJoEU, 2008: 94).

Directive 2009/28/EC was adopted as the main Directive for RES-E promotion before the Lisbon Treaty came into force and is based on two legal bases (Czeberkus, 2013: 28). The European Parliament and the Council adopted the Directive with regard to the Treaty establishing the European Community, and in particular Article 175(1) (now Article 192(1) TFEU) and Article 95 (now Article 114(1) TFEU) in relation to Articles 17,18 and 19 of the Directive (OJoEU, 2009b: 16; Czeberkus, 2013: 28). This means that the Directive is based on Title XIX 'Environment' of TEC (now Title XX, TFEU), except for Articles 17,18 and 19, where it refers to biofuels and bioliquids. These are based on Chapter 3 TEC (still Chapter 3, TFEU), since the measures provided for have an effect on the functioning of the internal market (OJoEU, 2009b: 26; OJoEU, 2008: 94 and 132; OJoEU, 2006b: 79 and 123; Czeberkus, 2013: 28-29). The fact that the Directive, whilst setting legally binding national overall targets for the share of energy from renewable sources, has been adopted through an ordinary legislative procedure means that this particular measure is not considered as *'significantly affecting a Member State's choice between different energy sources and the general structure of its energy supply'* (OJoEU, 2009a: 16 and 46; OJoEU, 2008: 133).

Whether legal measures concerning RES promotion could be adopted under the Lisbon Treaty, as has been the case for Directive 2009/28/EC under the TEC – that is, based on TFEU Title XX 'Environment' and pursuant to TFEU 'Chapter 3' – is questionable (Czeberkus, 2013: 29). With the entry into force of the Lisbon Treaty on 1st December 2009, the EU's authority was expanded by introducing a specific energy competence under TFEU Title XXI 'Energy' (German Advisory Council on the Environment, 2011: 179; Fouquet, Nysten, Held and Johnston, 2012: 15).

Text Box 3: Article 194 (OJoEU, 2008: 134-135)

Article 194:

1. In the context of the establishment and functioning of the internal market and with regard for the need to preserve and improve the environment, Union policy on energy shall aim, in a spirit of solidarity between Member States, to:

(a) ensure the functioning of the energy market;

(b) ensure security of energy supply in the Union;

(c) promote energy efficiency and energy saving and the development of new and renewable forms of energy; and

(d) promote the interconnection of energy networks.

2. Without prejudice to the application of other provisions of the Treaties, the European Parliament and the Council, acting in accordance with the ordinary legislative procedure, shall establish the measures necessary to achieve the

objectives in paragraph 1. Such measures shall be adopted after consultation of the Economic and Social Committee and the Committee of the Regions.

Such measures shall not affect a Member State's right to determine the conditions for exploiting its energy resources, its choice between different energy sources and the general structure of its energy supply, without prejudice to Article 192(2) (c).

3. By way of derogation from paragraph 2, the Council, acting in accordance with a special legislative procedure, shall unanimously and after consulting the European Parliament, establish the measures referred to therein when they are primarily of a fiscal nature.

Whilst the scope of Article 194 understands that the internal market should preserve and improve the environment, including the development of new and renewable forms of energy (194(1)), measures taken by the Union (194(2)) *'shall not affect a Member State's right to determine the conditions for exploiting its energy resources, its choice between different energy sources and the general structure of its energy supply'* (OJoEU, 2008: 134-135).

Based on this new energy chapter, some authors conclude that Article 194 applies as a lex specialis before all other provisions, as it appears to be the most obvious legal basis for any kind of energy-related measures (Fouquet, Nysten, Held and Johnston, 2012: 15; German Advisory Council on the Environment, 2011: 180). However, disagreement exists on the meaning of Article 194(2), which does not provide for a special decision-making procedure with unanimity voting in the Council (Fouquet, Nysten, Held and Johnston, 2012: 16, Ballesteros, 2010: 11, OJoEU, 2008: 135). Whilst Article 192(1) of the environment chapter does provide for a special legislative procedure for measures *significantly affecting* a Member State's choice between different energy sources and the general structure of its energy supply, 194(2) prohibits the adoption of measures affecting a Member State's rights to determine the conditions for exploiting its energy resources. In other words, whilst pursuant to Article 192(1), EU measures *affecting* Member States can be adopted in accordance with the ordinary legislative procedure (as was the case with the RES Directive), pursuant to Article 194(1), measures *affecting* Member States cannot be adopted (Ballesteros, 2010: 12; OJoEU, 2008: 132 and 135). Whilst it is hard to imagine that Member States, with the introduction of a special energy chapter in the Lisbon Treaty, established a competence limit that would ex post take away the legal basis for pre-existing measures (e.g. Third Energy Package, Climate Package), the clause mentioned above could serve as an absolute limit on EU competence, since it remains questionable whether there is any matter in the field of energy that can be regulated without effecting Member States' rights pursuant to Article 194(2) (Czeberkus, 2013: 26; Fouquet, Nysten, Held and Johnston, 2012: 16-18; Ballesteros, 2010: 12; OJoEU: 133 and 135).

Some authors argue, however, that it remains unclear as to whether future RES promotion is to be based on Article 192(2) – as has been the case with Directive 2009/28/EC – or on Article 194(1). The precise wording of Article 194(1) refers to the *'development of new and renewable forms of energy'* as opposed to the promotion of established RES. Furthermore, measures under Article 194(1) shall be without prejudice to Article 192(2) (c), which, following the German Advisory Council on the Environment (2011: 180), only makes sense if Article 192 applies in conjunction with Article 194. In consequence *'the EU's newfound authority (under Article 194, <A/N>) [...] solely empowers it to promote the technological development of renewable energies, whereby any economically or ecologically motivated support henceforth is governed by environmental regulations* (under Article 192, <A/N>)' (German Advisory Council on the Environment, 2011: 180). *'Consequently, the EU has no authority over non-environmental energy policy measures that fall within the competence of the member states',* but can adopt RES measures pursuant to environmental regulations (German Advisory Council on the Environment, 2011: 189).

Whether or not Article 194(2) constitutes an absolute competence limit on the EU or grants Member States the opportunity to opt out of EU legislation in the context of energy, *'it is clearly possible that this provision may have an impact upon the behaviour of any Member State during discussions of Commission (legislative) proposals on various aspects of energy and environmental policy as well as on the application of EU legislation adopted in this are'* (Fouquet, Nysten, Held and Johnston, 2012: 18). It should be clearly emphasized at this point, since the European Court of Justice has not yet had the opportunity to comment on the matter, any argument with respect to Article 194(2) and Article 192(2) remains a mere interpretation of the law (Czeberkus, 2013: 26; Ballesteros, 2010: 12).

3.5.1 Summary

Based on the jurisdiction of the European Court of Justice (ECJ), a recognised constitutional principle of the European Union is reflected in the sui generis nature of EU law, its supremacy, direct effect, pre-emption, and protection of fundamental rights (Tatham, 2013: 15). As established in the 17th Declaration of the Final Act of the Treaty of Lisbon, *'in accordance with well settled case law of the Court of Justice of the European Union, the Treaties and the law adopted by the Union on the basis of the Treaties have primacy over the law of Member States, under the conditions laid down by the said case law'* (OJoEU, 2007a: 256).

With regard to energy policy competencies, energy and environmental policies are inextricably bound up to each other, which *'raises a number of issues concerning horizontal competency overlaps and the attendant issue of vertical competency delimitation in terms of leeway allowed Member States to set their own energy policies'* (German Advisory Council on the Environment, 2011: 175). Directive 2009/28/EC was adopted as the main Directive for RES-E promotion before the Lisbon Treaty came into force and is based on two legal bases (Czeberkus, 2013: 28). The European Parliament and the Council have adopted the Directive with regard to the Treaty establishing the European Community, and in particular Article 175(1) (now Article 192(1) TFEU) and Article 95 (now Article 114(1) TFEU) in relation to Articles 17,18 and 19 of the Directive (OJoEU, 2009b: 16; Czeberkus, 2013: 28).

With the entry into force of the Lisbon Treaty on 1st December 2009, the EU's authority was expanded by introducing a specific energy competence pursuant to TFEU Title XXI 'Energy' (German Advisory Council on the Environment, 2011: 179; Fouquet, Nysten, Held and Johnston, 2012: 15). It remains questionable as to whether, under the Lisbon Treaty, measures concerning RES promotion can be grounded in the same legal basis (Czeberkus, 2013: 29). Whether or not Article 194(2) TFEU constitutes an absolute competence limit on the EU or grants Member States the opportunity to opt out of EU legislation in the context of energy, *'it is clearly possible that this provision may have an impact upon the behaviour of any Member State during discussions of Commission (legislative) proposals on various aspects of energy and environmental policy as well as on the application of EU legislation adopted in this area'* (Fouquet, Nysten, Held and Johnston, 2012: 18). Nevertheless, since the European Court of Justice has not yet had the opportunity to comment on the matter, any argument with respect to Article 194(2) and Article 192(2) remains a mere interpretation of the law (Czeberkus, 2013: 26; Ballesteros, 2010: 12).

3.6 Implementation of Directive 2009/28/EC in light of RES-E

Notwithstanding the Lisbon Treaty–based uncertainties regarding the legal grounds for the EU's authority on energy, Directive 2009/28/EC on the promotion of renewable energies was adopted as part of the climate and energy legislative package on April 6th, 2009 (see also Tab. 1) (OJoEU, 2009a-f: 11-145). Until now (December 2013) the legality of the Directive has not been questioned in front of the ECJ. As has been mentioned above, it is hard to imagine that Member States, with the introduction of a special energy chapter in the Lisbon Treaty, established a competence limit that would ex post take away the legal basis for pre-existing measures (Czeberkus, 2013: 26; Fouquet, Nysten, Held and Johnston, 2012: 16-18; Ballesteros, 2010: 12). The Directive thus has to be treated pursuant to the principles of the conferral of powers regulated in Article 3(6), 4, 5 and 13(2) TEU, as well as the Union's competencies, to adopt secondary measures pursuant to Article 288 TFEU and in accordance with the 17th Declaration of the Final Act of the Treaty of Lisbon (OJoEU, 2010; OJoEU, 2008: 171; OJoEU, 2007a: 256; Tobler and Beglinger, 2012: 82 and 91).

It is about time now to focus on the impact of the Directive 2998/28/EC on Member States' RES(-E) promotion. According to Article 4 of Directive 2009/28/EC, each Member State is obliged to adopt a National Renewable Energy Action Plan (NREAP) in order to communicate Member States' national targets for transport, electricity, heating and cooling (OJoEU, 2009b: 28). Complementary to the NREAPs, the Directive (here Article 4(3)) calls on Member States to prepare a forecast document for the Commission six months before the NREAP, and (Article 22) to prepare a report on the progress of the promotion of renewable energies by 31st December 2011, and every two years thereafter (OJoEU, 2009a: 28 and 41). Based on these documents, as well as energy figures provided by Eurostat and research carried out by and for the Commission, the Commission on 27th April 2013 published a renewable energy progress report (EC, 2013b). A 'Commission Staff Working Document' accompanies the report (EC, 2013c). Both documents refer to a 'renewable energy progress and biofuels sustainability'-report produced for the Commission by ECOFYS, Fraunhofer, Becker Bütter Held (BBH), Energy Economics Group and Windrock International (Ecofys,[9] 2012). The following chapter will be based on these documents.

3.6.1 Past Progress

Between the coming into force of Directive 2009/28/EC in 2009, and 2010, all countries but Latvia, Lithuania, the Netherlands, Austria, Portugal, Slovakia and Sweden increased their overall RES share (see Tab. 4) (Ecoyfys, 2010: 18). Malta, the Netherlands and Latvia are not on track for meeting their minimum trajectory. Ireland, Cyprus, Austria and Poland have a negative deviation of the actual 2010 share from the planned 2010 share, but exceeded their minimum trajectory for 2011/2012 (Ecoyfys, 2010: 18). The Commission concludes that, since 20 Member States achieved or exceeded the NREAP's indicative target for RES share in 2010, renewable energy growth in the EU shows a solid start (EC, 2013b: 2; EC, 2013c: 3). Since 23 countries achieved or exceeded their minimum trajectory for 2011/2012, the Commission feels, unless their shares decreased in 2011 and 2012, that the Member States in question can be seen as being on track with their trajectory for 2020. The same is not true for Luxemburg and the UK, for their minimum trajectory RES share for 2011/2012 is higher than the indicative target for 2010 (Ecofys, 2012: 19; EC, 2013c: 3). It is important to bear in mind, the Commission emphasizes, that, however encouraging the findings, they are based on data from the period 2008-2010, before the financial crisis of 2008 led to a significant change in the overall economic climate, whose possible effects on the overall prospects of Member States' meeting their overall targets are not thoroughly reflected in the datasets (EC, 2013c: 3).

---

[9] Since the consortium was headed by Ecofys, it will be referred to as 'Ecofys, 2012'.

Tab. 4: Actual and planned RES Shares 2009-2011/2012 (Ecofys, 2012: 18)

Member State	EUROSTAT actual RES share 2009 [%]	EUROSTAT actual RES share 2010 [%]	NREAP indicative target RES share 2010 [%]	NREAP minimum trajectory RES share 2011/2012 [%]	Deviation of actual 2010 share from planned 2010 share [%]	Deviation of actual 2010 share from minimum trajectory 2011/2012 [%]
Belgium	4,9	5,38	3,8	4,36	41,58	23,39
Bulgaria	11,88	13,79	10,06	10,72	37,11	28,67
Czech Republic	8,67	9,35	8,3	7,48	12,7	25,06
Denmark	20,23	22,22	21,9	19,4	16,36	25,36
Germany	9,54	11	10,1	8,24	8,94	33,53
Estonia	23,01	24,32	20,9	19,4	16,36	25,36
Ireland	5,31	5,83	6,6	5,68	-11,71	2,6
Greece	8,57	9,69	8	9,12	21,12	6,25
Spain	12,84	13,83	13,6	10,96	1,71	26,21
France	12,34	32,57	32,7	34,08	-0,39	-4,42
Italy	9,13	10,43	8,05	7,56	29,55	37,95
Cyprus	5,31	5,65	6,5	4,92	-13,01	14,92
Latvia	34,34	32,57	32,7	34,08	-0,39	-4,42
Lithuania	19,96	19,72	16	16,6	23,23	18,78
Luxembourg	2,93	2,95	2,2	2,92	33,98	0,94
Hungary	8,18	8,79	7,4	6,04	18,76	45,5
Malta	0,26	0,43	1,8	2	-76,37	-78,73
Netherlands	4,17	3,77	4,2	4,72	-10,32	-20,2
Austria	30,97	30,05	30,9	25,44	-2,75	18,12
Poland	8,93	9,49	9,58	8,76	-0,99	8,28
Portugal	24,59	24,57	24,1	22	1,96	11,69
Romania	22,58	23,64	17,5	19,04	35,07	24,14
Slovenia	18,99	19,9	17,7	17,8	12,41	11,78
Slovakia	10,42	9,8	9,5	8,16	3,13	20,07
Finland	31,95	33	28,7	30,4	14,97	8,54
Sweden	49,36	49,07	43,5	41,64	12,81	17,85
UK	2,99	3,26	3	4,04	8,8	-19,21

Concerning the past progress in RES-E promotion, based on Member States' NREAP's as well as Eurostat data on 'electricity generated from renewable sources', the sources paint a more differentiated picture (see Tab. 5). Whilst a similar analysis, based on the same sources, can be found in the Ecofys report, there is a moderate difference between the data presented based on Tab. 5 and the Ecofys data, concerning the actual deviations from NREAP 2010 indicative targets for the RES-E sector. In order to present a comprehensive fundament for later analysis, both figures will be given below.

Tab. 5: Actual and planned RES-E shares (NREAP_EU27, 2010; Eurostat, 2013b)

Member State	EUROSTAT actual RES-E share 2010 [%]	NREAP indicative target RES-E share 2010 [%]	Deviation of actual 2010 share from planned 2010 share [%]
Belgium	6,9	4,8	43,75%
Bulgaria	12,5	10,6	17,9%
Czech Republic	7,5	7,4	1,35%
Denmark	32,8	34,3	-4,37%
Germany	18,1	17,4	4,02%
Estonia	10,4	1,7	511,76%
Ireland	14,9	20,4	-26,96%
Greece	13,1	13,3	-1,50%
Spain	29,8	28,8	3,47%
France	14,9	15,5	-3,87%
Italy	20,1	18,71	7,43%
Cyprus	1,4	4,3	-67,44%
Latvia	42	44,7	-6,04%
Lithuania	7,4	8	-7,50%
Luxembourg	3,8	4	-5,00%
Hungary	7,1	6,7	5,97%
Malta	0,1	0,6	-83,33%
Netherlands	9,7	8,6	12,79%
Austria	65,7	69,3	-5,19%
Poland	6,7	7,53	-11,02%
Portugal	41,2	41,4	-0,48%
Romania	30,4	27,48	10,63%
Slovenia	32,1	32,4	-0,93%
Slovakia	18,4	19,1	-3,66%

Member State	EUROSTAT actual RES-E share 2010 [%]	NREAP indicative target RES-E share 2010 [%]	Deviation of actual 2010 share from planned 2010 share [%]
Finland	27,6	26	6,15%
Sweden	56	54,9	2,00%
UK	7,4	9	-17,78%

According to Ecofys sources, in 2010, twelve Member States were able to exceed their individual 2010 target for RES-E (Ecofys, 2012: 22). Leading the group are Estonia (511% above target), Belgium (49% above target) and Bulgaria (21% above target) (Ecofys, 2012: 22 and 27). Starting from the highest positive deviation downwards, the Netherlands (13%), Romania (11%), Italy (7%), Finland (6%), Hungary (6%), Germany (4%), Spain (2%), Sweden (2%), and the Czech Republic (2%) exceeded their 2010 RES-E targets as well (Ecofys, 2012: 22 and 27).

Fifteen Member States did not achieve their RES-E targets, out of which Malta (-86%), Cyprus (-68%), Ireland (-27%), the United Kingdom (-18%), Poland (-11%), and Greece (-10%) missed them by more then 10% (Ecofys, 2012: 22 and 27). Four Member States, Denmark (-4%), France (-2%), Slovenia (-0,5%) and Portugal (-0,5%) missed their RES-E targets for 2010 by less than 5% (Ecofys, 2012: 22 and 27).

According to Tab. 5, based on Eurostat data on 'electricity generated from renewable sources' (Eurostat, 2013b) for the actual RES-E share in 2010 and Member States' NREAP's for the indicative RES-E 2010 targets (NREAP_EU27, 2010), twelve Member States were able to exceed their individual 2010 target. Leading the group are Estonia (511% above target), Belgium (44% above target) and Bulgaria (18% above target). Starting from the highest positive deviation downwards, the Netherlands (13%), Romania (11%), Italy (7%), Finland (6%), Hungary (6%), Germany (4%), Spain (2%), Sweden (2%), and the Czech Republic (1%) exceeded their 2010 RES-E target as well (Eurostat, 2013b; NREAP_EU27: 2010).

Fifteen Member States did not achieve their RES-E targets, out of which Malta (-83%), Cyprus (-67%), Ireland (-27%), the United Kingdom (-18%), and Poland (-11%), missed it by more then 10% (Ecofys, 2012: 22 and 27). Six Member States, Denmark (-4%), France (-4%), Slovakia (-4%), Greece (-2%), Slovenia (-1%) and Portugal (-0,5%) missed their RES-E targets for 2010 by less than 5%. Luxembourg missed its RES-E target by 5% (Eurostat, 2013b; NREAP_EU27: 2010). All together, the data presented in Tab. 5 allow for a slightly more positive impression in terms of Member States' deviation from the 2010 target compared to the data presented in the Ecofys report (Ecofys, 2012: 22 and 27). However, less than half the Member States (12 out of 27) were able to reach or exceed their NREAP RES-E target, whilst twenty Member States achieved or exceeded the NREAP indicative target for overall RES share in 2010 Tab. 4)

(Ecofys, 2012: 22 and 27; Eurostat, 2013b; NREAP_EU27: 2010; EC, 2013c: 3). Whilst the Commission acknowledges that *'15 Member States failed to meet their legally agreed indicative 2010 targets'* (EC, 2013c: 4), it does not comment further on the matter.

3.6.2 Administrative Barriers

As mentioned before, administrative procedures concerning national rules for authorisation, certification and licensing procedures for plants producing electricity from RES, as well as transmission and distribution of RES-E, are outlined in Article 13 of Directive 2009/28/EC (OJoEU, 2009b: 32). All together the Article calls for a comprehensive, transparent, proportionate, discrimination-free and simple authorisation process, limited in time and facilitating smaller and decentralised projects (OJoEU, 2009b: 33; EC, 2013b: 7). In addition, Article 22(3) of the Directive demands that Member States outline in their first progress report whether they (a) intend to establish a single administrative body processing authorisation, certification and licensing for renewable energy application, (b) provide for an automatic approval of planning and permit applications, or (c) indicate geographical locations suitable for exploitation of energy from renewable sources (OJoEU, 2009b: 42). Implicitly, the legislator, although just asking for these questions to be addressed, pushes for Member States to implement a 'one stop shop' as a single agency for authorisation, certification and licensing procedures, with permission automatically granted in case respective authorities do not reply to an application within a certain time frame (Ecofys, 2012: 183; OJoEU, 2009b: 42).

As early as 2011, the Commission concludes in a 'Renewable Energy: Progress towards the 2020 target' report, primarily based on the pre-Directive 2009/28/ EC legislative framework (that is, in accordance with Article 3 of Directive 2001/77/EC and Article 4(2) of Directive 2003/30/EC) that the duration of procedures and the number of authorities involved in the authorisation represent a sever barrier to the growth of the renewable electricity sector (EC, 2011c: 2 and 4-5). In January 2011, an average of over nine different authorities need to be contacted for building RES-E plants. Overall, a *'lack of specific experience in renewables of the civil servants, the non-homogenous application of laws in different regions, or even different individual cases, the lack of clarity of the administrative framework, including problems such as legal uncertainty, contradictory or unclear legal provisions, opaque procedures, excessive margins of discretion of the administration and sheer extortion and corruption'* are the reasons for administrative barriers as listed by the Commission (EC, 2011c: 5-6). These obstacles, addressed in Article 13 of Directive 2009/28/EC, still remain a key challenge in 2013, for, as the Commission concludes, *'progress in removing the administrative barriers is still limited in slow'* (EC, 2013b: 7; EC, 2013c: 8). With respect to Article 22(3) of the Directive, many Member States do not even address the specificities asked for in their reports on the progress of the promotion of renewable energies (EC, 2013b: 7-8).

The Commission's conclusions are based on an evaluation undertaken in the 'renewable energy progress' report by the Ecofys consortium (Ecofys, 2012: 183; EC, 2013c: 8-9). In order to assess Member States' progress on implementing Article 13(2) of Directive 2009/28/EC, the following methodology has been applied (Ecofys, 2012: 183; OJoEU, 2009b: 32-33 and 42-43):

- Article 22(3)a of the Directive asks for a single administrative body, therefore the authors inquire as to whether there is a so called 'one-stop-shop'. In the assessment, this was considered the preferred option.

- A next step analysed whether a project could be implemented with one single permit, or needed more than one. The assumption was that fewer permits were preferable.

- In addition, the authors assessed whether a permit could be applied for online. The underlying argument is that, pursuant to Article 13(1) e, administrative costs must be proportionate, which can be achieved by online applications, supposedly further reducing costs.

- Based on Article 13(1) a, which specifically asks Member States to define time frames for determining planning and building procedures, the next indicator of the assessment is whether there are mandatory time frames within which a decision has to be taken.

- With respect to Article 22(3) b, the authors looked at whether – if the authorising body did not responded within set time limits – permission is granted 'automatically'.

- Furthermore, the existence of any kind of special procedure for smaller projects and decentralised devices pursuant to Article 13(1) f has been taken as an indicator.

- With respect to Article 22(3) c, the authors identify whether the Member States have established some sort of open database or plan for projectors to identify geographical sites for the deployment of renewables.

- A final indicator looks at whether an additional procedure is required to access renewable support schemes, or whether this happens automatically.

Tab. 6: Assessment of administrative procedures in the Member States (Ecofys, 2012: 186-187; EC, 2013c: 9)

Member State	"One Stop Shop"?	One permit? (Nr. of permits?)	Online application for permit?	Max time limit for procedure?	Automatic permission?	Facilitated procedure for small-scale?	Identification of geographic sites?	Automatic entry into financial support scheme?	Overall assessment
Beligium	No	No	n.a.	Partly (6 mths. - 1 yr.)	No	No	n.a.	n.a.	☹
Flanders	No	Partly	n.a.	Yes (15 days - 4 mths.)	No	Yes	Yes	No	☺
Wallonia	No	Partly	n.a.	Yes (90 - 140 days)	No	Yes	Yes	No	☺
Brussels	Yes	Partly	n.a.	Yes (20 - 450 days)	No	Yes	n.a.	n.a.	☺
Bulgaria	No	No	No	No	No	Yes	Yes	Yes	☹
Czech Republic	No	No	n.a.	Yes (60 days - 72 mths.)	No	Yes	No	n.a.	☹
Denmark	Yes	Yes	n.a.	No	n.a.	Yes	n.a.	Yes	☺
Germany	Partly	Partly	Partly	Partly (? - 10 mths.)	n.a.	Yes	Yes	Yes	☺
Estonia	No	No	No	No	No	No	Yes	No	☹
Ireland	No	No	No	Partly (6 - 8 weeks)	n.a.	Yes	Yes	No	☹
Greece	Yes	No	No	Yes (n.a.)	n.a.	Yes	n.a.	n.a.	☹
Spain	No	No	n.a.	Yes (3 mths.)	Yes	Partly	n.a.	No	☹
France	No	No	Partly	Partly (? - 1 yr.)	No	Yes	n.a.	No	☹
Italy	Yes	Yes	No	Yes (30 - 90/180 days)	Partly	Yes	n.a.	No	☺
Cyprus	Yes	No	No	Yes (2 - 3 mths.)	n.a.	Yes	Yes	n.a.	☺
Latvia	No	No	No	Partly (30 - 180 days)	n.a.	n.a.	n.a.	No	☹
Lithuania	Partly	No	n.a.	Partly (10 - 30 days)	Partly	Yes	n.a.	No	☹
Luxembourg	No	No	n.a.	Partly (3 - 5,5 mths.)	n.a.	Yes	n.a.	n.a.	☹
Hungary	Yes	Partly	Partly	Yes (n.a.)	n.a.	Yes	n.a.	No	☺
Malta	No	Partly	No	Partly (4 weeks)	n.a.	Yes	n.a.	No	☹
Netherlands	Yes	Yes	Yes	Partly (6 mths.)	n.a.	Yes	Yes	No	☺
Austria	Yes	No (?)	No	No	No	Yes	No	No	☹

Poland	No	No	No	Partly (30 - 65 days)	Partly	Yes	n.a.	n.a.	☹
Portugal	Yes	Partly	Partly	Yes (120 - 250 days + 30 days for connection)	n.a.	Yes	Yes	n.a.	☺
Romania	No	No	n.a.	Partly (30 days)	n.a.	No	n.a.	No	☹
Slovenia	No	No	n.a.	No	No	Yes	n.a.	n.a.	☹
Slovakia	No	No	No	Partly (n.a.)	n.a.	Yes	Yes	n.a.	☹
Finland	No	No	n.a.	n.a.	n.a.	Yes	Yes	n.a.	☹
Sweden	Partly	Partly	Partly	Partly (n.a.)	n.a.	Yes	Yes	No	☺
UK	No	No	n.a.	Partly (1 yr.)	n.a.	Yes	Partly	No	☹

The assessment of the administrative procedures in the Member States is summarised in Tab. 6. The Commission added an 'overall assessment' cell to the table, which is otherwise based on the Ecofys analysis, where ☺ means 'advanced', ☺ means 'fair', and ☹ means 'needs improvement' (EC, 2013c: 9, Ecofys, 2012: 186-187).

It becomes apparent based on Tab. 6 that most countries have room for improvement (Ecofys, 2012: 187). Following the Commission's conclusions, no Member State can be regarded as being 'advanced' with respect to administrative procedure reforms, whilst only a few countries have achieved a 'fair' rating (EC, 2013c: 8).

Although a detailed analysis of the assessment of the administrative procedures in the Member States can be found in the indicated reports (see Ecofys, 2012; EC, 2013b; EC, 2013c), the Commission concludes: *'Generally, the concreteness and completeness of the administrative simplification measures intended and reported is very slow in all Member State reports: the quantity of permits required is often not mentioned, neither the number of authorities involved in procedures. [...] This shows a lack of coordination and poor implementation of Article 13 of the Directive 2009/28/EC which explicitly asks for improved coordination of the administrative procedures'* (EC, 2013c: 8).

3.6.3 Electricity Grid Integration

According to Article 16 of Directive 2009/28/EC, access to and operation of electricity grids must meet the demands of increasing RES-E production. In consequence, Member States are responsible for taking appropriate measures for the development of transmission and distribution infrastructure, storage facilities and the secure operation of the electricity system (OJoEU, 2009b: 35). Of particular relevance for the promotion of RES-E are the transparent and non-discriminatory criteria (Article 16(2) a, b and c) leading to a guarantee of transmission and distribution of electricity produced from RES, priority access or guaranteed access to the grid , and priority for generating installations using RES, as long as the secure operation of the national electricity system permits it (OJoEU, 2009b: 35).

Based on the 'overall assessment' indicators established by the Commission, where ☺ means 'advanced', ☺ means 'fair', and ☺ means 'needs improvement' (EC, 2013c: 9), the findings of the Ecofys consortium (Ecofys, 2012: 177, 179 and 181) are summarized in Tab. 7. The table indicates that the majority of Member States perform well in the whole procedure of RES grid integration, with 14 out of 27 countries rated 'advanced' (Ecofys, 2012: 177). However, there are eight barriers that are directly relevant for grid capacity limitations. According to Ecofys (2012: 178), those are a lack of communication or conflicts between stakeholders with regard to grid connection (barrier 1), a lack of grid capacity for the integration of RES-E production (barrier 2), limited accessibility to and limited exchange of information with regard to the electricity sector in general, and concerning cost, time and procedure for grid connection (barrier 3), long lead times and delays with regard to RES-E facility connection requests (barrier 4), no obligation of Transmission System Operator (TSO) to optimize and expand the electricity infrastructure and to connect RES-E production (barrier 5), no clarification of the distribution costs for the general development of the grid and concerning the distribution of costs (barrier 6 and 7), and no cost compensation in case of curtailment (barrier 8) (Ecofys, 2012: 176, 178 and 180).

The fact that the bulk of Member States are affected by barrier 2 and barrier 4 (for more information, see: Ecofys, 2012: 171-177) leads to the conclusion that there is an urgent need for grid extension and shorter approval procedures (EC, 2013c: 10; Ecofys, 2012: 176). When it comes to recognizing these barriers – that is, addressing the grid capacity limitations – Member States are preforming well. Whilst seven countries 'need improvement', 14 countries are 'advanced' when it comes to addressing the grid capacity limitations (Ecofys, 2012: 179).

The adoption of effective cost regulation measures aiming at a clear distribution and level of cost, as well as setting incentives for investment (barrier 6 and 7) – in other words, rules for bearing and sharing the costs of grid development – seems to be less of a priority for Member States (EC, 2013c: 10; Ecofys 2012: 181). As can be seen in Tab. 7, the rating shows eight 'advanced' countries, six countries that 'need improvement', and 13 countries that are 'fair' (Ecofys: 2012: 181).

Tab. 7: Member State rating on progress in electricity grid integration (Ecofys, 2012: 177, 179 and 181)

Member State	Rating on overall RES grid integration	Rating on addressing grid capacity limitations	Rating on rules for bearing and sharing the costs of grid developmet
Belgium	☺	😐	☺
Bulgaria	😐	😐	😐
Czech Republic	😐	☺	😐
Denmark	😐	😐	😐
Germany	☺	☺	☺
Estonia	☺	☺	☹
Ireland	☺	☺	😐
Greece	☺	☺	😐
Spain	☺	☺	😐
France	☹	☹	☹
Italy	☺	☺	☺
Cyprus	☺	☺	😐
Latvia	😐	☹	☺
Lithuania	☹	☹	😐
Luxembourg	☹	☹	☹
Hungary	☹	☹	😐
Malta	☺	☺	☺
Netherlands	☹	☹	😐
Austria	☺	☺	😐
Poland	😐	😐	☺
Portugal	😐	☺	☹
Romania	☺	☺	☺
Slovenia	☺	☺	☹
Slovakia	😐	😐	😐
Finland	☺	☺	😐
Sweden	☹	☹	☹
UK	☺	😐	☺

The analysis by Ecofys (2012: 171-182) leads to the conclusion that the majority of Member States perform well in the whole procedure of RES grid integration, whilst there is an urgent need for grid extension and shorter approval procedures. When it comes to recognising these barriers – that is, addressing the grid capacity limitations – Member States are preforming well. However, rules for bearing and sharing the costs of grid development seem to be less of a priority for Member States. The Commission concludes (EC, 2013b: 8) that *'given the longer term expectations of the growing share of EU electricity from renewable energy sources, full implementation of Article 16 of the Directive is important. The current failure to modernise the grid [...] is causing problems for the development of the internal market, technical problems related to loop flows, grid stability and growing power curtailment, and investment bottlenecks resulting from delayed connection of new producers'.*

3.6.4 Expected RES shares in 2020

In order to perform a quantitative assessment of the future deployment of RES in the European Union, Ecofys (2012: 58-60) applied the Green-X model, developed by the Energy Economic Group at Vienna University of Technology, Institute of Power Systems and Energy Economic (EEG, 2007: 1).

Text Box 4: Short characterisation of the model Green-X by EEG (EEG, 2007: 3-4)

Green-X:

Within the model Green-X, the most important RES-E (e.g. biogas, biomass, bio-waste, wind on- & offshore, hydropower large- & small-scale, solar thermal electricity, photovoltaic, tidal & wave energy, geothermal electricity), RES-H technologies (e.g. biomass – subdivided into log wood, wood chips, pellets, district heating, geothermal and solar heat) and RES-T options (e.g. traditional biofuels such as biodiesel and bioethanol, advanced biofuels as well as the impact of biofuel imports) are described for each investigated country by means of dynamic cost-resource curves. Dynamic cost curves are characterised by the fact that the costs as well as the potential for electricity generation/demand reduction can change each year. The magnitude of these changes is given endogenously in the model, i.e. the difference in the values compared to the previous year depends on the outcome of this year and the (policy) framework conditions set for the simulation year. Based on the derivation of the dynamic cost curve, an economic assessment takes place considering scenario-specific conditions like selected policy strategies, investor and consumer behaviour as well as primary energy and demand forecasts.

Within this step, a transition takes place from generation and saving costs to bids, offers and switch prices. It is worth mentioning that the policy setting influences the effective support, e.g. the guaranteed duration and the stability of the planning horizon or the kind of policy instrument to be applied.

Policies that can be selected are the most important price-driven strategies (feed-in tariffs, tax incentives, investment subsidies, subsidies on fuel input) and demand-driven strategies (quota obligations based on tradable green certificates (including international trade), tendering schemes). All the instruments can be applied to all RES technologies (and conventional options within the EU-15) separately for the various energy sectors. In addition, general taxes can be adjusted and the effects simulated. These include energy taxes (to be applied to all primary energy carriers as well as to electricityand heat) and environmental taxes on CO2 -emission as well as policies supporting demand-side measures.

As Green-X is a dynamic simulation tool, the user has the possibility to change policy and parameter settings within a simulation run (i.e. by year). Furthermore, each instrument can be set for each country individually.
(see: www.green-x.at)

The general approach used by Ecofys (2012: 59) is to conduct a model-based quantitative analysis of Member States' expected future RES deployment, both in absolute (i.e. GWh produced, MW installed) and relative (i.e. RES shares on gross demands) terms. Key assumptions regarding potential costs for RES in the Member States are taken from the Green-X database. Since the Green-X model is a policy assessment tool, Ecofys (2012: 60) defines two policy tracks that are complemented by investigations of expected demand based on Member States' NREAPs, and contrasted by actual Eurostat data. The first policy scenario is defined as a 'business as usual' case, where currently implemented RES support policies are assumed to continue. It is labelled 'Current Policy Initiative (CPI)' (Ecofys, 2012: 60). The second policy scenario takes into account planned measures as proposed by the Member State's in their progress reports, including measures to either improve the support framework for RES promotion, or to mitigate applicable non-economic barriers. In consequence, this scenario is labelled 'Current Policy Initiatives complemented by Planned Policy Initiatives (CPI+PPI) (Ecofys, 2012: 60). Whilst a detailed description of the general approach and definitions for the projected future RES progress in the EU by 2020 can be found in the respective reports, the results for the expected, planned and required RES shares for the EU Member States in 2020 are illustrated in Tab. 8 (Ecofys, 2012: 66).

According to Tab. 8, only very few countries are expected to meet their targets for 2020, and only if complementary energy efficiency measures are implemented successfully (Ecofys, 2012: 69). Latvia, Portugal, the Netherlands, Lithuania, France, Ireland, the UK, Greece, Denmark and Spain may end up with a gap higher than 7% of gross final energy demand under the CPI scenario. Compared to 2012, twelve countries may even end up with a lower RES share by 2020 than in 2012. However, these are countries with a high RES share. Consequently, a strong overall energy demand growth would reduce the future RES shares (Ecofys, 2012: 65). Taking into account the CPI+PPI scenario, only Bulgaria, Estonia, Austria and Slovakia are expected to meet their 2020 RES targets (Ecofys, 2012: 65 and 66). If currently implemented policies are taken into account (CPI scenario), Estonia and Austria are top of the list, followed by Sweden, whose performance improves if the default uncertainty range of the Green-X model is taken into account (Ecofys, 2012: 69).

Member State	Expected RES share 2020 (CPI Scenario)		Expected RES-share 2020 (CPI+PPI Scenario)		NREAP indicative target RES-share 2020	NREAP minimum trajectory RES-share 2020	Deviation of expected from planned 2020 share (CPI and CPI+PPIScenario)		Deviation of expected from minimum trajectory 2020 share (CPI and CPI+PPI Scenario)	
	Min. [%]	Max. [%]	Min. [%]	Max. [%]	[%]	[%]	Min. [%]	Max. [%]	Min. [%]	Max. [%]
Belgium	7,9	8	8,2	8,2	13	13	-39,2	-36,6	39,2	-36,6
Bulgaria	12,5	14,5	15,3	17,8	16	16	-22	11,2	-22	11,2
Czech Republic	9,1	9,4	9,3	9,5	13,5	13	-32,4	-29,4	-29,8	-26,7
Denmark	22,5	24,7	22,8	25,9	30	30	-24,9	-16,7	-24,9	-16,7
Germany	15,4	16,4	15,8	16,8	19,6	18	-21,3	-14,5	-14,3	-6,9
Estonia	24,3	25,2	24,2	25,1	25	25	-3,1	0,9	-3,1	0,9
Ireland	8,1	8,7	8,6	9,1	16	16	-49,2	-49,2	-49,2	-49,2
Greece	10,3	10,6	10,3	10,6	18	18	-42,8	-41,4	-42,8	-41,4
Spain	12,6	13,8	15,4	17,1	22,7	22	-44,6	-24,6	-37,2	-14,4
France	14,9	16,9	15,8	17,9	23	23	-35,1	-22,1	-35,1	-22,1
Italy	13,1	14	13,1	13,9	17	17	-23,1	-17,8	-25,1	-17,8
Cyprus	8,7	9,1	8,9	9,3	13	13	-33	-28,5	-33	-28,5
Latvia	21,6	24,2	26,1	29,2	40	40	-46	-26,9	-46	-26,9
Lithuania	14,1	14,6	15,6	16	24	23	-41,4	-33,2	-38,8	-30,3
Luxembourg	5,8	5,9	5,9	6,1	11	11	-47,7	-44,4	-47,7	-44,4
Hungary	8,6	9	9,2	9,6	14,7	13	-41,1	-34,7	-33,6	-26,4
Malta	4	4	4	4	10,2	10	-61	-60,6	-60,2	-59,8
Netherlands	5	5	5,1	5,1	14,5	14	-65,3	-64,5	-64,1	-63,2
Austria	32	35,7	33,2	36,6	34,2	34	-6,5	7	-5,9	7,6
Poland	8,2	9,6	9	10,7	15,5	15	-47,1	-30,7	-45,4	-28,4
Portugal	20,5	21,1	21,7	22,3	31	31	-33,8	-28	33,8	-28
Romania	19,5	21	21,2	22,7	24	24	-18,9	-5,5	-18,9	-5,5
Slovenia	21,8	21,8	22	22	25,3	25	-13,8	-12,9	-12,8	-11,9
Slovakia	11,7	12,9	13,1	14,3	14	14	-16,1	1,8	-16,1	1,8
Finland	34,8	34,9	34,8	24,8	38	38	-8,5	-8,2	-8,5	-8,2
Sweden	46,1	49,4	46,2	49,5	50,2	49	-8,2	-1,5	-6	1
UK	7,3	7,5	11,1	11,5	15	15	-51,5	-23,4	-51,5	-23,4

The negative ranking is headed by the Netherlands and Malta (deviation >60%), followed by the UK, Ireland, Luxembourg, Poland, Latvia, Spain, Greece, Lithuania and Hungary (deviation >40%). The ranking changes if the CPI+PPI scenario is taken into account. This is significantly true for the UK, Latvia, Spain and Poland (Ecofys, 2012: 69)

Tab. 9: Expected and planned RES deployment at EU level by 2012 and 2020 (Ecofys, 2012: 71)

Renewable Energy Commodities and Technology-specific deployment	Status Quo 2010	Expected deployment 2012 (CPI)	NREAP indicative target 2012	Expected deployment 2020 (CPI)		Expected deployment 2020 (CPI+PPI)		NREAP indicative target 2012	Deviation of expected from planned deployment		
				Min.	Max.	Min.	Max.		2012	2020	
										Min.	Max.
Categories	[Mtoe]	[Mtoe]	[Mtoe]	[Mtoe]	[Mtoe]	[Mtoe]	[Mtoe]	[Mtoe]	[%]	[%]	[%]
RES Electricity	56,2	62,5	64,3	77,3	77,9	87,1	87,9	104,5	-2,8	-26,1	-15,9
Biomass	8,53	9,6	8,73	12,04	12,15	14,24	4,53	14,45	9,9	-16,7	0,6
Biogas	2,1	2,53	2,92	4,68	4,68	5,5	5,24	5,5	-13,3	-15	-4,8
Geothermal	0,48	0,51	0,55	0,79	0,79	0,87	1,9	0,94	-6,9	-15,5	6,7
Hydro large-scale	25,96	26,15	25,87	26,8	26,9	26,84	26,95	27,12	1,1	1,2	0,6
Hydro small-scale	3,83	3,91	4,14	4,57	4,59	4,64	4,67	4,66	-5,5	-1,9	0,1
Photovoltaic	1,94	3,01	3	6,82	6,82	6,97	6,98	7,17	0,2	-4,9	-2,7
Concentrated Solar Power	0,06	0,08	0,4	0,12	0,12	0,73	0,74	1,72	-79,8	-92,9	-56,8
Wind onshore	12,76	15,89	17,05	17,55	17,91	18,28	18,47	30,45	-6,8	-42,4	-39,3
Wind offshore	0,53	0,81	1,61	3,66	3,66	9,09	9,1	12	-49,8	-69,5	-24,1
Tidal/Wave/Ocean	0,04	0,04	0,05	0,23	0,24	0,18	0,2	0,52	-14,2	-64,7	-54,4
RES Heating & Cooling	80,6	81	70,6	84,3	84,6	88,1	88,9	104,7	14,7	-19,5	-15,1
Biomass	72,24	72,18	60,86	74,78	75,12	77,14	77,92	81,56	18,6	-8,3	-4,5
Biogas	2,01	2,32	1,87	2,75	2,75	3	3,03	4,45	25	-38,1	-31,9
Geothermal	0,53	0,59	0,86	1,13	0,13	1,32	1,32	2,55	-32,1	-55,5	-48,4
Heat Pumps	4,3	4,24	5,12	2,87	2,88	3,15	3,17	9,88	-17,2	-70,9	-67,9

Renewable Energy Commodities and Technology-specific deployment	Status Quo 2010	Expected deployment 2012 (CPI)	NREAP indicative target 2012	Expected deployment 2020 (CPI)		Expected deployment 2020 (CPI+PPT)		NREAP indicative target 2012	Deviation of expected from planned deployment 2012	Deviation of expected from planned deployment 2020	
				Min.	Max.	Min.	Max.		2012	2020 Min.	Max.
Categories	[Mtoe]	[Mtoe]	[Mtoe]	[Mtoe]	[Mtoe]	[Mtoe]	[Mtoe]	[Mtoe]	[%]	[%]	[%]
Solar Thermal	1,49	1,67	1,91	2,73	2,73	3,47	3,47	6,28	-12,2	-56,6	-44,7
RES Transport	13,6	15	16,2	18,9	20,6	19,1	20,8	28,9	-7,8	-34,8	-28
Biofuels (1st generation)	13,57	14,97	15,36	16,71	18,41	16,91	18,64	26,43	-2,6	-26,8	-29,5
Biofuels (2nd generation)	0	0	0,88	2,16	2,18	2,16	2,18	2,5	100	-13,7	-12,7
RES Total	150,4	158,5	151,2	180,4	183,1	194,2	197,6	238,2	4,8	-24,3	-17

At the EU level and based on 'Million Tonnes of Oil Equivalent (Mtoe)' with respect to RES total or gross final energy consumption from RES, as defined pursuant to Article 2 f of Directive 2009/28/EC, the expected deviations are -24,3% to -17% in 2020 (see Tab. 9) (Ecofys, 2012: 71; OJoEU, 2009b: 27). The deviations from planned deployment in electricity produced from RES are expected to be in between -26,1% and -15,9% by 2020. The range indicated for 2020 is related to energy demand and to the different framework conditions assumed in the CPI and CPI+PPT scenario (Ecofys, 2012: 69). A positive trend can be observed for the 2012 period. Herein, the heat sector appears to be the most advanced amongst all energy sectors. Nonetheless, at the technological level, heat pumps, solar thermal collectors and mid to large-scale geothermal heating systems may need additional initiatives to meet the 2020 obligations (Ecofys, 2012: 69). The same is true for the electricity sector, where the highest improvement demands arise for concentrated solar power and ocean technologies (incl. tidal stream and wave power). The authors of the report highlight, however, that 'most important for achieving RES targets appears to improve support and in particular framework conditions for wind energy' (Ecofys, 2012: 69 and 70).

Overall, Tab. 9 paints a rather pessimistic picture of planned and/or required RES deployment in comparison with expected RES deployment in 2020. Whilst Member States are almost all expected to reach and significantly exceed their minimum trajectory targets in 2012 (see Tab. 4), almost all Member States will fail to achieve their required RES targets for 2020. This is true if no further measures or adaptations are provided for by the Member States (Ecofys, 2012:

64 and 65). Only four out of 27 countries – Bulgaria, Estonia, Austria and Slovakia – are expected to achieve a positive deviation from the planned RES share under the CPI+PPI scenario by 2020. Taking into account positive deviations from planned minimum trajectory RES share under the CPI+PPI scenario, Sweden joins the group (Ecofys, 2012: 66). However, for the majority of countries, the currently implemented RES policies appear insufficient to promote the required RES deployment. Following Ecofys (2012: 65), *'this reflects deficits in both the financial support for RES and the required mitigation steps related to non-economic barriers that hinder an accelerated RES diffusion'* (Ecofys, 2012: 65).

3.6.5 Summary and Addition

Since 23 countries achieved or exceeded their minimum trajectory for the overall RES share in gross final consumption of energy 2011/2012 (see Tab. 4), the Commission feels that, unless their shares decreased in 2011 and 2012, the Member States in question can be seen as 'on track' with their 2020 trajectories (EC, 2013c: 3). According to Tab. 5 – based on Eurostat data on 'electricity generated from renewable sources' (Eurostat, 2013b) for the actual RES-E share in 2010 and Member States' NREAPs for the indicative RES-E 2010 targets (NREAP_EU27, 2010) – less than half the Member States (12 out of 27) were able to reach or exceed their NREAP RES-E target (Ecofys, 2012: 22 and 27; Eurostat, 2013b; NREAP_EU27: 2010; EC, 2013c: 3). Indeed, the Commission acknowledges that *'15 Member States failed to meet their legally agreed indicative 2010 targets'* (EC, 2013c: 4) but does not further comment on the matter, contradicting the Commissions intention that the bulk of Member States can be seen *'on track with their trajectory towards 2020'* (EC, 2013c: 3).

What becomes apparent based on Tab. 6 is that no Member State can be regarded as being 'advanced' with respect to administrative procedure reforms, whilst only a few countries have achieved a 'fair' rating (Ecofys, 2012: 187; EC, 2013c: 8). *'Generally, the concreteness and completeness of the administrative simplification measures intended and reported is very slow in all Member State reports: the quantity of permits required is often not mentioned, neither the number of authorities involved in procedures. [...] This shows a lack of coordination and poor implementation of Article 13 of the Directive 2009/28/EC which explicitly asks for improved coordination of the administrative procedures'* (EC, 2013c: 8).

With respect to Tab. 7, the analysis by Ecofys (2012: 171-182) leads to the conclusion that the majority of Member States perform well in the whole procedure of RES grid integration, whilst there is an urgent need for grid extension and shorter approval procedures. When it comes to addressing the grid capacity limitations, Member States are preforming well. However, rules for bearing and sharing the costs of grid development seem to be less of a priority for Member States. The Commission concludes (EC, 2013b: 8) that *'given the longer term expectations of the growing share of EU electricity from renewable energy sources, full implementation of Article 16 of the Directive is important. The cur-*

rent failure to modernise the grid [...] is causing problems for the development of the internal market, technical problems related to loop flows, grid stability and growing power curtailment, and investment bottlenecks resulting from delayed connection of new producers'.

According to Tab. 8, only very few countries are expected to meet their trajectories for 2020, and only if complementary energy efficiency measures are implemented successfully (Ecofys, 2012: 69). Other than the Commission, the Ecofys consortium concludes that, for the majority of countries, the currently implemented RES policies aimed at 2020 appear insufficient for promoting the required RES deployment (Ecofys, 2012: 65). 'Generally this reflects deficits in both the financial support for RES and the required mitigation steps related to non-economic barriers that hinder an accelerated RES diffusion. [...] [M]any countries may end up with a significant gap in their 2020 RES target' (Ecofys, 2012: 65).

Another aspect the Commission highlights with respect to the given subject is the impact of the changed economic climate following the financial crisis of 2008. Whilst evaluating the possible impact of the crisis on the promotion of renewable energies in the EU lies behind the scope of this thesis, the generally increased cost of capital, especially for southern European countries, must be included in the overall picture (EC, 2013b: 13). The financial crisis also affects the growth predictions for RES deployment, as reflected in the Ecofys report based on the Green-X model (Ecofys, 2012: 65). Whilst EU countries face a different risk rating today (2012/2013) that also has a significant impact on investment in RES, the above analysis mainly focuses on non-economic barriers for accelerated RES diffusion (Ecofys, 2012: 65). It becomes apparent, however, that the consequences of the on-going economic crisis – along with budgetary problems of Member States and businesses who have difficulties mobilising funds for long term investments (EC, 2013a) – add to the challenge of a cost-effective deployment of RES through market-oriented, efficient and effective support schemes (EC, 2013b: 2 and 9).

Overall, and notwithstanding the different appreciation of the future prospects for RES deployment with regard to the Ecofys report (EC, 2013b: 15; Ecofys, 2012: 65), the Commission acknowledges that further measures by the Member States will be needed in order to stay on the RES trajectories (EC, 2013b: 2). The Commission identifies four elements that should be part of a reform for the promotion and integration of renewable energies in the EU. These are strong growth, cost control, market integration and Europeanization – that is, developing common approaches to support renewable energy based on joint projects in the field of specific renewables projects, technologies, support schemes and regional markets, where consumers physically profit from renewable energy capacity installed in a neighbouring country (EC, 2013b: 9-10). To foster this agenda, the Commission plans on releasing a guidance document for its Member States (EC, 2013b: 10).

With respect to the overall evaluation of the impact of Directive 2009/28/EC in the European Member States, the Commission underlines its intention to continue to examine Member States' implementation of the respective Directive and will take legal measures wherever necessary (EC, 2013b: 7-8 and 13). Pursuant to Article 258 TFEU, the Commission of the European Union is responsible for ensuring that EU law is correctly implemented by the Member States (OJoEU, 2008: 160): '*If the Commission considers that a Member State has failed to fulfil an obligation under the Treaties, it shall deliver a Reasoned Opinion on the matter after giving the State concerned the opportunity to submit its observations. If the State concerned does not comply with the opinion within the period laid down by the Commission, the latter may bring the matter Before the Court of Justice of the European Union.*' In line with this mandate, cases have been opened and Reasoned Opinions have been sent to Austria, Bulgaria, Cyprus, Czech Republic, Finland, Hungary, Ireland, Latvia, Luxembourg, the Netherlands, Poland and Slovenia (EC, 2013b: 13). Asked for further information on the infringement procedures launched, based on a mail sent to the Commission on 26th June 2013, the contact person approached failed to reply.

3.7 Analysis

Based on the overall hypothesis in light of the multi-level policy dependence theory applied here, three subordinate hypotheses have been identified in the theory section. To identify constraints and dependencies of the European Integration of RES-(E) promotion, the explanatory factors implicitly describe three different influencing factors that follow a top-down, vertical logic, in turn leading to a horizontal logic. The first subordinate hypothesis (i) claims supranational renewable energy measures and policies produce compliance at the level of Member States (Radaelli, 2003: 19). Subordinate hypothesis (ii) counterpoints the first by claiming the European Union lacks the executive authority to expect compliance at the Member States level. Instead, it is national preferences that dominate the Member States' policy outcomes. Subordinate hypothesis (iii) is not relevant for the analysis based on Chapter 3.

As has been stressed, the overall framework includes theories derived from Liberal Intergovernmentalism and Europeanization. They are reflected in subordinate hypothesis (i) (Europeanization) and subordinate hypothesis (ii) (Liberal Intergovernmentalism) and read as follows:

(i) The supranational RES-(E) measures and policies of the European Union, when expressed in legislation, excite change in Member States' RES-(E) policies only when there is a degree of misfit between EU and domestic RES-(E) policies that causes adaptational pressure, leading to the national promotion of RES-(E) in accordance with EU level implications.

(ii) The RES-E policy of a Member State is an exclusive result of the domestic policy process that reflects energy-specific economic interests of the dominant domestic group, and is determined by geopolitical and economic interest.

Since any progress in the promotion of RES-(E) can only be measured at the Member States level, a model has been derived where the EU Member States are the focal point. In the applied model, the dependent variable are the Member State RES-E policies of Poland and Germany, with and ultimate focus being on the vertical and horizontal impacts on the RES-E policy of Poland. It has been stressed that a mono-causal approach cannot explain the impact on the dependent variable 'Member State RES-E'. Therefore there cannot be a single independent variable (e.g. defined at EU level) (Radaelli, 2003: 27-28). Instead, there is a set of three decisive variables, occurring at the EU level, the national level and the inter-state level. Fig. 2 illustrates their interrelations. Since all three influencing factors for the promotion of RES-E in Member States (supranational extrinsic, national intrinsic, inter-state interdependencies) will be considered, the analysis aims at identifying general insights about constraints and dependencies of European RES-E promotion mechanisms.

'Supranational extrinsic logic' has been the analytical focus of this chapter. EU RES-E policies, as well as the underlying European polity reflect it. Note that at this point the dependent variable is 'Member State RES-E'. Germany and Poland will be the exclusive analysis-focus of the following chapters. Fig. illustrates the analytical focus of chapter three.

Fig. 5: Applied model of adaptational pressure for RES-E policies with top-down analytical focus

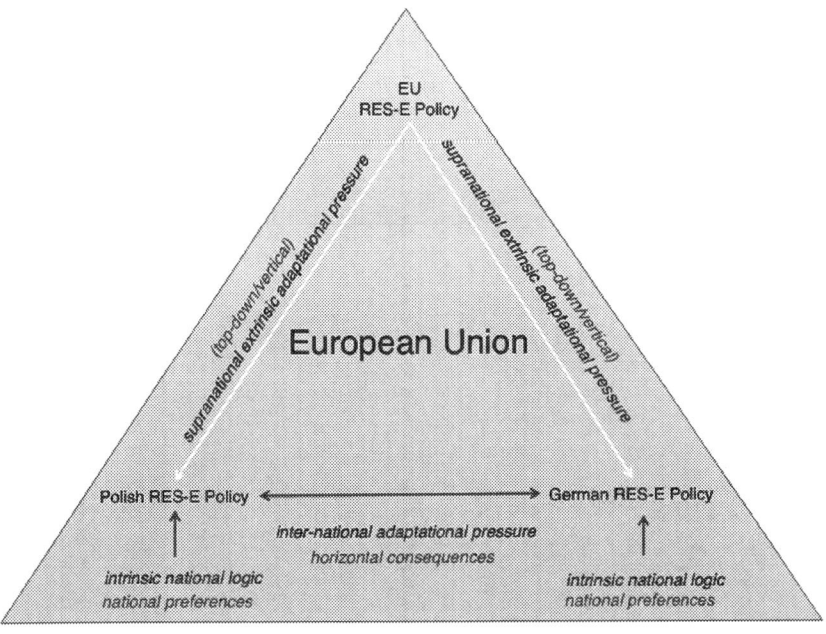

If subordinate hypothesis (i) should be accepted as a valid explanatory factor, the questions are: On which judicial basis does the European Union adopt secondary provisions? What is the nature of these secondary provisions in light of the subject matter? Did the secondary provision in question excite change in Member States' RES-E policies? And, if change has been excited, was it because of a misfit between EU and domestic RES-E policies?

If – in order to consider alternative reasons for empirical evidence presented in this chapter (Ladrech, 2010: 40-41) – subordinate hypothesis (ii) should serve as a valid explanatory factor, the question is: Can the RES-E policy of the EU Member States be considered an exclusive result of the domestic policy process?

Based on the arguments in light of the EU constitutional context, it can be concluded that there is a recognised constitutional principle of the European Union reflected in the sui generis nature of EU law, its supremacy, direct effect, preemption and protection of fundamental rights (Tatham, 2013: 15). With respect to the judicial basis for executive authority at the EU-level, attention has been given to the principles of the conferral of powers, which are regulated, *inter alia,* in Article 3(6), 4, 5 and 13(2) of the Treaty on European Union (OJoEU, 2010: 17-22; Tobler and Beglinger, 2012: 82):

- The Union shall only act where it enjoys powers given to it by the Member States (Art. 5(2); TEU, OJoEU, 2010: 18)

- In accordance with Art. 5, all competencies not conferred upon the Union remain with the Member States (Art. 4; OJoEU, 2010: 18)

- The only field entirely excluded from EU law is national security, which remains the sole responsibility of each Member State (Art. 4(2); OJoEU, 2010: 18).

There are three different types of Union competencies, which are listed in the Treaty on the Functioning of the European Union (TFEU) (Tobler and Beglinger, 2012: 83):

- Exclusive competences of the Union, where Member States shall no longer act, except for the implementation of Union acts or where they are empowered by the Union (Art. 2(1);OJoEU, 2008: 50).

- Shared competences between the Union and the Member States, where Member States exercise competence to the extent that the Union has not exercised its own competence. Member States have lost their competence if the field is occupied by Union law (pre-emption) (Art. 2(2); OJoEU, 2008: 50; Tobler and Beglinger, 2012: 83)

- Supporting, coordinating or supplementing competencies of the Union, where Union action does not supersede Member States' competencies (Art. 2(5); OJoEU, 2008: 50)

In order to facilitate the attainment of the primary goals of the Treaties (Art. 288 TFEU) – that is, to exercise the Union's competencies – the Union can adopt secondary measures identical to regulations, directives, decisions, recommendations and opinions (Tobler and Beglinger, 2012: 91; OJoEU, 2008: 171). The Lisbon Treaty distinguishes between the ordinary legislative procedure and the special legislative procedure (Tobler and Beglinger, 2012: 95). Union legislative acts may, unless the Treaties provide otherwise, only be adopted on the basis of a Commission proposal (OJoEU, 2010: 25). The ordinary legislative procedure demands both the European Parliament and the Council to adopt a Commission proposal (Tobler and Beglinger, 2012: 94; OJoEU, 2008: 172).

The conferral of powers from the European Member States to the European Institutions has been a conscious act by the representatives of the governments of the Member States, ultimately confirmed by common accord during the adoption of the Lisbon Treaty in 2007 (OJoEU, 2007a: 232-248), for which *'[...] the states have limited their sovereign rights, albeit within limited fields'* (ECJ, 1963: II B, Franzius, 2010: 39). In a case law judgement, the ECJ defines the interrelations of the Union and Member States' legal order by stating that *'the relationship between provisions of the treaty and directly applicable measures of the institutions on the on hand and the national law of the Member States on the other is such that those provisions and measures not only by their entry into force render automatically inapplicable any conflicting provisions of current national law but – in so far as they are an integral part of, and take precedence in, the legal order applicable in the territory of each of the Member States – also preclude the valid adoption of new national legislative measures to the extent to which they would be incompatible with the community provisions'* (ECJ, 1978: Summary 3.; Franzius, 2010: 43). The judgement of the ECJ was confirmed by the representatives of the governments of the Member States in the 17th Declaration of the Final Act of the Treaty of Lisbon (OJoEU, 2007a: 256). To underline the self-evidence of the primacy of EU law, the conference of the representatives of the governments of the Member States *'has also decided to attach as an Annex to this Final Act the Opinion of the Council Legal Service on the primacy of EC law as set out in 11197/07 (JUR 260): [...] It results from the case-law of the Court of Justice that primacy of EC law is a cornerstone principle of Community law. According to the Court, this principle is inherent to the specific nature of the European Community. At the time of the first judgment of this established case law (Costa/ENEL, 15 July 1964, Case 6/641 (1)) there was no mention of primacy in the treaty. It is still the case today. The fact that the principle of primacy will not be included in the future treaty shall not in any way change the existence of the principle and the existing case-law of the Court of Justice'* (OJoEU, 2007: 256).

> It follows in principal, and notwithstanding the specific nature of the promotion of renewable energies in the EU, that, pursuant to Art. 5(2) and Art. 4 TEU (OJoEU, 2010: 18), Member State policies cannot a priori be considered an exclusive result of the domestic policy process, except for natio-

nal security policies, which remain the sole responsibility of each Member State (Art. 4. (2); OJoEU, 2010: 18). It must be concluded, furthermore, that this not only results from the treaties in question, but also reflects the intention of the representatives of the governments of the Member States who signed the treaties. In terms of explanatory factors (i) and (ii), this means, with the conferral of powers by European Member States to the European Institutions, that the signatory parties intentionally legislated the prerequisite for EU measures to excite change in Member States, and voluntarily limited their sovereign rights. Whilst the treaty negotiations are an exclusive result of Member States' policies, these policies created a supranational polity, which with *'the entry into force of the treaty, became an integral part of the legal systems of the Member States and which their courts are bound to apply'* (ECJ, 1964: Summary 3.).

Whilst these findings reflect a polity principle of the European Union, it is now time to focus on the consequences it has on the secondary provisions for the promotion of renewable energies on the EU Level, and its consequences for the EU Member States. With respect to top-down and bottom-up implications of the proposed framework, and the theories therein, EU-level supranational policy for the promotion of renewable energies – as reflected in the Directive 2009/28/EC and as part of the overall Climate and Energy Package of the European Union (EC, 2008), as well as the relevant aspects of the third legislative package for an internal EU gas an electricity market (EC, 2011b: 3) – have been described. Based on six legislative proposals, the President of the European Parliament and the President of the Council signed each legislative act of the Climate Change Package – including Directive 2009/28/EC – on 23rd April 2009 (see also Tab. 1) (OJoEU, 2009a-f: 11-145). Each of the five legislative proposals of the Third Energy Package were signed by the President of the European Parliament and the President of the Council on 13th July 2009 (see Tab. 2) (OJoEU, 2009i-k: 14-132). All secondary provisions were established following the ordinary legislative procedure (see Text Box 1), making the European Parliament and the Council equal co-legislators (EP, 2008: 1).

Since Directive 2009/28/EC can be considered the key piece of legislation in the context of the promotion of renewable energies in the EU, the analysis will now focus on the polity framework for RES promotion. The fact that legislation described above was adopted through the ordinary legislative procedure is important in the on-going discussion. As for Directive 2009/28/EC: it was adopted before the Lisbon Treaty came into force, and is founded on two legal bases (Czeberkus, 2013: 28). The European Parliament and the Council adopted the Directive with regard to the Treaty establishing the European Community, and in particular Article 175(1) (now Article 192(1) TFEU) and Article 95 (now Article 114(1) TFEU), and in relation to Articles 17,18 and 19 of the Directive (OJoEU, 2009b: 16; Czeberkus, 2013: 28). This means that the Directive is based on Title XIX 'Environment' of TEC (now Title XX, TFEU), except for Articles 17, 18 and 19, where it refers to biofuels and bioliquids. These are based on Chapter 3 TEC (still Chapter 3, TFEU) (OJoEU, 2009b: 26; OJoEU, 2008: 94 and 132;

OJoEU, 2006b: 79 and 123; Czeberkus, 2013: 28-29). Since in the pre–Lisbon Treaty EU there were no provisions for regulatory authority over the energy sector, the EU shaped its energy policies through competencies in the field of the environment (ex Article 174, 175, 176 Treaty establishing the European Community, TEC), the internal market (ex Article 94, 95, 96, 97 TEC) and trans-European Networks (ex Article 154, 155, 156 TEC) (German Advisory Council on the Environment, 2011: 175; Czeberkus, 2013: 23; OJoEU, 2006b: 79-81, 116-117 and 123-125). However, with respect to the environment, and pursuant to Article 192(2) (ex Article 175 TEC), (a) provisions primarily of fiscal nature, (b) measure affecting town and country planning, quantitative management of water resources, land use (except waste management), and *'measures significantly affecting a Member State's choice between different energy sources and the general structure of its energy supply'* are subject to a special legislative procedure, where the Council acts unanimously after consulting the Parliament, the Economic and Social Committee, and the Committee of the Regions (OJoEU, 2008: 133; German Advisory Council on the Environment, 2011: 175). In other words, EU measures significantly affecting Member States' choice can be adopted only if unanimity is guaranteed in the Council (Ballesteros, 2010: 11). The fact that Directive 2009/28/EC, although setting legally binding national overall targets for the share of energy from renewable sources, was adopted through an ordinary legislative procedure implies that this particular measure is not considered as *'significantly affecting a Member State's choice between different energy sources and the general structure of its energy supply'* (OJoEU, 2009a: 16 and 46; OJoEU, 2008: 133).

> It follows –, based on the executive rights of the European bodies pursuant to the pre–Lisbon Treaty EU – that the legislative provisions of the Climate Change Package, and the third legislative package for an internal EU gas and electricity market, reflect non-exclusive rights of the Member States in the given policy context. Each Member State, however, is represented in the Council by way of the ordinary legislative procedure and can non-exclusively influence the policy outcome. With respect to Article 192(2), which can be considered a safeguard for exclusive Member State influence (because measures falling under that article are subject to Council unanimity), it is important to emphasize, that the implementation of Directive 2009/28/EC is not considered a measure *'significantly affecting a Member State's choice between different energy sources and the general structure of its energy supply'* (OJoEU, 2008: 133; German Advisory Council on the Environment, 2011: 175). The legislative packages in question, and particularly Directive 2009/28/EC, were adopted through the ordinary legislative procedure and therefore under non-exclusive participation of the Member States, seeking to excite change pursuant to each provision's content. Whilst subordinate hypothesis (ii) can be falsified under the very narrow conditions of the so primary and secondary legislative European sources cited so far, subordinate hypothesis (i) can only be accepted at this point as expressing the intention of the European legislator.

It remains questionable as to whether legal measures concerning RES promotion could be adopted under the Lisbon Treaty, as was the case for Directive 2009/28/EC under the TEC – that is, based on TFEU Title XX 'Environment' and pursuant to TFEU 'Chapter 3' is questionable (Czeberkus, 2013: 29). With the entry into force of the Lisbon Treaty on 1st December 2009, the EU's authority was expanded by introducing a specific energy competence under TFEU Title XXI 'Energy' (German Advisory Council on the Environment, 2011: 179; Fouquet, Nysten, Held and Johnston, 2012: 15). Based on this new energy chapter, some authors conclude that Article 194 applies as a lex specialis before all other provisions, as it appears to be the most obvious legal basis for any kind of energy-related measures (Fouquet, Nysten, Held and Johnston, 2012: 15; German Advisory Council on the Environment, 2011: 180). However, disagreement exists on the meaning of Article 194(2), which does not provide for a special decision-making procedure with unanimity voting in the Council (Fouquet, Nysten, Held and Johnston, 2012: 16, Ballesteros, 2010: 11, OJoEU, 2008: 135). Whilst Article 192(1) of the environment chapter does provide for a special legislative procedure concerning measures *significantly affecting* a Member State's choice between different energy sources and the general structure of its energy supply, 194(2) prohibits the adoption of measures affecting Member States' rights to determine the conditions for exploiting their energy resources. In other words, whilst pursuant to Article 192(1), EU measures *affecting* Member States can be adopted in accordance with the ordinary legislative procedure (as was the case with the RES Directive), pursuant to Article 194(1) measures affecting Member States cannot be adopted (Ballesteros, 2010: 12; OJoEU, 2008: 132 and 135).

> These new realities lead to the conclusion that Article 194 TFEU theoretically has the potential to withdraw all executive authorities from the European bodies that affect a *'Member State's right to determine the conditions for exploiting its energy resources, its choice between different energy sources and the general structure of its energy supply'* (OJoEU, 2008: 134-135). Whilst it is hard to imagine that Member States, with the introduction of a special energy chapter in the Lisbon Treaty, established a competence limit that would ex post take away the legal basis for pre-existing measures (e.g. Third Energy Package, Climate Package), the clause mentioned here could serve as an absolute limit on EU competence. It remains questionable as to whether there is any matter in the field of energy that can be regulated without effecting Member States' rights pursuant to Article 194(2) (Czeberkus, 2013: 26; Fouquet, Nysten, Held and Johnston, 2012: 16-18; Ballesteros, 2010: 12; OJoEU: 133 and 135). Some authors argue, however, that it remains unclear whether future RES promotion should be based on Article 192(2) – as was the case with Directive 2009/28/EC – or on Article 194(1). The precise wording of Article 194(1) refers to the *'development of new and renewable forms of energy'* as opposed to the promotion of established RES. Furthermore, measures under Article 194(1) shall be without prejudice to Article 192(2) (c), which, following the German

Advisory Council on the Environment (2011: 180), only makes sense if Article 192 applies in conjunction with Article 194. In consequence *'the EU's newfound authority (under Article 194, <A/N>) [...] solely empowers it to promote the technological development of renewable energies, where by any economically or ecologically motivated support henceforth is governed by environmental regulations* (under Article 192, <A/N>)' (German Advisory Council on the Environment, 2011: 180). *'Consequently, the EU has no authority over non-environmental energy policy measures that fall within the competence of the member states'*, but can adopt RES measures pursuant to environmental regulations (German Advisory Council on the Environment, 2011: 189). Whether or not Article 194(2) constitutes an absolute competence limit on the EU, or grants Member States the opportunity to opt out of EU legislation in the context of energy, *'it is clearly possible that this provision may have an impact upon the behaviour of any Member State during discussions of Commission (legislative) proposals on various aspects of energy and environmental policy as well as on the application of EU legislation adopted in this area'* (Fouquet, Nysten, Held and Johnston, 2012: 18). It should be clearly emphasized at this point, since the European Court of Justice has not yet had the opportunity to comment on the matter, that any argument with respect to Article 194(2) and Article 192(2) remains a mere interpretation of the law (Czeberkus, 2013: 26; Ballesteros, 2010: 12). If, however, the ECJ would follow the argument that Article 194(2) serves as an absolute limit on EU competencies, explanatory factor (ii) would be verified in its entirety. This interpretation highly unlikely, however, given the fact that the existence of Directive 2009/28/EC is not considered a measure *'significantly affecting a Member State's choice between different energy sources and the general structure of its energy supply'* pursuant to Article 192(2) TFEU (OJoEU, 2008: 133) – since that the legality of the European measures as reflected in the Climate and Energy Package and the Third Energy Package have not been questioned in front of the ECJ – and also as a result of an on-going policy cycle for a European strategy for the promotion of renewable energies beyond 2020 (EC, 2013a).

The bulk of arguments so far lead to the conclusion that subordinate hypothesis (ii) has to be revised when taking into account a strictly European perspective. Of course, by the very nature of the European polity design, the European perspective is in parts always shaped by preferences of the governments of the Member State countries. These preferences are integrated into the European policy process through the Council. With respect to the promotion of renewable energies, the RES-E policy of a Member State would only be an exclusive result of the domestic policy process if Article 194(2) serves as an absolute limit on EU competencies. For the purpose of precision, it remains the exclusive 'right' of the Member States to implement Union acts (Art. 2(1); OJoEU, 2008: 50). With respect to subordinate hypothesis (i), the aforementioned arguments did not give rise to useful insights. The question remains: Did secondary provision Directive 2009/28/EC excite change in the RES-E policies of the European Member States? To answer this question with respect to the scope of this thesis, insights on the implementation effectiveness of Directive 2009/28/EC are based on Member States' NREAPs and the respective forecast documents (NREAP_EU27: 2010), energy figures provided by Eurostat, a renewable energy progress report, and an accompanying working staff document by the Commission (EC, 2013b; EC, 2013c), as well as a 'renewable energy progress and biofuels sustainability' report produced for the Commission by ECOFYS, Fraunhofer, Becker Bütter Held (BBH), Energy Economics Group and Windrock International (Ecofys, 2012).

Quoted in Chapter 3 are evaluations on the following subjects:

- Past progress of EU 27 RES shares in between the coming into force of Directive 2009/28/EC in 2009 and 2010, as reflected in 'Tab. 4: Actual and planned RES Shares 2009-2011/2012 (Ecofys, 2012: 18)'.

- Past progress in RES-E promotion, based on Member States' NREAPs, as well as Eurostat data on 'electricity generated from renewable sources' with respect to the individual 2010 RES-E Member State targets, as reflected in 'Tab. 5: Actual and planned RES-E shares (NREAP_EU27, 2010; Eurostat, 2013b)'.

- Evaluation of the reduction of administrative barriers pursuant to Article 13 of Directive 2009/28/EC regarding administrative procedures concerning national rules for authorisation, certification and licensing procedures for plants producing electricity from RES and transmission and distribution of RES-E, as reflected in 'Tab. 6: Assessment of administrative procedures in the Member States (Ecofys, 2012: 186-187; EC, 2013c: 9)'.

- Rating of Member States' progress in electricity grid integration, including ratings of overall RES grid integration, of addressing grid capacity limitations, and of rules for bearing and sharing the costs of grid development, as reflected in 'Tab. 7: Member State rating on progress in electricity grid integration (Ecofys, 2012: 177, 179 and 181)'.

- Quantitative assessment of the future deployment of RES in the European Union based on the Green-X model developed by the Energy Economic Group at Vienna University of Technology, Institute of Power Systems and Energy Economic (EEG, 2007: 1), as reflected in 'Tab. 8: Expected, planned and required RES share in 2020 (Ecofys, 2012: 66)' and 'Tab. 9: Expected and planned RES deployment at EU level by 2012 and 2020 (Ecofys, 2012: 71)'.

The results of the implementation-effectiveness evaluation of Directive 2009/28/EC have been described in detail in the respective paragraph, and in the summary and addition section. In short what can be concluded is:

23 countries achieved or exceeded their minimum trajectory for the overall RES share in gross final consumption of energy 2011/2012 (see Tab. 4). According

to Tab. 5 – based on Eurostat data on 'electricity generated from renewable sources' (Eurostat, 2013b) for the actual RES-E share in 2010, and Member States' NREAPs for the indicative RES-E 2010 targets (NREAP_EU27, 2010) – less than half the Member States (12 out of 27) were able to reach or exceed their NREAP RES-E target (Ecofys, 2012: 22 and 27; Eurostat, 2013b; NREAP_EU27: 2010; EC, 2013c: 3). What becomes apparent based on Tab. 6 is that no Member State can be regarded as being 'advanced' with respect to administrative procedure reforms, Whilst only a few countries achieved a 'fair' rating (Ecofys, 2012: 187; EC, 2013c: 8), with respect Tab. 7, the analysis by Ecofys (2012: 171-182) leads to the conclusion that the majority of Member States perform well in the whole procedure of RES grid integration, although there is an urgent need for grid extension and shorter approval procedures. When it comes to addressing the grid capacity limitations, Member States are preforming well. However, rules for bearing and sharing the costs of grid development seem to be less of a priority for Member States. According to Tab. 8, only very few countries are expected to meet their trajectories for 2020, and only if complementary energy efficiency measures are implemented successfully (Ecofys, 2012: 69). The Commission has launched infringement cases for Member States' non-transposition of Directive 2009/28/EC and has sent Reasoned Opinions to Austria, Bulgaria, Cyprus, Czech Republic, Finland, Hungary, Ireland, Latvia, Luxembourg, the Netherlands, Poland and Slovenia (EC, 2013b: 13).

> The outlined findings have an obvious impact on the validity of the explanatory capacity of subordinate hypothesis (i). All Member States fulfilled their obligation pursuant to Article 4 of Directive 2009/28/EC to adopt a National Renewable Energy Action Plan (NREAP) in order to communicate Member States' national targets for transport, electricity, heating and cooling (OJoEU, 2009b: 28; NREAP_EU27, 2010). It becomes apparent from Tab. 4 and Tab. 5 that, with the entry into force of Directive 2009/28/EC, the overall RES share in gross final consumption of energy, and the RES-E share, evidently rose in the bulk of the EU Member States. Whilst the possibility cannot be entirely eliminated that the measured change in RES and RES-(E) shares would have occurred without the implementation of Directive 2009/28/EC, meaning that the measured changes would be a result of an independent domestic policy process simultaneously occurring in at least 23 European Countries, such a possibility seems – in light of the circumstances laid out in this chapter – highly unlikely. Against this backdrop, it seems very likely that the supranational RES-(E) measures and policies of the European Union excited change in Member States' RES-(E) policies. However, it also be comes apparent that the change excited is not in its entirety in accordance with the change intended. This fact is equally important. The majority of Member States to date (Jan. 2014) did not transpose Directive 2009/28/EC in its entirety. Reasons for this result will be the subject of the on-going discussion. Accepting the first aspect of subordinate hypothesis (i) to be validated under usual reserve leads to the converse argument that the evident

promotion of renewable energies in the European Member States cannot be result of an independent domestic policy process (subordinate hypothesis (ii)).

Tab. 10: Evaluation Matrix for Chapter 3

	European Union	GER	PL	GER and PL
Subordinate Hypothesis (I)	☺	X	X	X
Subordinate Hypotehsis (II)	☺	X	X	X
Subordinate Hypothesis (III)	X	X	X	X

What follows from the analysis above? The constitutional principle of the European Union designed and agreed upon by the representatives of the governments of the Member States was designed for the purpose of having an effect on the national policy process originating in the supranational arena of the European Union, according to which *'the states have limited their sovereign rights'* (ECJ, 1963: II B). Within the boundaries of the polity defined by the European treaties and their validity period, the actors and institutions legitimised therein adopted renewable energy policies to be transposed into national policies, aiming to trigger domestic change as intended in the supranational policy. Only if the ECJ would rigidly interpret the indistinct implications of Article 194(2) TFEU as a lex specialis, ultimately leading to an absolute competence limit in defined cases resembling the particular conditions described above, would there be no treaty-based legal grounds for the EU to adopt (renewable) energy policies. Empirical evidence for Directive 2009/28/EC suggests that European renewable energy policies and measures excite change in Member States' policies, but the change excited is not in its entirety in accordance with the change intended. With respect to subordinate hypothesis (i), the first aspect (that the supranational RES-(E) measures and policies of the European Union excite change in Member States' RES-(E) policies) can be verified under usual reserve. With regard to the second aspect (if there is a degree of misfit between EU and domestic RES-(E) policies that causes adaptational pressure leading to the national promotion of RES-(E) in accordance with EU level implications), no conclusions can be drawn based on this chapter.

Subordinate hypothesis (ii) cannot be falsified in its entirety. Based on both pre–Lisbon Treaty realties and Directive 2009/28/EC implementation-effectiveness evidence, it must be falsified. Based on Lisbon Treaty provisions in interaction with the possibility of a rigid ECJ judgment with respect to Article 194(2) TFEU, it holds the potential to be verified at some indefinite future date. Whilst the latter thought certainly bears too many speculative variables to be the basis for scientific deductions, the principle of completeness demands that this possibility be accounted for. For an overview of the results visualized in a basic evaluation matrix, see Tab. 10.

4. Germany

4.1 German Constitutional Context

What was true for the European constitutional context is equally important for the German constitutional context. In order to understand the likelihood that the EU Climate Change Package and Third Energy Package have an impact on Germany, it is necessary to understand the German judicial framework for the adaptation of secondary provisions at the supranational level. Again, the set of problems in question should be linked to the overall framework of this study, namely Liberal Intergovernmentalism (Moravcsik, 1998) and Europeanization (Börzel and Risse, 2000; Radaelli, 2003). It furthermore should be linked to Grimm's (2001: 13-22) definitions of the relation of law and politics (see, *inter alia*, 2.2 Trans-, Supra-, and International Bound Policy Analysis Frameworks). The argument of the following chapter follows the descriptive design of an analysis by Alan F. Tatham (2013) on Central European constitutional courts in the face of EU Membership.

In light of the transfer of the exercise of sovereignty based on the 1949 Constitution of the Federal Republic of Germany, two provisions are relevant for European Integration (Tatham, 2013: 77; Pernice, 1998: 3-4): (a) the Preamble affirms the will of the people to *'promote world peace as an equal partner in a united Europe'* (BMJ, 2012: 1); and (b) correspondingly Art. 24(1) of the Constitution states that the *'Federation may by law transfer sovereign powers to international organisations'* (BMJ, 2012: 9). As Pernice states (1998: 4) and Tatham emphasizes (2013: 78), Germany was, based on lessons taken by the German people from the historical failure of the classical concept of the nation state, constituted to be a member of a broader political system from the very beginning. Consequently, constitutional law in Germany cannot be based on the Constitution alone, but must be construed in its constitutional and legislative European context (Pernice, 1998: 4). However, with the so called 'eternity clause', the Constitution holds a counterpoint to the above-mentioned provisions. Pursuant to Article 79(3) of the Constitution, *'Amendments to this Basic Law affecting the division of the Federation into Länder, their participation on principle in the legislative process, or the principles laid down in Articles 1 and 20 shall be inadmissible'* (BMJ, 2012: 25). Integration of Germany into a European federal state, including a comprehensive transfer of sovereignty to the European level, would therefore require the German people to adopt freely a new Constitution (Tatham, 2013: 83-84).

Based on Article 24(1) and Art 59(2), which requires the consent and/or participation of bodies responsible for the enactment of federal law (BMJ, 2012: 18), Germany, by the passing of a Ratification Act, adopted the European Economic Community Treaty, the European Atomic Energy Community and the Single European Act (Tatham, 2013: 78). However, Art 24 *'was not regarded as providing a sufficient constitutional basis for the continued progress of European integration under the terms of the Maastricht Treaty since the Treaty touched the core of German sovereignty more than the previous European Community Treaties'* (Talham, 2013: 80). A firmer constitutional basis for EU integration under the Maastricht Treaty was provided with the adoption of a new Article 23, which, before its amendment, served as an accession clause for the East-German states that were not in the field of the application of the Constitution of the Federal Republic of Germany (BPB, 2013). After reunification, the new Article 23 of the Constitution borrows the idea of the Preamble and translates it into obligations with legally binding effects and remarkable provisions, as Pernice (2008: 9-19) concludes.

Text Box 5: New Integration Clause, Article 23 Basic Law (BMJ, 2012: 8-9)

Article 23
[European Union – Protection of basic rights – Principle of subsidiarity]

(1) With a view to establishing a united Europe, the Federal Republic of Germany shall participate in the development of the European Union that is committed to democratic, social and federal principles, to the rule of law, and to the principle of subsidiarity, and that guarantees a level of protection of basic rights essentially comparable to that afforded by this Basic Law. To this end the Federation may transfer sovereign powers by a law with the consent of the Bundesrat. The establishment of the European Union, as well as changes in its treaty foundations and comparable regulations that amend or supplement this Basic Law, or make such amendments or supplements possible, shall be subject to paragraphs (2) and (3) of Article 79.

(1a) The Bundestag and the Bundesrat shall have the right to bring an action before the Court of Justice of the European Union to challenge a legislative act of the European Union for infringing the principle of subsidiarity. The Bundestag is obliged to initiate such an action at the request of one fourth of its Members. By a statute requiring the consent of the Bundesrat, exceptions from the first sentence of paragraph (2) of Article 42, and the first sentence of paragraph (2) of Article 52, may be authorised for the exercise of the rights granted to the Bundestag and the Bundesrat under the contractual foundations of the European Union.

(2) The Bundestag and, through the Bundesrat, the Länder shall participate in matters concerning the European Union. The Federal Government shall keep the Bundestag and the Bundesrat informed, comprehensively and at the earliest possible time.

(3) Before participating in legislative acts of the European Union, the Federal Government shall provide the Bundestag with an opportunity to state its position. The Federal Government shall take the position of the Bundestag into account during the negotiations. Details shall be regulated by a law.

(4) The Bundesrat shall participate in the decision-making process of the Federation insofar as it would have been competent to do so in a comparable domestic matter, or insofar as the subject falls within the domestic competence of the Länder.

(5) Insofar as, in an area within the exclusive competence of the Federation, interests of the Länder are affected, and in other matters, insofar as the Federation has legislative power, the Federal Government shall take the position of the Bundesrat into account. To the extent that the legislative powers of the Länder, the structure of Land authorities, or Land administrative procedures are primarily affected, the position of the Bundesrat shall be given the greatest possible respect in determining the Federation's position consistent with the responsibility of the Federation for the nation as a whole. In matters that may result in increased expenditures or reduced revenues for the Federation, the consent of the Federal Government shall be required.

(6) When legislative powers exclusive to the Länder concerning matters of school education, culture or broadcasting are primarily affected, the exercise of the rights belonging to the Federal Republic of Germany as a member state of the European Union shall be delegated by the Federation to a representative of the Länder designated by the Bundesrat. These rights shall be exercised with the participation of, and in coordination with, the Federal Government; their exercise shall be consistent with the responsibility of the Federation for the nation as a whole.

(7) Details regarding paragraphs (4) to (6) of this Article shall be regulated by a law requiring the consent of the Bundesrat.

Article 23 guarantees a complete parliamentary process and the full participation of the 'Länder', as well as the observation of fundamental principles of the State before any transfer of powers to the European level can occur (Tatham, 2013: 81). Nonetheless, the rather general language of the provision in question allows for flexibility on how these principles are put into practice (Pernice, 1998: 11, Pernice, 2006: 18-19). Flexibility, however, leaves room for interpretation, which in terms of the Constitution falls within the scope of the German Federal Constitutional Court (FCC). What is true for the ECJ, its case law and its effect on European constitutionalism – based on Article 19 TEU and the 17th Declaration of the Final Act of the Treaty of Lisbon (OJoEU, 2010: 83; OJoEU, 2007a: 256) – is also true, based on Article 93 of the German Constitution, for the FFC (BMJ, 2012: 33). The relationship of the ECJ and the FFC can be described as rather tempestuous (Tatham, 2013: 84).

As opposed to the position confirmed by the representatives of the governments of the Member States in the 17th Declaration with respect to the primacy of EU law (OJoEU, 2007a: 256), the FCC decided in a judgement of the Second Senate of 30 June 2009, known as the 'Lisbon case', inter alia, that 'with Declaration no.

17 Concerning Primacy Annexed to the Treaty of Lisbon, the Federal Republic of Germany does not recognise an absolute primacy of application of Union law, which would meet with constitutional objections, but merely confirms the legal situation as interpreted by the Federal Constitutional Court' (BverfG, 2009: at 331). In order to understand this judgement, clarification is needed on 'the legal situation as interpreted by the Federal Constitutional Court' (BVerfGE, 2009: 123 (267) at 331).

Based on case law between 1967 and 1971, the FCC accepted the autonomous non-state sovereign authority of treaty-based European legislation that, following ECJ interpretation, led to the conclusion that supranational provisions overlaid inconsistent national law (Pernice, 2006: 28; Tatham, 2013: 86; BVerfGE, 1971: 31 (145); BVerfGE 31(145) quoted in IfIL, 2007a: 6). The FCC states in the so-called 'Milk-powder' case: 'To decide whether a domestic norm of ordinary law is incompatible with a paramount provision of European Community Law and ought therefore to be denied validity, the Federal Constitutional Court is not competent; the solution to this conflict of norms is therefore left to the comprehensive review and dismissal powers of the competent courts' (BVerfGE 31(145) quoted in IfIL, 2007a: 6; BVerfGE, 1971: 31 (145) at 97). However, the FCC recognized the primacy of European law over national law in the form of 'priority in application', as opposed to 'priority of validity' (Tatham, 2013: 86).

A change of this attitude and potential conflict between the case law of the ECJ and the FCC is reflected in the so-called 'Solange I' judgement of the FFC in 1974. In this case, the court denied European supremacy because fundamental rights enshrined in the German Constitution were not protected to the same extent in European law (Pernice, 2006: 28; Tatham, 2013: 100, Grimm, 2001: 287). 'As long as the integration process has not progressed so far that Community law receives a catalogue of fundamental rights decided on by a parliament and of settled validity, which is adequate in comparison with the catalogue of fundamental rights contained in the Basic Law, a reference by a court of the Federal Republic of Germany to the Federal Constitutional Court in judicial review proceedings, following the obtaining of a ruling of the European Court under Article 177 of the Treaty, is admissible and necessary if the German court regards the rule of Community law which is relevant to its decision as inapplicable in the interpretation given by the European Court, because and in so far as it conflicts with one of the fundamental rights of the Basic Law' (BVerfGE 37 (271) quoted in IfIL, 2007b: 1; BVerfGE, 1974: 37 (271)). This conflict was resolved in 1986 in the so-called 'Solange II' judgment. Here the Court states that, as long as the general level of protection of human rights remains adequate by German standards, the FCC should no longer examine secondary European legislation for compliance with constitutional provisions relating to fundamental rights in the German Constitution (Tatham, 2013: 101; Pernice, 2006: 28-29; Grimm, 2001: 288). 'The power conferred by Article 24 (1) of the Basic Law, however, is not without limits under constitutional law. The provision does not confer a power to surrender by way of ceding sovereign rights to international institutions the identity of the prevailing constitutional order of the Federal Republic by breaking into its basic framework, that is, into its very structure' (BVerfGE 73 (339) quoted in IfIL, 2007c: 8; BVerfGE, 1986: 73 (339)). Whilst, according to this judgment, the FCC 'will no longer exercise its jurisdiction to decide on the applicability of secondary Community legislation cited as the legal basis for any acts of German courts or authorities within the sovereign jurisdiction of the Federal Republic of Germany' (BVerfGE 73 (339) quoted in IfIL, 2007c: 1; BVerfGE, 1986: 73 (339)), it nonetheless preserved its final authority in applying national provisions over European provisions where problems concerning the protection of fundamental rights in European law arise (Tatham, 2013: 102).

A new perspective on the interrelation of domestic primary German law and supranational European law was introduced by the so-called 'Maastricht' judgement of the FCC in 1993. The FCC argues that *"[e]ven after the Maastricht Treaty has entered into force, the Federal Republic of Germany remains a member of a compound of States, the authority of which is derived from the Member States and has binding effect in German sovereign territory only by virtue of the German command to apply the law [Rechtsanwendungsbefehl]. Germany is one of the 'High contracting parties' which have given as the reason for their commitment to the Maastricht Treaty, concluded 'for an unlimited period' (Art. Q), their desire to be members of the European Union for a lengthy period; such membership may, however, be terminated by means of an appropriate act being passed'* (BVerfGE 89 (155) quoted in University of Nevada, 1994: 21; BVerfGE, 1993: 89 (155) at 112). Consequently, the Member States remain the 'Masters of the Treaties' who have conferred, as an example of international organisations, the exercise of some of their powers to the EU, whose primary Treaty law merely remains a derivative legal order. Not only can Germany terminate its membership, but the FFC will also examine whether instruments of the European Union exceed the limits of the sovereign rights accorded to them (Tatham, 2013: 83; Franzius, 2010: 25 and 34 and 43; Pernice, 2006: 30-31; Grimm, 2001: 288; BVerfGE 89 (155) quoted in University of Nevada, 1994: 2; BVerfGE, 1993: 89 (155) at 112). On the other hand, the FFC acknowledges a relation of co-operation with the ECJ, where secondary EU legislation infringes constitutional rights, by stating that *'the Federal Constitutional Court exercises its jurisdiction regarding the applicability of derivative Community law in Germany in a 'co-operative relationship' with the European Court of Justice'* (BVerfGE 89 (155) quoted in University of Nevada, 1994: 2; BVerfGE, 1993: 89 (155) 7.; Tatham, 2013: 92; Pernice, 2006: 30-31).

Altogether, as Pernice (2006: 27-28) concludes, there are three development phases regarding FFC case law in light of European integration and the primacy of EU legislation. In the early days of European Integration, before 1974, FCC

case law was principally in accord with ECJ case law ('Milk-powder' decision). This phase was followed by FCC judgements leaning towards a more state-dominated interpretation ('Solange I' and 'Solange II' decisions), complemented later on (third phase) with a sense of co-operative acknowledgment, whilst re-affirming the ultimately binding competence of the European Member States ('Maastricht' decision). The last stance was confirmed in the 'Lisbon' decision, where the FCC refers to *'the legal situation as interpreted by the Federal Constitutional Court'* (BVerfGE, 2009: 123 (267) at 331). In short, the FCC claims ultimate authority to protect fundamental rights in the national system but refrains *'from striking down secondary European legislation per se'* (Tatham, 2013: 212). It reserves the right to review national legal rules that transpose European provisions with regard to whether they occurred on the basis of manifest transgressions of EU competence pursuant to the Lisbon treaty. The FFC rarely invokes this power (as long as fundamental rights are not infringed by EU secondary legislation) and would only do so in a co-operative manner with ECJ jurisdiction (Tatham, 2013: 126). Nevertheless, as Tatham (2013: 99) concludes, *'[t]he FCC [...] has not accepted without question the constitutionalisation of the European legal order, as created and managed by the ECJ. [...] [A] number of aspects have been challenged by the FCC but, so far, they have not led to open conflict with their European counterpart although [...] its ruling in the Lisbon case could render an eventual clash more likely'.*
For completeness and comparability, it should be noted that the constitution of Germany does not explicitly include provisions with respect to energy, as is the case with Article 194 TFEU (BMJ, 2012: 1-58; OJoEU, 2008: 134-135).

4.1.1 Summary

In light of the transfer of the exercise of sovereignty based on the 1949 Constitution of the Federal Republic of Germany, two provisions are relevant for European Integration (Tatham, 2013: 77; Pernice, 1998: 3-4): (a) the Preamble affirms the will of the people to *'promote world peace as an equal partner in a united Europe'* (BMJ, 2012: 1); and (b) correspondingly, Art. 24(1) of the Constitution states that the *'Federation may by law transfer sovereign powers to international organisations'* (BMJ, 2012: 9). However, with the so-called 'eternity clause', the Constitution forms a counterpoint to the above-mentioned provisions. Pursuant to Article 79(3) of the Constitution, *'[a]mendments to this Basic Law affecting the division of the Federation into Länder, their participation on principle in the legislative process, or the principles laid down in Articles 1 and 20 shall be inadmissible'* (BMJ, 2012: 25). After reunification, the new Article 23 of the Constitution borrows the idea of the Preamble and translates it into obligations with legally binding effects and with remarkable provisions, as Pernice (2008: 9-19) concludes. Article 23 guarantees a complete parliamentary process and the full participation of the 'Länder', as well as the observation of fundamental principles of the State, before any transfer of powers to the European level can occur (Tatham, 2013: 81). Nonetheless, the rather general language of the provision in question allows for flexibility on how these principles are put into practice (Pernice, 1998: 11, Pernice, 2006: 18-19). Flexibility, however, leaves room for interpretation, which in terms of the Constitution falls within the scope of the German Federal Constitutional Court (FCC).

Altogether there are three development phases regarding FFC case law in light of European integration and the primacy of EU legislation. In the early days of European Integration, before 1974, FCC case law was principally in accord with EJC case law ('Milk-powder' decision). This phase was followed by FCC judgements leaning towards a more state-dominated interpretation ('Solange I' and 'Solange II' decisions), complemented later on (third phase) with a sense of co-operative acknowledgment, whilst re-affirming the ultimately binding competence of the European Member States ('Maastricht' decision) (Pernice, 2006: 27-28). The last stance was confirmed in the 'Lisbon' decision, where the FCC refers to *'the legal situation as interpreted by the Federal Constitutional Court'* (BVerfGE, 2009: 123 (267) at 331). In short, the FCC claims ultimate authority for protecting fundamental rights in the national system but refrains *'from striking down secondary European legislation per se'* (Tatham, 2013: 212). It reserves the right to review national legal rules that transpose European provisions with regard to whether they occurred on the basis of manifest transgressions of EU competence pursuant to the Lisbon treaty. The FFC rarely invokes this power (as long as fundamental rights are not infringed by EU secondary legislation) and would only do so in a co-operative manner with ECJ jurisdiction (Tatham, 2013: 126). Nevertheless, as Tatham (2013: 99) concludes, *'[t]he FCC [...] has not accepted without question the constitutionalisation of the European legal order, as created an managed by the ECJ. [...] [A] number of aspects have been challenged by the FCC but, so far, they have not led to open conflict with their European counterpart although [...] its ruling in the Lisbon case could render an eventual clash more likely'.*
The constitution of Germany does not explicitly include provisions with respect to energy, as is the case with Article 194 TFEU (BMJ, 2012: 1-58; OJoEU, 2008: 134-135).

4.2 German Transposition of Directive 2009/28/EC and NREAP

Directive 2009/28/EC, as part of the Climate and Energy Package, was signed by the President of the European Parliament and the President of the Council on 23rd April 2009 and published in the Union's Official Journal on 5th June 2009 (see also Tab. 1) (OJoEU, 2009a-f: 11-145).

Pursuant to Article 4(3) of Directive 2009/28/EC (OJoEU, 2009b: 28-29), the German government on 21st December 2009 published a forecast document indicating the estimated potential for joint projects with respect to estimated excess production of energy from renewable sources compared to indicative trajectories in Germany up until 2020 (BMU, 2009: 1; OJoEU, 2009b: 28). One year later, on 4th August 2010, the German government adopted a national action plan on the promotion of the use of energy from renewable sources (BMU, 2010a: 1). Based on the action plan, Germany expects to achieve its national overall target for the share of energy from RES in gross final energy consumption of 18% in 2020 (see Tab. 3) and expects to exceed the overall target by 1,6%, without making use of statistical transfers or joint projects (BMU, 2010a: 11 and 14; OJoEU, 2009b: 30 and 46).

On 29th September 2010, the German government adopted a proposal for a law for the transposition of Directive 2009/28/EC on the promotion of the use of energy from renewable sources (BMU, 2010b: 1). Based on the legislative proposal, the Bundestag (German Federal Parliament, lower house of the German parliament), in a first reading on 11th November 2010, assigned the proposal for further proceedings to the Committee of Environment, Nature Conservation, and Nuclear Safety (in charge), the Committee of Internal Affairs, the Committee of Economics and Technology, the Committee for Food, Agriculture and Consumer Protection, the Committee for Transport, Building and Urban Development, and the Committee for Finance (DIP, 2010a: 7744). Correspondingly, on 26th November 2010, the Bundesrat (German Federal Council, upper house of the German parliament) delivered an opinion on the draft proposal by the German government and called upon the government to consolidate technical and functional specificities for energy efficiency demands for real estate as part of the proposed amendments of German provisions for the transposition of Directive 2009/28/EC (DIP, 2010b: 15).

On 15th December 2010, the German government communicated the opinion of the Bundesrat to the Bundestag including a counterstatement (DIP, 2010c: 7). According to the latter, the law proposed by the government reflected a straight transposition of implementation requirements pursuant to Directive 2009/28/ EC, whereas amendments to German provisions as envisioned in the opinion of the Bundesrat would risk not being in accordance with Directive 2009/28/ EC, and could ultimately cause the danger of being unlawful with regard to European law (DIP, 2010c: 7).

On 24th February 2011, the Bundestag debated the legislative proposal in a second and third reading (DIP, 2011a: 10566-10573), and adopted it based on a draft version of the Committee of Environment, Nature Conservation, and Nuclear Safety from 23rd February 2011 (DIP, 2011b: 3-11). On 18th March 2011, the Bundesrat adopted the draft version of the Committee of Environment, Nature

Conservation, and Nuclear Safety of a law for the transposition of Directive 2009/28/EC on the promotion of the use of energy from renewable sources from 23rd February 2011, as previously accepted by the Bundestag (DIP, 2011c: 111). Approval by the Bundesrat was based on a recommendation of its Committee on Environment, Nature Protection and Reactor Safety not to call upon the joint Mediation Committee of the Bundestag and Bundesrat to resolve differences of opinion between both chambers in terms of adopted legislation and/or legislative proposals (DIP, 2011d: 1; Bundesrat, 2014: 1-2). On 12th April 2011, the President, the Chancellor and the Minister for Environment, Nature Conversation and Reactor Safety of the Federal Republic of Germany signed the 'Law for the Transposition of Directive 2009/28/EC on the Promotion of the Use of Energy from Renewable Sources' (BGBL, 2011a: 635). For an overview of the general legislative procedure, see Tab. 11.

Tab. 11: Legislative Observatory – Law for the Transposition of Directive 2009/28/EC (DIP, 2014)

Legislative Act	President	Bundes-regierung (Government)	Bundestag (Federal Parliament)	Parliament Committe	Bundesrat (Federal Council)	Council Committe
Legislative proposal published	-	15.10.2010	-	-	-	-
1st reading Parliament, Committee referral	-	-	11.11.2010	-	-	-
Recommen-dation Council Committees	-	-	-	-	-	15.11.2010
1st reading Council	-	-	-	-	26.11.2010	-
German government communicates opinion of Council including a counterstatement	-	15.12.2010	-	-	-	-
Recommendation and legislative proposal of the Committee of Environment, Nature Conser-vation, and Nuclear Safety	-	-	-	23.02.2011	-	-
2nd and 3rd reading in Parliament, adoption	-	-	24.02.2011	-	-	-
Recommen-dation of the Committee of Environment, Nature Protection, and Reactor Safety	-	-	-	-	-	07.03.2011
2nd reading Council, adoption	-	-	-	-	18.03.2011	-
Final act signed	12.04.2011	12.04.2011	-	-	-	-
Final act published in Official Journal	15.04.2011	15.04.2011	15.04.2011	15.04.2011	15.04.2011	15.04.2011

The 'Law for the Transposition of Directive 2009/28/EC on the Promotion of the Use of Energy from Renewable Sources' contains several amendments to the following national provisions:

- Renewable Energy Act, 11 overall amendments pursuant to Article 1 of transposition law (BGBL, 2011a: 619-623).

- Renewable Energies Heat Act, 22 overall amendments pursuant to Article 2 of transposition law (BGBL, 2011a: 623-633).

- Act on Energy Statistics, three amendments pursuant to Article 3 of transposition law (BGBL, 2011a: 633).

- Town and Country Planning Code, two amendments pursuant to Article 4 of transposition law (BGBL, 2011a: 633-634).

- Electricity from Biomass Sustainability Regulation, six amendments pursuant to Article 5 of transposition law (BGBL, 2011a: 634).

- Statistical Law for Building Construction, one amendment pursuant to Article 5(a) of transposition law (BGBL, 2011a: 635).

A more differentiated contextualisation of the Eu ropean impact on domestic legal provisions and overall policy measures, notably for the national Renewable Energy Act, is reflected in the National Renewable Energy Action Plan of Germany. As has been mentioned before, according to Article 4 of Directive 2009/28/EC, each Member State is obliged to adopt a National Renewable Energy Action Plan (NREAP) in order to communicate Member States' national targets for transport, electricity, heating and cooling (OJoEU, 2009b: 28). The document was due on 30th June 2010 (OJoEU, 2009b: 28).

Note that the following section reflects the information contained in the NREAP of the year 2010 and does not reflect domestic policy developments and energy statistics since. Those will be subject to the 'national preferences' subsection of this chapter.

In its NREAP (NREAP Germany, 2010: 8) the Federal Republic of Germany drafted a scenario for expected final consumption of energy from 2010–2020, based on energy efficiency and energy savings measures taken before 2009. A second scenario anticipates the gross final energy consumption of Germany, if additional measures be taken into account beyond the status quo of 2009. The developed scenarios, the authors emphasize, cannot be understood as a prediction of the development of energy consumption and the contribution of renewable energies to energy supply, but rather as a plausible and consistent development path that is to be used as a guideline (NREAP Germany, 2010: 8).

Tab. 12: German overall RES targets as defined in Annex I of Directive 2009/28/EC (NREAP Germany, 2010: 13)

National over all target based on Annex I of Directive 2009/28/EC	
Share of energy from renewable sources in gross final consumption of energy in 2005 (%)	5,8%
Target for share of energy from renewable sources in gross final consumption of energy 2005 (%)	18%
Expected total consumption in 2020 after adjustment (ktoe)	197.178
Expected amount of energy from renewable sources in accordance with the target 2020 (ktoe)	35.492

Based on Annex I of Directive 2009/28/EC (see Tab. 3 and Tab. 12), the share of energy from RES in gross final energy consumption was 5,8% in 2005. The legally binding target for Germany to be reached by 2020 is at least 18%. Germany assumes that it will achieve the target by 2020 without the benefit of surpluses from other countries. In addition, it expects to surpass the national target based on a 'scenario with additional energy efficiency measures' (EFF). This surplus could potentially be made available for flexible cooperation mechanisms (NREAP Germany, 2010: 8-9 and 13). In absolute terms, the expected percentages translate to an expected amount of energy from renewable energy sources of 35.492 ktoe with respect to an overall energy consumption of 197.178 ktoe in 2020 (NREAP Germany, 2010: 13).

Tab. 13 gives an overview of the expected trajectory of the RES share in the sectors of heating and cooling, electricity, transport, and in total gross final consumption of energy (NREAP Germany, 2010: 14). Trajectories in Tab. 13 reflect, based on the German NREAP, expected developments that surpass the overall national targets. Since targets by sector are not mandatory according to Directive 2009/28/EC, but must be stated pursuant to Article 4 and Annex VI of the Directive (OJoEU, 2009b: 28 and 60-61), the Federal government regards them as non-binding estimates (NREAP Germany, 2010: 14). With respect to these non-binding estimates, Germany expects a share of RES in the electricity sector of well beyond 30% (38,6%), a share of 15,5% RES in the heating and cooling sector, and a share of 13,2% RES in the transport sector by 2020 (NREAP Germany, 2010: 14).

In the electricity sector the total installed capacity amounts to 27.898 MW, reflecting an electricity production of 61.653 GWh of electricity in 2005 (see Tab. 15). The largest contributor has been wind energy (43%), followed by hydropower (32%) and biomass (23%) (NREAP Germany, 2010: 107). Following 'the estimate of the total contribution (installed capacity, gross electricity consumption) anticipated in Germany of each technology using renewable energy sources with regard to the binding targets for 2020 and the indicative trajectories for the share of energy from renewable sources in the electricity sector [...]' (NREAP Germany, 2010: 113), the installed capacity of all plants producing electricity will quadruple by 2020, with approximately 110.934 MW of installed capacity (NREAP Germany, 2010: 107). The anticipated energy consumption amounts to 216.935 GWh by 2020, reflecting an increase by a factor of 3,5 compared to 2005 (see Tab. 15 and Tab. 14). The share of RES-E will amount to approximately 38,6 % in 2020. In the same year, the share of wind energy will amount to 48% , whilst electricity produced from biomass, photovoltaic, and hydropower follow respectively with 23 %, 19 %, and 9% (NREAP Germany, 2010: 107-108).

Tab. 13: National target for 2020 and expected path for energy from renewable sources in the sectors of heating and cooling, electricity, and transport (NREAP Germany, 2010: 17)

	2005	2010	2011	2012	2013	2014	2015	2016	2017	2018	2019	2020
Renewable sources of energy - heating and cooling (%)	6,6	9	9,4	10	10,5	11,1	11,7	12,4	13,1	13,9	14,7	15,5
Renewable energy sources - electricity (%)	10,2	17,4	19,3	20,9	22,7	24,7	26,8	28,8	31	33,3	35,9	38,6
Renewable energy sources - transport (%)	3,9	7,3	7,5	7,6	7	7	7	7,1	9,3	9,4	9,7	13,2
Renewable energy sources, total (%)	6,5	10,1	10,8	11,4	12	12,8	13,5	14,4	15,7	16,7	17,7	19,6
Of which through cooperation mechanism (%)	0	0	0	0	0	0	0	0	0	0	0	0
Surplus for cooperation mechanism (%)	0	0	2,6	3,2	2,5	3,3	2,2	3,1	2	2,9	0	1,6

Tab. 14: Estimate of the total contribution (installed capacity, gross electricity consumption) anticipated in Germany of each technology using renewable energy sources with regard to the binding targets for 2020 and the indicative trajectories for the share of energy from renewable sources in the electricity sector in the period 2010-2014 including operating hours in h/a (NREAP Germany, 2010: 113)

	2005			2010			2011		
	MW	GWh	h/a	MW	GWh	h/a	MW	GWh	h/a
Hydropower:	4.329	19.687	4.548	4.052	18.000	4.442	4.068	18.000	4.425
< 1 MW	641	3.157	4.925	507	2.300	4.536	511	2.300	4.501
1 MW - 10 MW	1.073	3.560	3.318	987	4.050	4.103	991	4.050	4.087
> 10 MW	2.615	12.615	4.960	2.558	11.650	4.554	2.567	11.650	4.538
from pumped storage power plant	4.012	7.789	1.941	6.494	6.989	1.076	6.494	6.989	1.076
Geothermal power:	0	0	1.000	10	27	2.700	17	53	3.118
Solar energy	1.980	1.282	647	15.784	9.499	602	20.284	13.967	689
photovoltaics	1.980	1.282	647	15.784	9.499	602	20.284	13.967	689
concentrated solar energy	0	0	0	0	0	0	0	0	0
Tides, waves, other ocean energy:	0	0	0	0	0	0	0	0	0
Wind energy:	18.415	26.658	1.448	27.676	44.668	1.614	29.606	49.420	1.669
land-based	18.451	26.658	1.445	27.526	44.397	1.613	29.175	48.461	1.661
offshore	0	0	0	150	271	1.807	432	959	2.220
Biomass:	3.174	14.025	4.419	6.312	32.778	5.193	6.620	34.682	5.239
solid	2.427	10.044	4.138	3.707	17.498	4.720	3.860	18.298	4.740
biogas	693	3.652	5.270	2.368	13.829	5.840	2.523	14.933	5.919
liquid biofuels	54	329	6.093	237	1.450	6.118	237	1.450	6.118
Overall:	27.898	61.653	2.210	53.834	104.972	1.950	60.596	116.122	1.916
from combined heat and power	-	-	0	1.067	5.328	4.993	1.280	6.453	5.041

2012			2013			2014		
MW	GWh	h/a	MW	GWh	h/a	MW	GWh	h/a
4.088	18.000	4.403	4.111	19.000	4.622	4.137	19.000	4.593
515	2.300	4.466	521	2.450	4.702	527	2.450	4.649
995	4.050	4.070	1.000	4.250	4.250	1.005	4.250	4.229
2.557	11.650	4.556	2.590	12.300	4.749	2.604	12.300	4.724
6.494	6.989	1.076	6.494	6.989	1.076	6.494	6.989	1.076
27	97	3.593	40	164	4.100	57	257	4.509
23.783	17.397	731	27.282	20.293	744	30.781	23.218	754
23.783	17.397	731	27.282	20.293	744	30.781	23.218	754
0	0	0	0	0	0	0	0	0
0	0	0	0	0	0	0	0	0
31.357	53.055	1.692	32.973	57.314	1.738	34.802	63.657	1.829
30.566	51.152	1.673	31.672	54.064	1.707	32.763	58.420	1.783
792	1.903	2.403	1.302	3.250	2.496	2.040	5.237	2.567
6.934	36.710	5.294	7.214	38.562	5.345	7.475	40.359	5.399
4.017	19.294	4.803	4.140	20.114	4.858	4.253	20.901	4.914
2.680	15.966	5.957	2.837	16.998	5.992	2.985	18.008	6.033
237	1.450	6.118	237	1.450	6.118	237	1.450	6.118
66.189	125.258	1.892	71.621	135.333	1.890	77.251	146.490	1.896
1.503	7.681	5.110	1.740	9.002	5.174	1.990	10.424	5.238

Tab. 15: Estimate of the total contribution (installed capacity, gross electricity consumption) anticipated in Germany of each technology using renewable energy sources with regard to the binding targets for 2020 and the indicative trajectories for the share of energy from renewable sources in the electricity sector in the period 2015-2020 including operating hours in h/a (NREAP Germany, 2010: 114)

	2015			2016			2017		
	MW	GWh	h/a	MW	GWh	h/a	MW	GWh	h/a
Hydropower:	4.165	19.000	4.562	4.196	19.000	4.528	4.228	19.000	4.494
< 1 MW	534	2.450	4.588	539	2.450	4.545	546	2.450	4.487
1 MW - 10 MW	1.012	4.250	4.200	1.019	4.250	4.171	1.026	4.250	4.142
> 10 MW	2.620	12.300	4.695	2.638	12.300	4.663	2.657	12.300	4.629
from pumped storage power plant	6.494	6.989	1.076	6.494	6.989	1.076	6.494	6.989	1.076
Geothermal power:	79	377	4.772	107	534	4.991	142	730	5.141
Solar energy	34.279	26.161	763	37.777	29.148	772	41.274	32.132	779
photovoltaics	34.279	26.161	763	37.777	29.148	772	41.274	32.132	779
concentrated solar energy	0	0	0	0	0	0	0	0	0
Tides, waves, other ocean energy:	0	0	0	0	0	0	0	0	0
Wind energy:	36.647	69.994	1.910	38.470	76.067	1.977	40.415	82.466	2.040
land-based	33.647	61.990	1.842	34.371	64.583	1.879	34.815	66.873	1.921
offshore	3.000	8.004	2.668	4.100	11.484	2.801	5.340	15.592	2.920
Biomass:	7.721	42.090	5.451	7.976	43.729	5.483	8.211	45.299	5.517
solid	4.358	21.695	4.978	4.472	22.369	5.008	4.575	23.050	5.038
biogas	3.126	18.946	6.061	3.267	19.884	6.086	3.399	20.798	6.119
liquid biofuels	237	1.450	6.118	237	1.450	6.118	237	1.450	6.118
Overall:	82.891	157.623	1.902	88.526	168.479	1.903	94.009	179.626	1.911
from combined heat and power	2.250	11.937	5.305	2.530	13.533	5.349	2.823	15.220	5.391

2018			2019			2020		
MW	GWh	h/a	MW	GWh	h/a	MW	GWh	h/a
4.258	19.000	4.462	4.286	20.000	4.666	4.309	20.000	4.641
552	2.450	4.438	558	2.550	4.570	564	2.550	4.521
1.032	4.250	4.118	1.038	4.500	4.335	1.043	4.500	4.314
2.674	12.300	4.600	2.689	12.950	4.816	2.702	12.950	4.793
7.900	8.395	1.063	7.900	8.395	1.063	7.900	8.395	1.063
185	976	5.276	236	1.281	5.428	298	1.654	5.550
44.768	35.144	785	48.262	38.243	792	51.753	41.389	800
44.768	35.144	785	48.262	38.243	792	51.753	41.389	800
0	0	0	0	0	0	0	0	0
0	0	0	0	0	0	0	0	0
41.909	89.210	2.129	43.751	96.359	2.202	45.750	104.435	2.283
35.188	68.913	1.958	35.479	70.694	1.993	35.750	72.664	2.033
6.722	20.297	3.019	8.272	25.666	3.103	10.000	31.771	3.177
8.440	46.761	5.540	8.648	48.133	5.566	8.825	49.457	5.604
4.672	23.633	5.058	4.750	24.139	5.082	4.792	24.569	5.127
3.531	21.678	6.139	3.660	22.543	6.159	3.796	23.438	6.174
237	1.450	6.118	237	1.450	6.118	237	1.450	6.118
99.561	191.092	1.919	105.183	204.016	1.940	110.934	216.935	1.956
3.129	16.986	5.429	3.444	18.837	5.470	3.765	20.791	5.522

Tab. 16: Overview of all policies and measures according to NREAP Germany (NREAP Germany, 2010: 20)

Name and reference of the measure	Type of the measure	Expected result	Target group and/or activity	Exists/ planned	Date of beginning and end of the measure
Renewable Energy Act (EEG)	Legislative	Increased share of renewable energies in electricity	Investors, private households	Exists	Start: April 2000 (as a follow-up regulation to the Electricity Feed Act of 1991); amendments 2004 and 2009, next revision in 2011; the law is not limited in time
Renewable Energies Heat Act (EEWärmeG)	Legislative	Increased share of renewable energies in the heating of buildings (focus on new buildings)	Building owners (private and public)	Exists	Start: Jan 2009; first revision 2011
Market Incentive Programme (MAP)	Financial	Investments in renewable energy in heating	Private households, investors	Exists	Start: 1999 financed from funds established in EEWärmeG; until 2012
KfW-funding-programs (e.g. CO2 renovation-program)	Financial	Energy efficiency measures and investments in renewable energy in buildings	Private households, investors, building owners, municipalities, social services	Exists	e.g. Start: 1996, End of measures 2011
Combined Heat and Power Act (KWKG)	Legislative	New construction, modernization and operation of CHP-plants and heating networks	Power plant operators, energy suppliers, investors	Exists	Start: April 2002, amendment in January 2009
Energy Saving Ordinance (EnEV)	Legislative	Compliance with minimum standards for energy efficiency in buildings and heating/cooling systems in new construction and renovation of residential and non-residential buildings	Building owners (private and public)	Exists	Start (current version dated 01.10.2009): October 2007, Basis: Energy Saving Ordinance of 28.03.2009; next amendment 2011/2012
Biofuels Quota Act (BioKraftQuG)	Legislative	Minimum share of biofuels of total fuel put into circulation, and tax incentive for certain biofuels	Companies that bring fuels on the market	Exists	Start: January 2007, Duration: beyond 2020 / tax incentive until the end of 2015

Following the general introduction of expected scenarios for RES share in Germany by 2020, the NREAP outlines a detailed description of national measures to implement the overall domestic target of Directive 2009/28/EC (NREAP Germany, 2010: 20). The outline is in accordance with the Decision of the Commission to establish a template for National Renewable Energy Action Plans (EC, 2009: 1).

Article 13 of Directive 2009/28/EC calls for a comprehensive, transparent, proportionate, discrimination-free and simple authorisation process, limited in time and facilitating smaller and decentralised projects (OJoEU, 2009b: 33; EC, 2013b: 7). General administrative procedures concerning national rules for authorisation, certification and licensing procedures for plants producing electricity from RES, and transmission and distribution of RES-E, are outlined in this Article (OJoEU, 2009b: 32). Germany addresses the specific measures for meeting the requirements pursuant to Article 13 of Directive 2009/28/EC (OJoEU, 2009b: 32) with respect to administrative procedures and spatial planning (Article 13(1)), technical specifications (Article 13(2)) and buildings (Article 13(3)) in the respective NREAP chapters and in accordance with the Commission's template (NREAP Germany, 2010: 20-36; EC, 2009: 4).

According to Article 16 of Directive 2009/28/EC, access to and operation of electricity grids (Article 16) must meet the demands of increasing RES-E production. In consequence, Member States are responsible for taking appropriate measures for the development of transmission and distribution infrastructure, storage facilities, and the secure operation of the electricity system (OJoEU, 2009b: 35). Of particular relevance for the promotion of RES-E are the transparent and non-discriminatory criteria (Article 16(2) a, b and c) leading to a guarantee of transmission and distribution of electricity produced from RES, priority access or guaranteed access to the grid-system for electricity produced from RES, and priority for generating installations using RES, as long as the secure operation of the national electricity system permits it (OJoEU, 2009b: 35). Again, Germany addresses specific measures taken regarding electricity infrastructure development (Article 16(1), 16(3) to (6)) and electricity grid operation (Article 16(7), 16(9), and 16(10)) in the respective NREAP chapters and in accordance with the Commission's template (NREAP Germany, 2010: 45-56; EC, 2009: 4).

Unfortunately, the NREAP does not contain information about planned or implemented measures with respect to administrative barriers as addressed in Article 22(3) of Directive 2009/28/EC. This is because Article 22(3) of the Directive solely demands Member States to address in their first progress reports (rather than in the NREAPs) whether they (a) intend to establish a single administrative body processing authorisation, certification and licensing for renewable energy application, (b) provide for an automatic approval of planning and permit applications, or (c) indicate geographical locations suitable for exploitation of energy from renewable sources (OJoEU, 2009b: 42).

A detailed presentation and/or evaluation of German legislation, the authors of the NREAP conclude, would be too extensive (NREAP Germany, 2010: 21). Whilst substantial information with respect to the specific requirements in terms of overall national measures to implement the national targets of Directive 2009/28/EC, as outlined in the Commission's template, can be found in the respective NREAP sections and, an overview of all policies and measures can be found in Tab. 16 (NREAP Germany, 2010: 20-54; EC, 2009: 4).

With the rather timely adoption of a 'Law for the Transposition of Directive 2009/28/EC on the Promotion of the Use of Energy from Renewable Sources' (BGBL, 2011a: 635), and with the release of an NREAP (NREAP Germany, 2010: 1-182) in accordance with the template for National Renewable Energy Action Plans of the European Commission (EC, 2009: 1-40), it seems that the German transposition of Directive 2009/28/EC has begun solidly. To which extent this impression is supported by a more detailed evaluation of the nation's progress in the promotion of renewable energies will be the subject of the following section.

4.2.1 Summary

On 12th April 2011, the President, the Chancellor and the Minister for Environment, Nature Conversation and Reactor Safety of the Federal Republic of Germany signed the 'Law for the Transposition of Directive 2009/28/EC on the Promotion of the Use of Energy from Renewable Sources' (BGBL, 2011a: 635). For an overview of the general legislative procedure, see Tab. 11.

In its NREAP (NREAP Germany, 2010: 8), the Federal Republic of Germany drafted a scenario for expected final consumption of energy 2010-2020 based on energy efficiency and energy savings measures taken before 2009. Based on Annex I of Directive 2009/28/EC (see Tab. 3 and Tab. 12) ,the share of energy from RES in gross final energy consumption was 5,8% in 2005. The legally binding target for Germany to be reached by 2020 is at least 18% . Germany assumes that it will achieve the target by 2020 without the benefit of surpluses from other countries. In addition, it expects to surpass the national target based on a 'scenario with additional energy efficiency measures' (EFF). This surplus could potentially be made available for flexible cooperation mechanisms (NREAP Germany, 2010: 8-9 and 13). Following *'the estimate of the total contribution (installed capacity, gross electricity consumption) anticipated in Germany of each technology using renewable energy sources with regard to the binding targets for 2020 and the indicative trajectories for the share of energy from renewable sources in the electricity sector'* (NREAP Germany, 2010: 113), the installed capacity of all plants producing electricity will quadruple by 2020, with approximately 110.934 MW of installed capacity (NREAP Germany, 2010: 107). The share of RES-E will amount to approximately 38,6% in 2020. The information contained in the NREAP of the year 2010 does not reflect domestic policy developments and energy statistics after 2010. Those will be subject to the National Preferences' subsection of this chapter.

With the rather timely adoption of a 'Law for the Transposition of Directive 2009/28/EC on the Promotion of the Use of Energy from Renewable Sources' (BGBL, 2011a: 635), and with the release of a NREAP (NREAP Germany, 2010:

1-182) in accordance with the template for National Renewable Energy Action Plans of the European Commission (EC, 2009: 1-40), it seems that the German transposition of Directive 2009/28/EC has got off to a good start. To which extent this impression is supported by a more detailed evaluation of the nation's progress in the promotion of renewable energies will be subject of the following section.

4.3 Implementation Evaluation of Directive 2009/28/EC in Germany

In accordance with the section '3.6 Implementation of Directive 2009/28/EC in light of RES-E', evaluation of the implementation of Directive 2009/28/EC in Germany will be predominantly based on a 'renewable energy progress and biofuels sustainability' report produced for the Commission by ECOFYS, Fraunhofer, Becker Bütter Held (BBH), Energy Economics Group and Windrock International (Ecofys, 2012). Whilst this might cause criticism in terms of a lack of counterfactual reasoning – given there is only one major source – it is deemed adequate for the following reasons: (i) Empirical information about the implementation effectiveness serves as one aspect in the overall thesis design. It lies beyond the scope of this thesis to conduct a new comprehensive implementation analysis for all European Member States to counterpoint the study in question. (ii) Whilst additional information on certain aspects reflected in the evaluation report might be available from different sources (e.g. statistical agencies from the Member States in question), they do not apply the same evaluation methodology, which makes a data synopsis illegitimate. (iii) The research in question has been carried out for the Commission and informs its renewable energy progress report (EC, 2013b), and the 'commission staff working' document (EC, 2013c). It therefore not only reflects the basis for official evaluation statements in the political process, but also guarantees a common analysis basis for inter-state comparability. It furthermore provides the elements of evidence for further decisions in the context of infringement procedure for non-communication of national transposition measures and/or non-transposition of supranational provisions.

4.3.1 Past Progress Evaluation

As can be seen from 'Tab. 4: Actual and planned RES Shares 2009-2011/2012 (Ecofys, 2012: 18)', in conjunction with 'Tab. 13: National target for 2020 and expected path for energy from renewable sources in the sectors of heating and cooling, electricity, and transport (NREAP Germany, 2010: 17)', Germany reached its NREAP indicative target RES share in 2010 and exceeded it by 0,9 percentage points, which equals a positive deviation of 8,94% (Ecofys, 2012: 18; NREAP Germany, 2010: 17). Therefore, according to Ecofys (2012: 20), since Germany had a growth rate in 2009/2010 above the required average, it is in a good situation to reach its overall RES target in 2020. With respect to the non–legally binding RES-E target for 2010, Germany exceeded its target by 4% (see 'Tab. 5: Actual and planned RES-E shares (NREAP_EU27, 2010; Eurostat, 2013b), and 'Tab. 13: National target for 2020 and expected path for energy from renewable sources in the sectors of heating and cooling, electricity, and transport (NREAP Germany, 2010: 17)') (Ecofys, 2012: 21-22 and 27; NREAP Germany, 2010: 14; EC, 2013c: 4). Germany is among the sixteen Member States that achieved a growth rate in the RES-E sector that was higher than the average growth rate needed to move from the NREAP 2010 RES-E target to the 2020 RES-E target (Ecofys, 2012: 23).

The growth of RES-E technologies in between 2009-2010 are described below. Please regard the data cited in conjunction with Tab. 14: Estimate of the total contribution (installed capacity, gross electricity consumption) anticipated in Germany of each technology using renewable energy sources with regard to the binding targets for 2020 and the indicative trajectories for the share of energy from renewable sources in the electricity sector in the period 2010-2014 including operating hours in h/a (NREAP Germany, 2010: 113)':

- Hydropower: According to Ecofys (2012: 25), Germany achieved a growth rate of electricity produced from hydropower of > 10 MW (7,1%) in between 2009 and 2010. In the category small hydro, which combines installations of < 1 MW and 1–10 MW, Germany (7.900 GWh) together with Italy (9.321 GWh) and France (7.241 GWh) was among the leading producers in Europe (Ecofys, 2012: 25 and 33). Note: The latter category is in objection to the categories indicated in the NREAP, since it combines two subcategories used in the NREAP (see Tab. 14) (NREAP Germany, 2010: 113; Ecofys, 2012: 33).

- Photovoltaic: With 11.700 GWh of electricity produced from photovoltaic in 2010, Germany was the most important producer of electricity from photovoltaic in Europe. Ecofys (2012: 25) indicates a growth rate between 2009 and 2010 of 43,59% . Given an estimate of 9.499 GWh of electricity produced from photovoltaic in 2010 (see Tab. 14), an actual production of 11.700 GWh electricity from photovoltaic in 2010 (Ecofys, 2012: 32) shows an actual deviation from planned trajectories of 23,17% (own calculation).

- Wind: In 2010 Germany, together with Spain, was the biggest producer of electricity from onshore windmills (42.900 GWh and 42,732 GWh, respectively). The growth rate between 2009 and 2010 was 3,65% . Installation of new capacities, however, slowed down in 2010 (Ecofys, 2012: 25 and 29). Germany is one of only seven European Member States that report any offshore wind. Owing to lengthy administrative procedures and uncertainties with respect to grid connection procedures, however, German offshore wind development in 2010 was lagging behind its indicated development path. The growth rate in between 2009 and 2910 was 80,95% (Ecofys, 2012: 25 and 28; NREAP Germany, 2010: 113-114).

- Biomass: In 2010, Germany led the EU in electricity generation from solid biomass (16.000 GWh), with a growth from 2009-2010 of 1,88% . The same is true for electricity produced from biogas (16.200 GWh), with a growth rate of 22,22% (Ecofys, 2012: 25 and 30-31).

Overall, Germany was among the sixteen Member states that had a RES growth rate from 2009 to 2010 above required average (Ecofys, 2012: 20).
The assessment of overall policies and measures for the promotion of renewable energies in Germany, as indicted in the NREAP, leads the authors of the Ecofys-report (2012: 101) to conclude that Germany did fulfil its overall NREAP policy commitments, and its NREAP policy commitments for the promotion of RES-E. With respect to the latter, support levels for each technology are *'good'*, as is the long-term security of support (Ecofys, 2012: 101). Indeed, as will be described later on, Germany did *'well'* in adopting new RES-E measures between its NREAP and its Progress Report, implementing measures it had not planned in its NREAP. In August 2011, Germany shut down eight of its oldest nuclear power facilities, with the remaining nine reactors gradually to be switched off by 2022. Under what is referred to as the 'Energiewende', the lost capacities are envisioned to be replaced by RES-E plants (Ecofys, 2012: 113). Germany implemented a technology-specific feed-in tariff and an optional feed-in premium for the promotion of RES-E guaranteed for 20 years (Ecofys, 2012: 113).
With respect to the overall assessment of administrative procedures (see Tab. 6: Assessment of administrative procedures in the Member States (Ecofys, 2012: 186-187; EC, 2013c: 9)), Germany receives a *'fair'* rating. Following the Ecofys-report (2012: 194), no major non-cost barriers were reported in Germany. On the federal level, privileges are given to renewable energy projects, in particular with respect to the German Federal Building Code. Most other matters, however, lie within the competencies of the 'Bundesländer' (federal states). In some states, there are one-stop shops in place, and there may be only one single permit needed. In the federal state of Brandenburg, online application is possible. In many federal states, in particular with respect to wind, plans identifying suitable geographical sites are available (Ecofys, 2012: 194).
With respect to Germany's progress in electricity grid integration (see Tab. 7: Member State rating on progress in electricity grid integration (Ecofys, 2012: 177, 179 and 181)'), it receives an *'advanced'* rating on overall RES grid integration, as well as for addressing grid capacity limitations, and for rules for bearing and sharing the costs of grid development (Ecofys, 2012: 177, 179 and 181).
The quantitative assessment of the future deployment of RES in Germany, based on the Green-X model (see Text Box 4: Short characterisation of the model Green-X by EEG (EEG, 2007: 3-4) leads the authors of the Ecofys report (2012: 66) to predict an expected share of RES in Germany under the CPI Scenario of max. 16,4% , and 16,8% under the CPI+PPI Scenario in 2020 (see also section '3.6.4 Expected RES shares in 2020'). With respect to the latter

evaluation, and contrary to the overall positive assessment of the German implementation effectiveness of measures and policies for the promotion of renewable energies based on Directive 2009/28/EC, the authors of the Ecofys report (2012: 66) do not expect Germany to reach its overall RES target by 2020 based on measures and policies referred to in its NREAP.

4.3.2 Summary

As can be seen from 'Tab. 4: Actual and planned RES Shares 2009-2011/2012 (Ecofys, 2012: 18))', in conjunction with 'Tab. 13: National target for 2020 and expected path for energy from renewable sources in the sectors of heating and cooling, electricity, and transport (NREAP Germany, 2010: 17)', Germany reached its NREAP indicative target RES share in 2010 and exceeded it by 0,9 percentage points, which equals a positive deviation of 8,94% (Ecofys, 2012: 18; NREAP Germany, 2010: 17). Therefore, according to Ecofys (2012: 20), since Germany had a growth rate in 2009/2010 above the required average annual, it is in a good situation to reach its overall RES target in 2020. Germany is among the sixteen Member States that achieved a growth rate in the RES-E sector in between 2009 and 2010 higher than the average growth rate needed to move from the NREAP 2010 RES-E target to the 2020 RES-E target (Ecofys, 2012: 23).

According to the administrative assessment carried out in the Ecofys report (2012: 186-187), Germany receives a 'fair' rating. With respect to Germany's progress in electricity grid integration (see Tab. 7: Member State rating on progress in electricity grid integration (Ecofys, 2012: 177, 179 and 181)'), it receives an 'advanced' rating for overall RES grid integration, as well as for addressing grid capacity limitations, and for rules for bearing and sharing the costs of grid development (Ecofys, 2012: 92).

With respect to the quantitative assessment of the future deployment of RES in Germany, and contrary to the overall positive assessment of the German implementation effectiveness of measure and policies for the promotion of renewable energies based on Directive 2009/28/EC, the authors of the Ecofys report (2012: 66) do not expect Germany to reach its overall RES target by 2020, based on measures and policies referred to in its NREAP. On the one hand, empirically, Germany had a growth rate in 2009/2010, which is above the average annual growth rate required, leading to the conclusion that the country is in a good situation to reach its overall RES target in 2020 (Ecofys, 2012: 20). On the other hand, based on Green-X model projections (EEG, 2007: 3-4), Germany is not expected to reach its overall RES target of 18% (Ecofys, 2012: 66). Again, the last section reflects information contained in the German NREAP of from 2010 and so does not reflect domestic policy developments and energy statistics since. Those will be subject to the 'National Preferences' subsection of this chapter.

4.4 German National Preferences

So far, the supranational energy policy context for the promotion of renewable energies in the Union has been the focal point of analysis. Emphasis has been given to the broader legislative context leading, in particular, to the adoption of Directive 2009/28/EC. Additional emphasis has been given to the European polity framework based on the respective European treaties, the interpretation of the treaties in light of national primary law (based on case law of the ECJ), and the treaty provisions that enable the adoption of renewable energy policies. The aforementioned aspects have been described, linked together and analysed. Furthermore a first evaluation of the implementation effectiveness of Directive 2009/28/EC in the European Union has been sketched.

In a second step, the domestic response to the European stimuli has been described. This is the case for the German polity with respect to the European polity, the interpretation of the German constitution in light of European primary law (based on case law by the FCC), and the implementation of Directive 2009/28/EC both in legal and empirical terms. With respect to explanatory factor (ii) (see '2.3.2 Applied Theory'), it is now time to focus on the domestic policy process, which reflects issue-specific interest determined by geopolitical and economic factors through national preferences.

As has been stated, the overall framework for the applied theory of this thesis, illustrated in Fig. 2, includes theories derived from Liberal Intergovernmentalism (see: '2.3 Applied Framework, Theory and Model'). The following aspect should be re-emphasized: the nation state in Liberal Intergovernmentalism is considered to be a unitary and rational actor. This assumption is not to be mistaken with the idea that states are unitary in their internal politics. The unitary actor assumption, which maintains the idea that each nation state acts in international negotiations 'as if' with a single voice, subsumes the idea that 'once particular objectives arise out of [...] domestic competition, states strategize as unitary actors vis-à-vis other states in an effort to realize them' (Moravcsik, 1998: 22). The same holds for the assumption that nation states act rationally. The idea is that states make internal decisions 'as if' efficiently conducting a weighted choice based on a stable set of underlying principles. Both assumptions, however, should not be over emphasised (Moravcsik, 1998: 23). Fundamental goals of states are neither uniform nor fixed. State preferences vary over time and in response to changes in, for example, economic, ideological, or geopolitical environments (Moravcsik, 1998: 22-23). It is the goal of this section to summarise the particular objectives that arise out of domestic competition (Moravcsik, 1998: 22). Thus, the following analysis is not to be mistaken for a nation-bound policy analysis (see '2.1 Nation-Bound Policy Analysis Frameworks'). For an extensive nation-bound, multi-level policy analysis with respect to German RES-E promotion, see Bernd Hirschl (2008). However, features described in the nation-bound policy analysis theory approaches inform the 'metatheoretical language' (Ostrom, 2007: 25) reflected in the overall theory framework, and is to be understood in conjunction with, in particular, the applied Liberal Intergorvenmentalism approaches underlying the on-going discussion.

4.4.1 German Energy Mix

The German primary energy mix is – in comparison with other industrialised nations – dominated by fossil fuels (Röhrkasten and Westphal, 2012: 329). As can be seen from Fig. 6:, oil has a share of 33% in the primary energy mix, natural gas 21% , hard coal 13% , lignite 12% , renewables 12% , and nuclear 8% (BMWi, 2014a, based on AGEB, 2013).[10]
With regard to energy dependency, Fig. 7 illustrates that Germany is heavily dependent on net energy imports. It imports more than 98% of its oil consumption, over 85% of its gas consumption, and covers more than 80% of its hard coal demands through external supply (BMWi, 2014b, based on AGEB, 2013). All of Germany's uranium needs are obtained from foreign sources (Röhrkasten and Westphal, 2012: 329; BMWi, 2014b, based on AGEB, 2013). Whilst lignite is the only energy source where Germany is self-sufficient, it relies overall for about 68% of its primary energy supply on imports (BMWi, 2014b, based on AGEB, 2013).

Fig. 6: Primary Energy Consumption Germany in 2012 (BMWi, 2014a, based on AGEB, 2013)

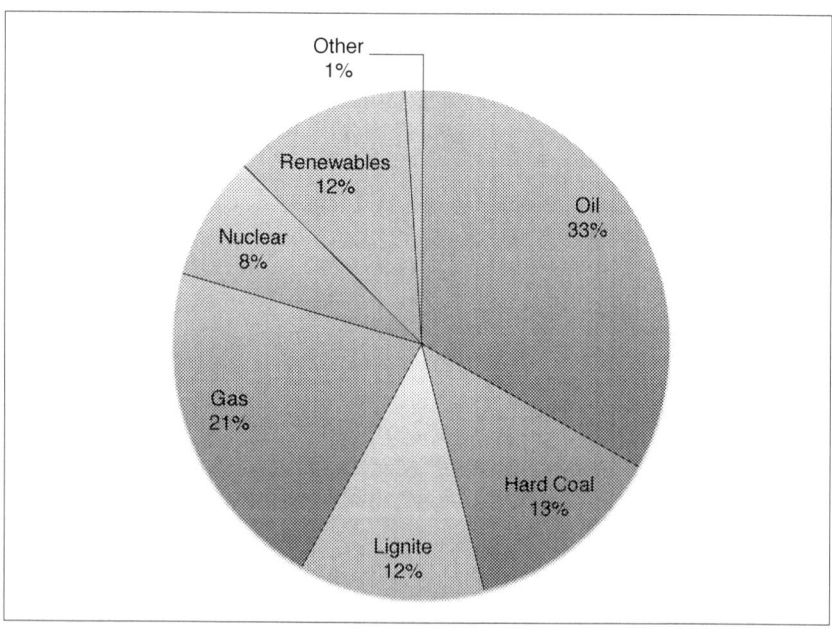

---

[10] Primary energy shall be defined as an energy carrier directly withdrawn from nature, that has not been subject to transformation (Pelte, 2010: 34).

Major energy suppliers for oil and gas to Germany are Russia (37% for oil, and 36% for gas in 2011), as well as Great Britain (14% for oil in 2011), Norway (31% for gas in 2011), and the Netherlands (20% for gas in 2011) (Röhrkasten and Westphal, 2012: 330-331).

Fig. 7: Net Energy Imports Germany in 2012 (BMWi, 2014b, based on AGEB, 2013)

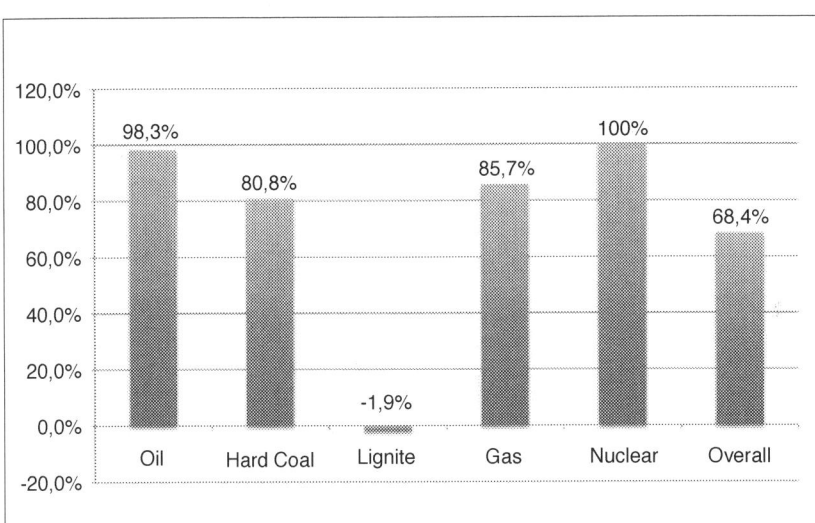

Concerning the conversion of energy into final energy, in particular electricity, about 26 % of gross electricity generation in 2012 was based on lignite (BMWi, 2014c, based on AGEB, 2013 and BDEW, undated; Pelte, 2010: 34-35). In addition, roughly 19% of gross electricity generation in 2010 was based on hard coal, 22% on renewable energy sources, 16% on nuclear, and about 11% on gas (BMWi, 2014c, based on AGEB, 2013 and BDEW, undated).

Fig. 8: Gross Electricity Generation in 2012 (BMWi, 2014c, based on AGEB, 2013 and BDEW, undated)

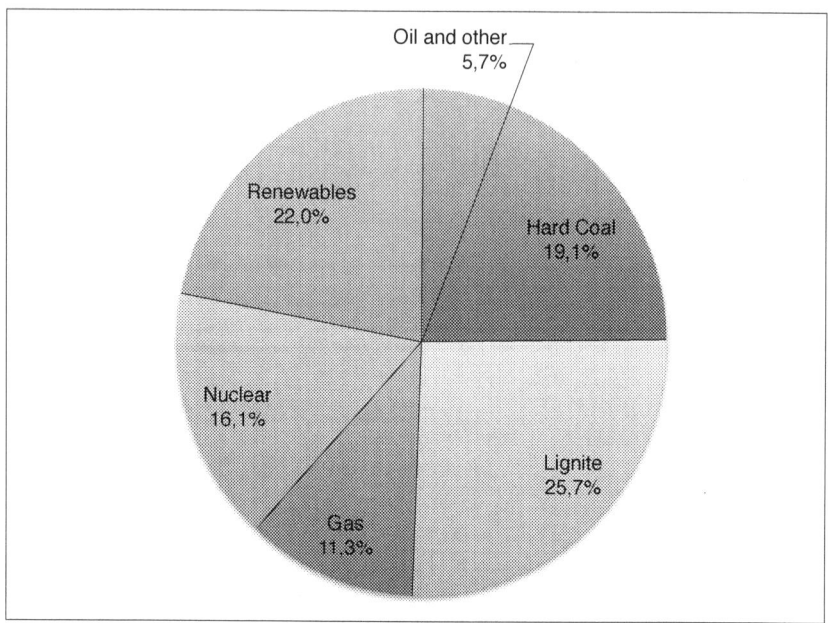

4.4.2 Domestic Energy Policies as Indicator of National Objectives

Following a seaquake on 11th March 2011, Japan was hit by a tsunami leading to what is referred to as the Fukushima nuclear disaster (IPPNW, 2014). In a government declaration of 17th March 2011, the Chancellor of Germany, Angela Merkel, declared the incident to be a *'catastrophe'* of *'apocalyptic size'* (Bundesregierung, 2011a: 2). In consequence, the Federal Government and the Federal State Governments agreed, based on the Atomic Energy Act, to shut down immediately all German nuclear reactors in operation since 1980, or earlier, for a three-month period (Bundesregierung, 2011a: 6). The goal of this three-month moratorium, affecting seven nuclear reactors, was to reassess the safety of the reactors and reanalyse the energy policy of the Federal Republic with respect to the implications of Fukushima (Bundesregierung, 2011a: 6-10). In a government declaration of 9th June 2011, the Chancellor declared the beginning of consultations for a new *'architecture of the German energy supply'* and presented eight legislative proposals previously adopted by the German government (Bundesregierung, 2011b: 1 and 3). The Chancellor offered the following reasons for a new energy infrastructure: *'I made a new assessment, only he can accept the remaining risk of nuclear power, who is convinced, within the boundaries of human conceivability, it will not occur. If it occurs nonetheless, the consequences are, both in geographical and temporal terms, so devastating and far-reaching, they by far surpass the risk of all alternative energy carriers. Before Fukushima I accepted the remaining risk of nuclear power, for I was convinced it would not occur, within the boundaries of human conceivability, in a high-technology country with high security standards such as Japan. But it occurred'* (Bundesregierung, 2011b: 2).

In consequence, on 30th June 2011, the Bundestag adopted, based on a draft proposal by the Conservative and Liberal parliamentary group, the 13th law on the amendment of the Atomic Energy Act (DIP, 2011e: 13404; DIP, 2011f: 1-8; BGBL, 2011b: 1704-1705). With the entry into force of this law on 31st July 2011, the seven oldest German nuclear power plants lost their operating license and the three newest facilities must end electricity production by 31st December 2022 (BGBL, 2011b: 1704; BMU, 2011: 1). The entire phase-out of nuclear energy, Merkel stated in her government declaration (Bundesregierung, 2011b: 4), is to start the *'era of renewable energies'*, with renewable energy being the foundation of the future German energy supply. Since the legally binding phase-out of nuclear energy by 2022 is considered a turning point in the strategy and energy policy of Germany, it is commonly referred to as the 'Energiewende', or Energy Transition (Morris and Pehnt, 2012: Preface). However, in order to understand the key elements underlying German energy policy, ,in particular with respect to the promotion of renewable energies and renewables energy electricity, a historical perspective cannot be limited to the 'Energiewende'. Instead, key elements of German renewable energy policies and measures, dating back to the legislative origins of renewable energy promotion, need to be taken into account.

The first successful initiative for a Feed-In Act for electricity from renewable energy sources (EEG-Clearingstelle, 2014a: 1) dates back to an unlikely alliance of Matthias Engelsberger (Member of Parliament, Conservative) and Wolfgang Daniels (Member of Parliament, Green Party) in the late 1980s (Hirschl, 2008: 131). Although inter-fractional cooperation between Conservatives (here Christian Social Union, CSU) and Greens (here Alliance 90/The Greens) was uncommon, owing to strong ideological reservations at the time, both worked together on a draft proposal for a Feed-In Act for RES-E (Hirsch, 2007: 131-132). On account of a general unwillingness on the part of Conservative and Liberal parliamentary group members, reflecting the majorities leading to the third Conservative–Liberal government coalition (1987-1991), the legislative procedure was not initiated with a joint draft proposal including the Greens parliamentary group as originator (Hirsch, 2007: 131-132; ZEIT Online, 2006: 1). The official draft proposal of the Conservative and Liberal parliamentary group, starting the legislative procedure for the first German Feed-In Act adopted by parliament on 5th October 1990, reflects, however, an initiative pushed forward by bipartisan co-operation of Conservative and Green parliamentarians (Hirsch, 2007: 131-132; ZEIT Online, 2006: 1; EEG-Clearingstelle, 2014a: 1). The subject matter of the provision was a guaranteed reimbursement for electricity produced from water, wind, solar, biogas, and biomass facilities with an installed capacity of < 5 MW, to be paid for by the electricity suppliers (EEG-Clearingstelle, 2014a: 2; BGBL, 1990: 2633).

It took another decade until a Renewable Energy Act replaced the Feed-in Act. Starting in autumn 1998, Germany was governed by a coalition of the Social Democrats and the Greens under Chancellor Gerhard Schröder (Morris and Pehnt, 2012: 49). Correspondingly, on 13th December 1999, a proposal for a Renewable Energy Act was drafted by the Social Democrats and the Greens parliamentary group (EEG-Clearingstelle, 2014b: 1). The act was published in the Official Journal of the Federal Republic of Germany (Bundesgesetzblatt) on 31st March 2000 (BGBL, 2000: 305). The subject matter and scope of the law was, as laid out in Article 1 (BGBL, 2000: 305), *'to significantly increase the contribution of electricity from renewable energy sources for the electricity supply, to, in accordance with the goals of the European Union and the Federal Republic of Germany, double the share of energy from renewable sources in gross final consumption of energy in 2010'*. It is interesting to note that the Renewable Energy Act explicitly refers to the European Union – namely, to a non-legally binding 1997 Commission communication 'Energy for the future: renewable sources of energy - White Paper for a Community strategy and action plan' – with the objective of attaining, and therefore doubling, a penetration of 12% renewable energy sources in the EU by 2010 (Europa.eu, 2001: 1). Furthermore, the Renewable Energy Act incorporates *'impulses'* of Directive 96/92/EC *'of the European Parliament and of the Council of 19 December 1996 concerning common rules for the internal market in electricity'* (EEG-Clearingstelle, 2014b: 1; OJoEU, 2007b: 1).

It was under the same government coalition that, in 2001, an agreement between the German government and nuclear power plant owners was reached concerning a nuclear phase-out, that is, a shut-down of all nuclear power plants by roughly 2022 (Morris and Pehnt, 2012: 49). The agreement includes fixed allowances for electricity production quantities for each nuclear facility starting on 1st January 2000. These define the remaining time-span of the operating license validity for each facility (BMU, 2001: 4). It furthermore includes tighter security standards for the remaining operating period of the power plants and specifications on transport of nuclear waste and nuclear waste repositories (BMU, 2001: 6-9). The signatories acknoweldge the fact that the government intends to enshrine the results of the agreement in an amended Atomic Energy Act. On behalf of the energy suppliers, the agreement was signed by Ulrich Hartmann for E.ON AG, Dr. Dietmar Kuhnt for RWE AG, Gerhard Goll for EnBW and Dr. Manfred Timm for Hamburger Electrizitäts-Werke AG (BMU, 2001: 14). On behalf of the German Government, Gerhard Schröder, Chancellor; Jürgen Trittin, Secretary of Environment, Nature Protection and Reactor Safety; and Dr. Werner Müller, Secretary for Economy and Technology, signed the agreement (BMU, 2001: 14).

On 14th December 2001, the Bundestag (lower house) adopted, based on a recommendation of the Committee of Environment, Nature Conservation, and Nuclear Safety, the draft proposal of the Social Democrats and the Greens parliamentary group on a law for the organised termination of nuclear power for commercial use (DIP, 2002: 1). Approval by the Bundesrat (upper house) was based on its decision of 1st February 2002 not to call upon the joint Mediation Committee of the Bundestag and Bundesrat to resolve differences of opinion between both chambers in terms of adopted legislation and/or legislative proposals (DIP, 2002: 2). The law amending the Atomic Energy Act was published on 26th April 2002 (BGBL, 2002: 1351). With its entry into force, the amendment act gave legal grounds to the framework implications of the agreement between the government and representatives of German energy suppliers for an organised termination of commercial nuclear power in Germany (BMU, 2002a: 1).

A next important aspect in the legislative history with respect to RES and RES-E promotion in Germany is the law on the readjustment of the Renewable Energy Act in 2004. As early as 1992, the representatives of the government of Germany signed the Agenda 21 of the United Nations at a 'United Nations on Environment & Development'conference in Rio, Brazil (UN, 1992: 1). As a result, in 2002, the government coalition under Chancellor Gerhard Schröder adopted a strategy for sustainable development – 'Perspectives for Germany: Our Strategy for Sustainable Development' – in which it set as its goal, inter alia, *'to increase the proportion of renewable energy sources to 4,2% of primary energy consumption and 12,5% of electricity consumption between the years 2000 and 2010'* (Bundesregierung, 2002: 97). Additionally, in 2001, the European Parliament and the Council adopted Directive 2001/77/EC *'on the promotion of electricity produced from renewable energy sources in the internal electricity market'* (OJoEU, 2001: 33). The Annex of the Directive defines Germany's national indicative target for the contribution of electricity produced from renewable energy sources to gross electricity consumption by 2010 as 12,5% (OJoEU, 2001: 39). Furthermore, in 2002, the German government released a field report on the development of RES-E promotion in light of the Renewable Energy Act, leading to the conclusion that, despite the positive effects of the legislative context for RES-E promotion, the further growth of renewable energy shares in final energy consumption and electricity production needs on-going state intervention (BMU, 2002b: 2).

The three aforementioned aspects are the reasons referred to in a draft proposal for a law to readjust the Renewable Energy Act, adopted by the Federal Government in January 2004 and assigned to the Committee of Environment, Nature Conservation, and Nuclear Safety for further proceedings (DIP, 2004a: 1 and 3). Additionally, in January 2004, the Social Democrats and the Greens parliamentary group adopted an identical draft proposal (DIP, 2004b: 1). On 2nd April 2004, the Bundestag, in a second reading, adopted, based on a recommendation of the Committee of Environment, Nature Conservation, and Nuclear Safety, the draft proposal of the Federal Government, and of the Social Democrats and the Greens parliamentary group, on a law to readjust the Renewable Energy Act (DIP, 2004c: 1). However, approval by the Bundesrat was denied based on a recommendation of its Committee on Environment, Nature Protection and Reactor Safety to call upon the joint Mediation Committee of the Bundestag and Bundesrat to resolve differences of opinion between both chambers in terms of the adopted legislation (DIP, 2004d: 1). On 17th June 2004, the Mediation Committee presented a recommendation for an amendment of the legislative act in question, to which both the Bundestag (on 18th June 2004) and the Bundesrat (on 9th July 2004) could agree (DIP, 2004e: 1; DIP, 2004f: 1; DIP, 2004g: 1). The law for the readjustment of the Renewable Energy Act entered into force on 21st July 2004 (BGBL, 2004: 1918).

The subject and scope of the amended 2004 Renewable Energy Act are highlighted in its first Article (BGBL, 2004: 1918). This reflects the intention of the legislator – in accordance with UN Agenda 21 (UN, 1992) and the strategy for sustainable development of the Federal Government (Bundesregierung, 2002: 97) – to enable the Federal Republic to have a sustainable energy supply that prevents international conflicts over fossil fuels, preserves the environment, averts climate change, and fosters renewable energy technologies (BGBL, 2004: 1918). In order to reach these goals, Article 1(2) calls for a legally binding share of 12,5% RES-E by 2010 and at least 20% by 2020 (BGBL, 2004: 1918). It is the latter aspect that the Bundesrat was opposed to, by declaring, *inter alia, that 'reaching the 2010 trajectory of 12,5% (double the value of the year 2000) asks for a substantial technical and financial effort already. An additional legally binding trajectory will, in light of the unsolved problems with respect to electricity network challenges, not serve the purpose of a sustainable energy supply'* (DIP, 2004d: 1). The Bundesrat could not, however, fulfill its intention to delete this aspect from the finally adopted provision. Instead, the joint Mediation Committee, inter alia, agreed on reduced remuneration tariffs for the promotion of wind energy (DIP, 2004e: 2).

With beginning of the 16th legislative period (2005-2009), the Government of the Federal Republic of Germany was formed from a coalition of the Conservatives and the Social Democrats under Chancellor Angela Merkel. During this period, the legislator once again agreed on an amendment of the Renewable Energy Act, increasing the RES-E goal for 2020 from 20% to 30% (BGBL, 2008: 2075). Overall, the topic of sustainable development, including a more integrated approach to energy and climate policies, was remarkably present on the political agenda. To understand the agenda-setting impulses of this period it is worth re-emphasizing the European context. In 2006 the Commission of the then European Communities published a Green Paper on a European strategy for sustainable, competitive and secure energy (EC, 2006). Based on seven findings, the paper claims that Europe has entered a new energy era. *'There is an urgent need for investment. [...] Our import dependency is rising. [...] Reserves are concentrated in a few countries. [...] Global demand for energy is increasing. [...] Oil and gas prices are rising. [...] Our climate is getting warmer. [...] Europe has not yet developed fully competitive internal markets'* (EC, 2006: 3). In consequence (see '3.1 Developing a European Energy Policy'), the Green Paper identifies six key areas, which are specified by follow-up questions, in order to identify whether there is agreement on the need for a new energy strategy (EC, 2006: 4-5).

In the first half of 2007, Germany held the Presidency of the Council of the European Union, which is held by each Member State in turn for a period of six months (EU2007.de, 2007a: 1). The order in which Member States hold the Presidency of the Council, during which they speak on behalf of all Member States and shape the Council agenda, is set out till 2020 in the Annex of Council Decision *'of 1 January 2007 determining the order in which the office of President of the Council shall be held'* (OJoEU, 2007c: 11-12; EU2007.de, 2007a: 1). With respect to the subject of this thesis, the Presidency Programme reads as follows: *'A secure, environmentally friendly and competitive energy supply is crucial if Europe is to experience positive economic development. [...] In view of these challenges, adoption of the European action plan on energy policy will be a priority of the European Council in spring 2007. [...] In achieving our trio of goals, namely security of supply, efficiency and environmental compatibility, we must reduce the need for energy imports by boosting energy efficiency, saving energy and making greater use of renewable energies (also in the field of heating/cooling), for example, by increasingly tapping the potential offered by biomass and biofuels. All EU Member States are called to meet the goals set by 2010. Germany will promote the development of clear medium and long-term goals for renewable energies'* (EU2007.de, 2007b: 8-9).

After a March 2007 meeting of the Council of the European Union, as stated in a Cover Note, the Council adopted a comprehensive energy action plan fully respecting *'Member States' choice of energy mix and sovereignty over primary energy sources'* (Council of the European Union, 2007: 11). Over all, the Council stresses the need to increase energy efficiency in the EU up to 20% compared to projections for 2020 (Council of the European Union, 2007: 20). It furthermore reaffirms the long-term commitment to renewable energy development by endorsing a binding target of 20% share of renewable energies in the EU's overall energy consumption by 2020, with a 10% binding target for biofuels in the EU's transport consumption by 2020. Furthermore, the Council calls for an overall framework for renewable energies by establishing a new comprehensive Directive on the use of all renewable energy resources (Council of the European Union, 2007: 22). Whilst the adopted energy action plan was *'based on the Commission's Communication "An Energy Policy for Europe"'* (Council of the European Union, 2007: 13), it is interesting to note that the document *'An Energy Policy for Europe'* was officially released seven months after the Brussels European Council (EC, 2007a).

Back to the national arena. In July 2007, the German Government held an energy summit with representatives of the German Energy Industry to discuss elements of an integrated energy and climate programme for Germany (Prognos and EWI, 2007: 1; SPIEGEL Online, 2007a: 1). During the summit, Chancellor Merkel underlined the need for ambitious climate protection goals for Germany and the European Union, including an increase of annual energy efficiency and an increase of the share of RES-E (Spiegel Online, 2007a: 1). At the same time, Chancellor Merkel was *'reportedly paving the way to make a nuclear energy renaissance part of her party's next election campaign if she is unable to sway the Social Democrats to abandon a 1999 deal to close the country's nuclear power plants negotiated between former chancellor Gerhard Schröder and his coalition partner, the Green Party'* (Spiegel Online, 2007a: 1).

In August of the same year, and after Chancellor Angela Merkel and Secretary of Environment, Nature Protection and Reactor Safety, Sigmar Gabriel, paid a visit to Grönland to *'observe the consequences of climate change'* (SPIEGEL Online, 2007b: 1), the German Cabinet adopted *'Key Elements of an Integrated Energy and Climate Programme'*, building on insights of the German Council Presidency and results of the energy summit (BMU, 2007: 3). With this document, as the authors state, Germany is implementing fundamental European policy decisions at national level 'by means of a concrete programme of measures' (BMU, 2007: 3 and 5). The expansion of renewable energies in electricity generation is reflected in the programme by expressing the goal of increasing the share of RES-E production to 25 to 30% by 2020, with further expansion by 2030 (BMU, 2007: 10). The highlighted measures include the revision of the Renewable Energy Act and better integration of renewable energies into the electricity grid, whilst maintaining security of supply (BMU, 2007: 10). Based on these goals and the intended measures, and seeing that *'realizing a sustainable energy supply is of utmost importance for the Federal Government'*, the Government in January 2008 adopted a draft proposal for a law on the readjustment of the Renewable Energy Act and corresponding legal provisions (DIP, 2008a: 1).

The Bundestag adopted the respective draft proposal based on a recommendation of the Committee of Environment, Nature Conservation, and Nuclear Safety on 6th June 2008 (DIP, 2008b: 1). Approval by the Bundesrat was based on its decision of 4th July 2008 not to call upon the joint Mediation Committee of the Bundestag and Bundesrat to resolve differences of opinion between both chambers in terms of the adopted legislation (DIP, 2008b: 2). The law for the readjustment of the Renewable Energy Act and corresponding legal provisions entered into force on 25th October 2008 (BGBL, 2008: 2074). In accordance with the *'Key Elements of an Integrated Energy and Climate Programme'* (BMU, 2007: 10) adopted by the German Cabinet, subject matter and scope of the 2008 Renewable Energy Act, as defined in Part 1, § 1(2), call for a 30% share of RES-E production by 2020 with a continuous, yet undefined, growth beyond 2020 (BGBL, 2008: 2075). It is important to re-emphasise that the amendment of the German Renewable Energy Act, as described above, happened in the same year in which, after 11 months of legislative work (see Tab. 1), the European Parliament adopted the Climate Change Package including Directive 2009/28/EC (EP, 2008: 1-14).

In 2009, results of the election to the Bundestag led to a Government coalition of the Conservatives and the Liberals. In the same year, Directive 2009/28/EC entered into force (OJoEU, 2009a-f: 11-145). Consequently, Germany adopted a National Renewable Energy Plan (see '4.2 German Transposition of Directive 2009/28/EC and NREAP'). Almost simultaneously, the Conservative–Liberal German Cabinet adopted a new *'Energy Concept for an Environmentally Sound, Reliable and Affordable Energy Supply'* (BMU, 2010c: 1). In terms of RES and RES-E promotion, the energy concept read as follows: *'By 2020 renewable energies are to account for 18% of gross final energy consumption. After that, the German government will seek to make renewable energies account for the following proportion of gross final energy consumption: 30% by 2030, 45% by 2040 and 60% by 2050. By 2020 electricity generated from renewable energy sources is to account for 35% of gross electricity consumption. Following this, the German government will seek to increase the proportion of gross electricity consumption contributed by electricity from renewable energy sources to: 50% by 2030, 65% by 2040 and 80% by 2050'* (BMU, 2010c: 8).

As opposed to the Social Democrat and the Green Federal Government under Chancellor Schröder, who succeeded in pushing forward a legally binding path for nuclear phase-out (BMU, 2002a: 1), the Government coalition of the Conservatives and the Liberals under Chancellor Angela Merkel considered nuclear power a 'bridging technology' to achieve its energy policy goals (BMU, 2010c: 15). In consequence, the energy concept envisions that *'[t]he operating lives of the 17 nuclear power plants in Germany will be extended by an average of 12 years. In the case of nuclear power plants commissioned up to and including 1980 there will be an extension of 8 years. For plants commissioned after 1980 there will be an extension of 14 years'* (BMU, 2010c: 15).

On 28th September 2010, the Conservatives and the Liberals parliamentary group adopted a draft proposal on the 11th law for the amendment of the Atomic Energy Act, translating the extension of the operating lives of nuclear power plants, as envisioned in the energy concept (BMU, 2010c: 14), into a legally binding form (DIP, 2010d: 1). Although in a second parliamentary reading in the Bundestag the Greens parliamentary group made 24 law amendment requests and a rules of procedure request to withdraw a decision on the draft proposal from the agenda – all of which were denied by the Conservative and the Liberals who held the Bundestag majority – the draft proposal was adopted after a third reading on 28th October 2010 (DIP, 2010e: 1; DIP, 2010f: 7159; DIP, 2010g: 1). Approval by the Bundesrat was based on its decision of 26th November 2010 not to call upon the joint Mediation Committee of the Bundestag and Bundesrat (DIP, 2010h: 1). The 11th law for the amendment of the Atomic Energy Act entered into force on 8th December 2010 (BGBL, 2010: 1814).

Only six months later, in June 2011, the same Government initiated the 'Energiewende' or Energy Transition (see above). An introductory statement newly attached to the Government's energy concept reads as follows: *'Against the backdrop of the nuclear meltdown at Fukushima in March 2011, the role assigned to nuclear power in the energy concept was reassessed and the seven oldest nuclear power plants and the one at Krümmel were shut down permanently. Furthermore, a decision was taken to phase out operation of the remaining nine nuclear power plants by 2022'* (BMU, 2010c: -2). On 30th June 2011, the Bundestag adopted, based on a draft proposal by the Conservative and Liberal parliamentary group, the 13th law on the amendment of the Atomic Energy Act (DIP, 2011e: 13404; DIP, 2011f: 1-8; BGBL, 2011b: 1704-1705). A detailed description of the legislative process can be found in the beginning of this section.

With respect to renewable energies, the German Government adopted a proposal for a law for the transposition of Directive 2009/28/EC on the promotion of the use of energy from renewable sources on 29th September 2010, already three months after the Directive had been published in the Union's Official Journal (see also Tab. 1) (OJoEU, 2009a-f: 11-145; BMU, 2010b: 1). The subject and scope of the law include detailed modifications in the national overall legislative context, tangential to requirements pursuant to Directive 2009/28/EC (see '4.2 German Transposition of Directive 2009/28/EC and NREAP') (DIP, 2011i: 1). It does not include, however, any adjustments in terms of overall RES-E trajectories (BGBL, 2011a: 619-635). The law for the transposition of Directive 2009/28/EC on the promotion of the use of energy from renewable sources was published on 15th April 2011 (BGBL, 2011a: 619). Eight months after the legislative process for the transposition law started, in June 2011, the Conservative and Liberal parliamentary group adopted a draft proposal on a law for the re-adjustment of the legal framework for the promotion of electricity from renewable energies (DIP, 2011g: 1). With this draft proposal, both parliamentary groups intended to translate the unchanged pre–Energy Transition goals of the energy concept with respect to RES-E promotion – 35% by 2020, 50% by 2030, 65% by 2040 and 80% by 2050 (BMU, 2010c: 8) – into a legal form (DIP, 2011g: 1). The Bundestag adopted the proposal on 29th June 2011 (DIP, 2011h: 1). Approval by the Bundesrat was given, following the described routine, on 8th July 2011 (DIP, 2011i: 1). The law for the re-adjustment of the legal framework for the promotion of electricity from renewable energies was announced on 4th August 2011 (BGBL, 2011c: 1634). Owing to Article 1 of the respective law, subject matter and scope of the 2008 Renewable Energy Act, as defined in Part 1, § 1(2) (BGBL, 2008: 2075), were amended by incorporating the RES-E goals of 35% by 2020, 50% by 2030, 65% by 2040 and 80% by 2050 (BGBL, 2011c: 1635). It is important to note that the RES-E goals referred to in the respective paragraph are a result of the domestic policy process as part of the overall legislative and policy processes described before, and not part of the law for the transposition of Directive 2009/28/EC on the promotion of the use of energy from renewable sources (BGBL, 2011a: 635).

4.4.3 Summary

The German primary energy mix is – in comparison with to other industrialised nations – dominated by fossil fuels (Röhrkasten and Westphal, 2012: 329). As can be seen from Fig. 6, oil has a share of 33% in the primary energy mix, natural gas 21% , hard coal 13% , lignite 12% , renewables 12% , and nuclear 8% (BMWi, 2014a, based on AGEB, 2013). With regard to energy dependency, Fig. 7 illustrates that Germany is heavily dependent on net energy imports (overall 68,4%). In 2010, 26% of gross electricity generation was based on lignite, 19% was based on hard coal, 22% on renewable energy sources, 16% on nuclear, and about 11% on gas (BMWi, 2014c, based on AGEB, 2013 and BDEW, undated).

Tab. 17: Legislative history German renewable energy policy (for sources see references text)

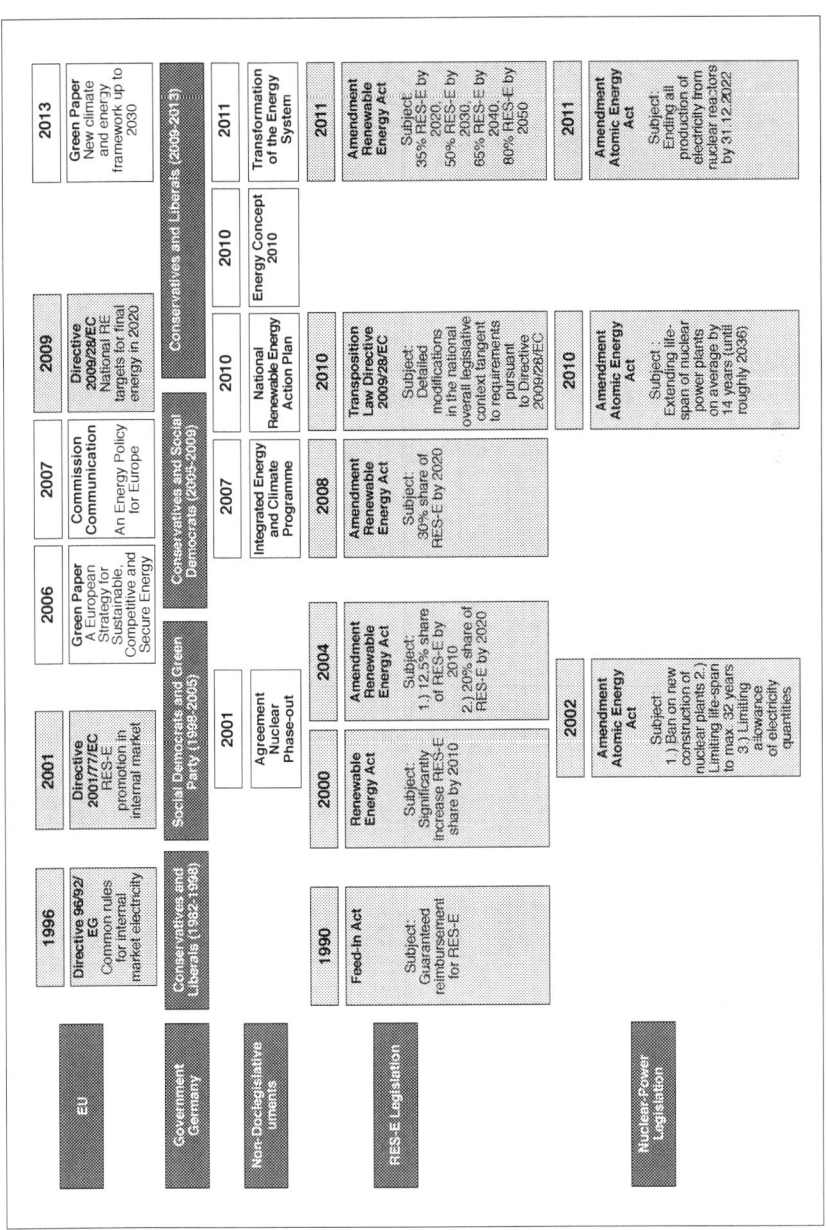

The overall legislative history of renewable energy promotion as part of the German energy policy as described in the previous section is summarised in Tab. 17. Note that the selection of non-legislative documents and legislative acts in this section reflects a substantial and adequate indication in light of the topic of this thesis, but should not be misunderstood as an exhaustive description of the German (renewable) energy policy process. Based on the indications presented here, it becomes apparent that after the adoption of a First Renewable Energy Act in 2000 (BGBL, 2000: 305), the legally enshrined trajectories for RES-E shares in the German electricity supply were constantly increased. In principle, either the acting Federal Government or those parliamentary groups representing the government majorities always adopted the draft proposals for Renewable Energy Act amendment laws (EEG-Clearingstelle, 2014b: 1; DIP, 2004a: 1 and 3; DIP, 2008a: 1; DIP, 2011g: 1). The most important adopted pieces of legislation with respect to RES-E promotion are the following:

- Feed-In Act (1990): The subject matter of the provision was a guaranteed reimbursement for electricity produced from water, wind, solar, biogas, and biomass facilities with an installed capacity of < 5 MW, to be paid for by the electricity suppliers (EEG-Clearingstelle, 2014a: 2; BGBL, 1990: 2633)

- Renewable Energy Act (2000): Subject matter and scope of the law were, as laid out in Article 1 (BGBL, 2000: 305), *to significantly increase the contribution of electricity from renewable energy sources for the electricity supply, to, in accordance with the goals of the European Union and the Federal Republic of Germany, double the share of energy from renewable sources in gross final consumption of energy in 2010'.*

- Amendment Renewable Energy Act (2004): Article 1(2) calls for a legally binding share of 12,5% RES-E by 2010 and at least 20% by 2020 (BGBL, 2004: 1918).

- Amendment Renewable Energy Act (2008): Part 1, § 1(2), calls for a 30% share of RES-E production by 2020 with a continuous, yet undefined, growth beyond 2020 (BGBL, 2008: 2075).

- Amendment Renewable Energy Act (2009): Owing to Article 1 of the amendment law, subject matter and scope of the 2008 Renewable Energy Act, as defined in Part 1, § 1(2) (BGBL, 2008: 2075) were amended by incorporating the RES-E goals of 35% by 2020, 50% by 2030, 65% by 2040 and 80% by 2050 (BGBL, 2011c: 1635).

Summarising the particular state objectives that arise out of domestic competition with respect to assumptions under Liberal Intergovernmentalism (Moravcsik, 1998: 22), it can be stated that each Federal Government, or the respective parliamentary groups, adopted legislative proposals significantly increasing the contribution of electricity from renewable energy sources in the period 1990-2013 (see Tab. 17).

In the first half of 2007, Germany held the Presidency of the Council of the European Union, which is held by each Member State in turn for a period of six month (EU2007.de, 2007a: 1). With respect to the subject of this thesis, the Presidency Programme reads as follows: *'A secure, environmentally friendly and competitive energy supply is crucial if Europe is to experience positive economic development. [...] In view of these challenges, adoption of the European action plan on energy policy will be a priority of the European Council in spring 2007. [...] In achieving our trio of goals, namely security of supply, efficiency and environmental compatibility, we must reduce the need for energy imports by boosting energy efficiency, saving energy and making greater use of renewable energies (also in the field of heating/cooling), for example, by increasingly tapping the potential offered by biomass and biofuels. All EU Member States are called to meet the goals set by 2010. Germany will promote the development of clear medium and long-term goals for renewable energies'* (EU2007.de, 2007b: 8-9). After a March 2007 meeting of the Council of the European Union (2007), as stated in a Cover Note, the Council adopted a comprehensive energy action plan (Council of the European Union, 2007: 11). To understand the national context for renewable energy promotion, in particular with respect to the Energy Transition (see above), additional focus was placed on the nuclear energy policy of the Federal Republic of Germany. On 14th December 2001 the Bundestag (lower house) adopted, based on a recommendation of the Committee of Environment, Nature Conservation, and Nuclear Safety, the draft proposal of the Social Democrats and the Greens parliamentary group of a law for the organised termination of commercial nuclear power (DIP, 2002: 1). The law amending the Atomic Energy Act was published on 26th April 2002 (BGBL, 2002: 1351). With its entry into force, the amendment act gave legal grounds for the framework implications of an agreement between the Federal Government and representatives of German energy suppliers for an organised termination of commercial nuclear power in Germany (BMU, 2002a: 1).

As opposed to the Social Democrats and the Greens Federal Government under Chancellor Schröder succeeded in pushing forward a legally binding path for a nuclear phase-out (BMU, 2002a: 1). The Government coalition of the Conservatives and the Liberals under Chancellor Angela Merkel considered nuclear power a *'bridging technology'* to achieve its energy policy goals (BMU, 2010c: 15). On 28th September 2010, the Conservatives and the Liberals parliamentary group adopted a draft proposal on the 11th law for the amendment of the Atomic Energy Act, translating the extension of the operating lives of nuclear power plants, as envisioned in the energy concept (BMU, 2010c: 14), into a legally binding form (DIP, 2010d: 1). Only six months later, in June 2011, the same Government initiated the 'Energiewende' or Energy Transition (see above). An introductory statement newly attached to the Government's energy concept reads as follows: *'Against the backdrop of the nuclear meltdown at Fukushima in March 2011, the role assigned to nuclear power in the energy concept was reassessed and the seven oldest nuclear power plants and the one at Krümmel were shut down permanently. Furthermore, a decision was taken to phase out operation of the remaining nine nuclear power plants by 2022'* (BMU, 2010c: -2). On 30th June 2011 the Bundestag adopted, based on a draft proposal by the Conservative and Liberal parliamentary group, the 13th law on the amendment of the Atomic Energy Act (DIP, 2011e: 13404; DIP, 2011f: 1-8; BGBL, 2011b: 1704-1705).

4.5 Analysis

Based on the overall hypothesis in light of the multi-level policy dependence theory applied here, three subordinate hypotheses have been identified in the theory section. To identify constraints and dependencies of the European Integration of RES-(E) promotion, the explanatory factors implicitly describe three different influencing factors that follow a top-down, vertical logic, in turn leading to a horizontal logic. The first subordinate hypothesis (i) claims supranational renewable energy measures and policies produce compliance at the level of Member States (Radaelli, 2003: 19). Subordinate hypothesis (ii) counterpoints the first by claiming the European Union lacks the executive authority to expect compliance at the Member States level. Instead, it is national preferences that dominate the Member States' policy outcomes. Subordinate hypothesis (iii) is not relevant for the analysis based on Chapter 4.

As has been stressed, the overall framework includes theories derived from Liberal Intergovernmentalism and Europeanization. They are reflected in subordinate hypothesis (i) (Europeanization) and subordinate hypothesis (ii) (Liberal Intergovernmentalism) and read as follows:

(i) The supranational RES-(E) measures and policies of the European Union, when expressed in legislation, excite change in Member States' RES-(E) policies only when there is a degree of misfit between EU and domestic RES-(E) policies that causes adaptational pressure, leading to the national promotion of RES-(E) in accordance with EU level implications.

(ii) The RES-E policy of a Member State is an exclusive result of the domestic policy process that reflects energy-specific economic interests of the dominant domestic group, and is determined by geopolitical and economic interest.

Since any progress in the promotion of RES-(E) can only be measured at the Member States level, a model has been derived where the EU Member States are the focal point. In the applied model, the dependent variable are the Member State RES-E policies of Poland and Germany, with and ultimate focus being on the vertical and horizontal impacts on the RES-E policy of Poland. It has been stressed that a mono-causal approach cannot explain the impact on the dependent variable 'Member State RES-E'. Therefore there cannot be a single independent variable (e.g. defined at EU level) (Radaelli, 2003: 27-28). Instead, there is a set of three decisive variables, occurring at the EU level, the national level and the inter-state level. Fig. 2. illustrates their interrelations. Since all three influencing factors for the promotion of RES-E in Member States (supranational extrinsic, national intrinsic, inter-state interdependencies) will be considered, the analysis aims at identifying general insights about constraints and dependencies of European RES-E promotion mechanisms.

The German reaction to 'supranational extrinsic adaptational pressure' and its 'intrinsic national logic' were the analytical focus of this chapter. German RES-E policies, as well as the underlying German polity with respect to supranational and in particular national implications, reflect it. Note that at this point the dependent variable is 'German RES-E policy'. Fig. 9 illustrates the analytical focus of this chapter. If subordinate hypothesis (i) should be accepted as a valid explanation, the questions are: On which judicial basis does Germany implement secondary provisions adopted by the European Union? Did the secondary provision in question excite change in Germany's RES-E policy? And: If change has been excited, was it because of a misfit between EU and German RES-E policies? If, in order to consider alternative explanations for the empirical evidence presented in this chapter (Ladrech, 2010: 40-41), subordinate hypothesis (ii) should serve as a valid explanation, then the question is: Can the RES-E policy of the EU Member State Germany be considered an exclusive result of the domestic policy process? What kinds of indicators verify or falsify subordinate hypothesis (ii)?

Fig. 9: Applied model of adaptational pressure for German RES-E policy

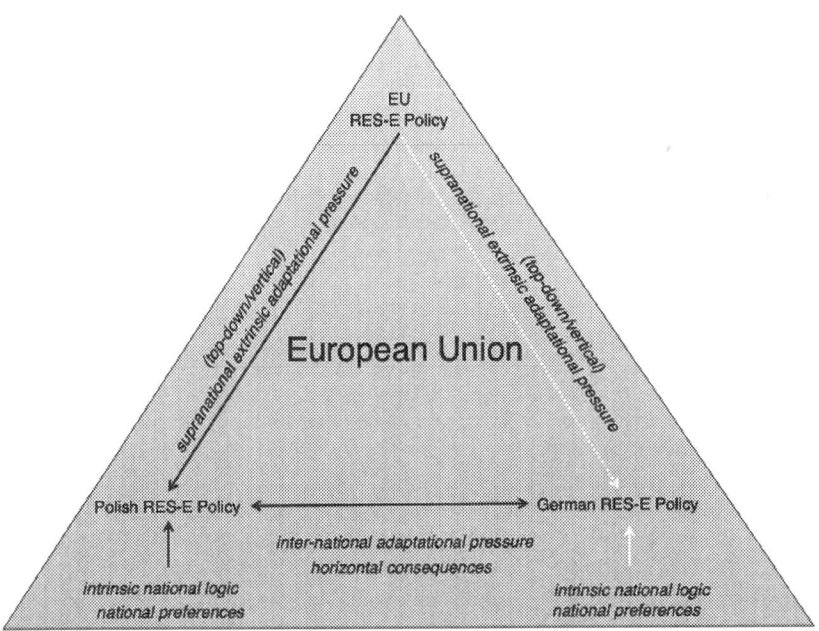

With respect to subordinate hypothesis (i), the German judicial framework for the adaptation of secondary provisions at the supranational level has been described. As a reminder it should be reiterated that, based on the arguments in light of the EU constitutional context, it has been concluded that there is a recognised constitutional principle of the European Union reflected in the *sui generis* nature of EU law, its supremacy, direct effect, pre-emption and protection of fundamental rights (Tatham, 2013: 15). But can the coherent logic of the supranational polity system of the EU be integrated into the pre-existing sovereign polity logic of the Member State Germany without constraints?

As has been described before, in light of the transfer of the exercise of sovereignty based on the 1949 Constitution of the Federal Republic of Germany, two provisions are relevant for European Integration (Tatham, 2013: 77; Pernice, 1998: 3-4): (a) the Preamble affirms the will of the people to '*promote world peace as an equal partner in a united Europe*' (BMJ, 2012: 1); and (b) correspondingly, Art. 24(1) of the constitution states that the '*Federation may by law transfer sovereign powers to international organisations*' (BMJ, 2012: 9). However, pursuant to Article 79(3) of the constitution '*[a]mendments to this Basic Law affecting the division of the Federation into Länder, their participation on principle in the legislative process, or the principles laid down in Articles 1 and 20 shall be inadmissible*' (BMJ, 2012: 25). A firmer constitutional basis for EU integration was provided with the adoption of a new Article 23, which before its amendment served as an accession clause for the East-German states that were covered by the constitution of the Federal Republic of Germany (BPB, 2013). After reunification, the new Article 23 (see Text Box 5) of the constitution borrows the idea of the preamble and translates it into obligations with legally binding effects (Pernice, 2008: 9-19). It guarantees a complete parliamentary process and the full participation of the 'Länder', as well as the observation of fundamental principles of the state, before any transfer of powers to the European level can occur (Tatham, 2013: 81). On the other hand, the rather general language of the provision in question allows for flexibility on how these principles are put into practice (Pernice, 1998: 11, Pernice, 2006: 18-19). Ultimately, however, integration of Germany into a federal-type Europe, including a comprehensive transfer of sovereignty to the European level, would require the German people to adopt a new constitution freely (Tatham, 2013: 83-84).

> As Pernice states (1998: 4) and Tatham emphasises (2013: 78), Germany was, based on lessons taken by the German people from the historical failure of the classical concept of the nation-state, constituted so as to be a member of a broader political system from the very beginning. Consequently, constitutional law in Germany cannot be based on the constitution alone, but must be construed in its constitutional and legislative European context (Pernice, 1998: 4). However, with the so called 'eternity clause', the constitution holds an absolute transfer-limit on German sovereignty with respect to the potential creation of a federal-type Europe (Tatham, 2013: 83-84). *Inter alia*, this is true for Article 23 of the constitution. It leaves room for

interpretation, which in terms of the constitution falls within the scope of the German Federal Constitutional Court (FCC). In this respect, what is true for the ECJ, its case law, and its effect on European constitutionalism – based on Article 19 TEU and the 17th Declaration of the Final Act of the Treaty of Lisbon (OJoEU, 2010: 83; OJoEU, 2007a: 256) – is also true, based on Article 93 of the German Constitution, for the FFC (BMJ, 2012: 33). The relationship of the ECJ and the FFC can be described as rather tempestuous (Tatham, 2012: 84).

Based on case law between 1967 and 1971, the FFC accepted the autonomous non-state sovereign authority of treaty-based European legislation, which, following ECJ interpretation, lead to the conclusion that supranational provisions overlaid inconsistent national law (Pernice, 2006: 28; Tatham, 2013: 86; BVerfGE, 1971: 31 (145); BVerfGE 31(145) quoted in IfIL, 2007a: 6). A change in this attitude and potential conflict between the case law of the ECJ and the FCC is reflected in the so-called 'Solange I' judgement of the FFC in 1974. In this case, the court denied European supremacy because fundamental rights enshrined in the German constitution where not protected to the same extent in European law (Pernice, 2006: 28; Tatham, 2013: 100, Grimm, 2001: 287). This conflict was resolved in 1986 in the so-called 'Solange II' judgment. Here, the Court stated that, as long as the general level of protection of human rights remained adequate by German standards, the FCC would no longer examine secondary European legislation for compliance with constitutional provisions relating to fundamental rights in the German constitution (Tatham, 2013: 101; Pernice, 2006: 28-29; Grimm, 2001: 288). Nevertheless, the FCC preserved its final authority to apply national provisions over European provisions where problems concerning the protection of fundamental rights in European law arose (Tatham, 2013: 102). A new perspective on the interrelation of domestic primary German law and supranational European law was introduced by the so-called 'Maastricht' judgement of the FCC in 1993 (BVerfGE 89 (155) quoted in University of Nevada, 1994: 21; BVerfGE, 1993: 89 (155) at 112). Following this judgment, the Member States remain the *'Masters of the Treaties'*, who have conferred, as an example of international organisations, the exercise of some of their powers to the EU, whose primary treaty law merely remains a derivative legal order. Not only can Germany terminate its membership, but the FFC can also examine whether instruments of the European Union exceed the limits of the sovereign rights accorded to them (Tatham, 2013: 83; Franzius, 2010: 25 and 34 and 43; Pernice, 2006: 30-31; Grimm, 2001: 288; BVerfGE 89 (155) quoted in University of Nevada, 1994: 2; BVerfGE, 1993: 89 (155) at 112). In short, as confirmed in the 'Lisbon' decision (BVerfGE, 2009: 123 (267) at 331), the FCC claims ultimate authority to protect fundamental rights in the national system, but refrains *'from striking down secondary European legislation per se'* (Tatham, 2013: 212).

> With respect to the implications of Europeanization as part of the overall theory framework, the FCC most definitely denies supremacy, direct effect,

and pre-emption of European law per se. It reserves for itself the right to review national legal rules that transpose European provisions with regard to whether they occurred on the basis of manifest transgressions of EU competence pursuant to the Lisbon treaty. The FCC rarely invokes this power (as longs as fundamental rights are not infringed by EU secondary legislation) and would do so in a co-operative manner with ECJ jurisdiction (Tatham, 2013: 126). Nevertheless, as Tatham (2013: 99) concludes, '[t]he FCC [...] has not accepted without question the constitutionalisation of the European legal order, as created an managed by the ECJ. [...] [A] number of aspects have been challenged by the FCC but, so far, they have not led to open conflict with their European counterpart although [...] its ruling in the Lisbon case could render an eventual clash more likely'. The German Basic Law, in its interpretation by the FCC, can definitely be regarded as a veto point in the domestic institutional setting to avoid constraints from supranational extrinsic adaptational pressure (Ladrech, 2010: 34). Assumptions, reflected in Europeanization, that EU law constitutes the law of the land can be disqualified as too simplistic (Börzel and Risse, 2000: 5).

With respect to the implications of Liberal Intergovernmentalism as part of the overall theory, it seems that the FCC incorporated the view of the state as dominant actor in European Integration. This becomes particularly apparent in the 'Maastricht' judgement, according to which Member States remain the 'Masters of the Treaties', who have conferred, as an example of international organisations, the exercise of some of their powers to the EU, whose primary Treaty law merely remains a derivative legal order. The FCC emphasizes that 'the Federal Republic of Germany remains a member of a compound of States, the authority of which is derived from the Member States and has binding effect in German sovereign territory only by virtue of the German command to apply the law [Rechtsanwendungsbefehl]' (BVerfGE 89 (155) quoted in University of Nevada, 1994: 21; BVerfGE, 1993: 89 (155) at 112). Not only, the FCC argues, can Germany terminate its membership, but the FFC can also examine whether instruments of the European Union exceed the limits of the sovereign rights accorded to them (Tatham, 2013: 83; Franzius, 2010: 25 and 34 and 43; Pernice, 2006: 30-31; Grimm, 2001: 288; BVerfGE 89 (155) quoted in University of Nevada, 1994: 2; BVerfGE, 1993: 89 (155) at 112). Overall, the FCC confirms assumptions of Liberal Intergovernmentalism whereby European Integration is not a preordained movement towards a federal type Europe, rather a series of pragmatic bargains amongst national governments based on concrete national interests, relative power, and a carefully calculated transfer of sovereignty (Moravcsik, 1998: 472; Moravcsik and Schimmelfennig, 2009: 73). A further indicator supporting this impression is the wording of the 'Solange II' judgement, in which the FCC refers to European Institutions as 'international institutions' (BVerfGE 73 (339) quoted in IfIL, 2007c: 8; BVerfGE, 1986: 73 (339)). What does this mean for the national implementation of the Climate and Energy Package, and Directive 2009/28/EC in particular? In the case of

Germany, the relation of the domestic and the supranational polity design, including interpretations of the FCC (see '4.1 German Constitutional Context ') and the ECJ (see '3.4 EU Constitutional Context in light of EU Renewable Energy Policy') has not lead to constraints regarding the transposition of Directive 2009/28/EC (see '4.2 German Transposition of Directive 2009/28/ EC and NREAP'). It is important to note that, whilst the FCC claims ultimate authority to protect fundamental rights in the national system, it refrains *'from striking down secondary European legislation per se'* (Tatham, 2013: 212). However, the uncertainties caused by Article 194 TFEU – prohibiting any measures that affect a *'Member State's right to determine the conditions for exploiting its energy resources, its choice between different energy sources and the general structure of its energy supply'* (OJoEU, 2008: 134-135) – might not only be subject to ECJ decisions, but might also be subject to the ultimate authority of the FCC to protect fundamental rights in the national system. Again, pre–Lisbon Treaty, Directive 2009/28/EC was not considered a measure *'significantly affecting a Member State's choice between different energy sources and the general structure of its energy supply'* (OJoEU, 2008: 133; German Advisory Council on the Environment, 2011: 175). The legislative packages in question, and particularly Directive 2009/28/EC, have been adopted through the ordinary legislative procedure, and therefore under non-exclusive participation of the Member States to excite change pursuant to each provision's content. It cannot be ruled out that, for the future adoption of energy related renewable energy policies, the German polity – by way of FCC decisions based on the German Constitution in light of an intrinsic national logic – might constitute a potential constraint. However, since the constitution of Germany does not explicitly include provisions with respect to energy, as is the case with Article 194 TFEU, this is not considered a likely scenario (BMJ, 2012: 1-58; OJoEU, 2008: 134-135).

On 12th April 2011, the President, the Chancellor, and the Minister for Environment, Nature Conversation and Reactor Safety of the Federal Republic of Germany signed the 'Law for the Transposition of Directive 2009/28/EC on the Promotion of the Use of Energy from Renewable Sources' (BGBL, 2011a: 635), roughly one year after Directive 2009/28/EC was published in the Union's Official Journal (see also Tab.1 and Tab. 11) (OJoEU, 2009a-f: 11-145). The 'Law for the Transposition of Directive 2009/28/EC on the Promotion of the Use of Energy from Renewable Sources' contains several amendments to the following national provisions:

- Renewable Energy Act, 11 overall amendments pursuant to Article 1 of transposition law (BGBL, 2011a: 619-623).

- Renewable Energies Heat Act, 22 overall amendments pursuant to Article 2 of transposition law (BGBL, 2011a: 623-633).

- Act on Energy Statistics, three amendments pursuant to Article 3 of transposition law (BGBL, 2011a: 633).

- Town and Country Planning Code, two amendments pursuant to Article 4 of transposition law (BGBL, 2011a: 633-634).

- Electricity from Biomass Sustainability Regulation, six amendments pursuant to Article 5 of transposition law (BGBL, 2011a: 634).

- Statistical Law for Building Construction, one amendment pursuant to Article 5 (a) of transposition law (BGBL, 2011a: 635).

In its NREAP (NREAP Germany, 2010: 8), the Federal Republic of Germany drafted a scenario for the expected final consumption of energy between 2010-2020, based on energy efficiency and energy savings measures taken before 2009. Based on Annex I of Directive 2009/28/EC (see Tab. 3 and Tab. 12), the share of energy from RES in gross final energy consumption was 5,8% in 2005. The legally binding target for Germany to be reached by 2020 is at least 18% . Germany assumes that it will achieve the target by 2020 without the benefit of surpluses from other countries. In addition, it expects to surpass the national target based on a 'scenario with additional energy efficiency measures' (EFF). An overview of all policies and measures with respect to renewable energy targets addressed in the German NREAP implemented before the supranational implications based on Directive 2009/28/EC can be found in Tab. 16 (NREAP Germany, 2010: 20-54; EC, 2009: 4).

> As mentioned before, with the rather timely adoption of a 'Law for the Trans-position of Directive 2009/28/EC on the Promotion of the Use of Energy from Renewable Sources' (BGBL, 2011a: 635) and the release of a NREAP (NREAP Germany, 2010: 1-182) in accordance with the template for National Renewable Energy Action Plans of the European Commission (EC, 2009: 1-40), it seems that the German transposition of Directive 2009/28/ EC has begun promisingly. However, the subject and scope of the German transposition law only include rather detailed modifications in the national overall legislative context, tangential to requirements pursuant to Directive 2009/28/EC (DIP, 2011i: 1). The fact that it does not include any major adjustments in terms of, for example, overall RES-E trajectories indicates the absence of fundamental adaptational necessities in light of supranational implications (BGBL, 2011a: 619-635). Furthermore, the overview of all policies and measures implemented in Germany before supranational implications based on Directive 2009/28/EC (see Tab. 16 (NREAP Germany, 2010: 20-54)) indicates an autonomous accordance of German renewable energy policy outcomes with European policy implications. In terms of these arguably non-comprehensive indicators, and in light of subordinate hypothesis (i), it seems that supranational RES-(E) measures and policies of the European Union excite change in Germany's RES-(E) policies because there is a limited degree of misfit between EU and German domestic RES-(E) policies, and because adaptational needs only include rather detailed modifications in the national overall legislative context. These rather

detailed modifications in the national overall legislative context, however, are with respect to the subject matter and scope an exclusive result of the supranational policy process.

These analytical deductions are supported by the results of the implementation evaluation of Directive 2009/28/EC. As can be seen from 'Tab. 4: Actual and planned RES Shares 2009-2011/2012 (Ecofys, 2012: 18)', in conjunction with 'Tab. 13: National target for 2020 and expected path for energy from renewable sources in the sectors of heating and cooling, electricity, and transport (NREAP Germany, 2010: 17)', Germany reached its NREAP indicative target RES share in 2010 and exceeded it by 0,9 percentage points, which equals a positive deviation of 8,94% (Ecofys, 2012: 18; NREAP Germany, 2010: 17). Therefore, according to Ecofys (2012: 20), since Germany had a growth rate in 2009/2010 above the required average, it is in a good situation to reach its overall RES target in 2020. Germany is among the sixteen Member States that achieved a growth rate in the RES-E sector between 2009 and 2010, which was higher than the average growth rate needed to move from the NREAP 2010 RES-E target to the 2020 RES-E target (Ecofys, 2012: 23).

According to the administrative assessment carried out in the Ecofys report (2012: 186-187), Germany receives a *'fair'* rating. With respect to Germany's progress in electricity grid integration – see Tab. 7: Member State rating on progress in electricity grid integration (Ecofys, 2012: 177, 179 and 181)' – it receives an *'advanced'* rating on overall RES grid integration, as well as on addressing grid capacity limitations, and on rules for bearing and sharing the costs of grid development (Ecofys, 2012: 92).

> According to the implementation evaluation, Germany is in a good situation to reach its overall RES target by 2020 (Ecofys, 2012: 20). However, the overall RES targets, whilst being in accord with Directive 2009/28/EC, have not been addressed in the transposition law, for they were already reflected in the existing national legislation. The 11 overall amendments of the Renewable Energy Act pursuant to Article 1 of the transposition law (BGBL, 2011a: 619-623) do mainly address administrative procedures such as the establishment of a national register of guarantees of origin. Whilst the national register of guarantees of origin is considered *'advanced'*, according to Ecofys (2012: 157), it is nevertheless the overall progress of Germany's administrative procedures that only receive a *'fair'* rating (Ecofys, 2012: 186-187). In order to validate these findings, it is important to pay closer attention to the national legislative context and its policy dependencies. At this point it is not possible to apply findings from the implementation evaluation with regard to subordinate hypothesis (i) or (ii).

With respect to the applied model of this thesis, and the need for the consideration of an 'intrinsic national logic' for Germany's RES-(E) policies, the respective overall legislative history has been shown to be an indicator. Seen from the perspective of Liberal Intergovernmentlism, it is the domestic policy outcome that subsumes the particular objectives that arise from domestic competition, and therefore reflect the 'intrinsic national logic' (Moravcsik, 1998: 22). The overall legislative history of renewable energy promotion as part of the German energy policy described in the previous section is summarized in Tab. 17. The most important adopted pieces of legislation with respect to RES-E promotion are the following:

- Feed-In Act (1990): The subject matter of the provision was a guaranteed reimbursement for electricity produced from water, wind, solar, biogas, and biomass facilities with an installed capacity of < 5 MW, to be paid for by the electricity suppliers (EEG-Clearingstelle, 2014a: 2; BGBL, 1990: 2633).

- Renewable Energy Act (2000): Subject matter and scope of the law were, as laid out in Article 1 (BGBL, 2000: 305), *to significantly increase the contribution of electricity from renewable energy sources for the electricity supply, to, in accordance with the goals of the European Union and the Federal Republic of Germany, double the share of energy from renewable sources in gross final consumption of energy in 2010'*.

- Amendment Renewable Energy Act (2004): Article 1(2) calls for a legally binding share of 12,5% RES-E by 2010 and at least 20% by 2020 (BGBL, 2004: 1918).

- Amendment Renewable Energy Act (2008): Part 1, § 1(2), calls for a 30% share of RES-E production by 2020 with a continuous, yet undefined, growth beyond 2020 (BGBL, 2008: 2075).

- Amendment Renewable Energy Act (2009): Owing to Article 1 of the amendment law, subject matter and scope of the 2008 Renewable Energy Act, as defined in Part 1, § 1(2) (BGBL, 2008: 2075) were amended by incorporating the RES-E goals of 35% by 2020, 50% by 2030, 65% by 2040 and 80% by 2050 (BGBL, 2011c: 1635).

> Besides the rather zigzag course in terms of German nuclear power policy in the last decade of the 20th and first decade of the 21st Century, it becomes apparent that after the adoption of a First Renewable Energy Act in 2000 (BGBL, 2000: 305) that the legally enshrined trajectories for RES-E shares in the German electricity supply were constantly increased. In principle, either the acting Federal Government or those parliamentary groups representing the government majorities adopted the draft proposals for Renewable Energy Act amendment laws (EEG-Clearingstelle, 2014b: 1; DIP, 2004a: 1 and 3; DIP, 2008a: 1; DIP, 2011g: 1). Starting with the first successful initiative for a Feed-In Act for electricity from renewable energy sources (EEG-Clearingstelle, 2014a: 1) whose informal origins were based in an inter-fractional cooperation between Conservative and Green Members of Parliament (Hirschl, 2008: 131), all political parties but Die Linke (left-wing

party) have been involved in fostering incrementally rising trajectories for RES-E in Germany between 1990 and 2011. Having stated this, however, it must be acknowledged that studies applying a nation-bound policy analysis, such as that by Bernd Hirschl (2008: 185-196), focus exclusively on domestic German developments in multi-level realities and naturally derive a more differentiated picture with regard to national advocacy-coalitions in light of RES-E promotion in Germany. Nonetheless, this study confirms the impression that the overall German RES-E policy can be labelled a success (Hirschl, 2008: 185), in so far as incrementally rising trajectories for RES-E, and implementation effectiveness, reflect the successful translation of issue-specific policy intentions of the dominant domestic group (Leuffen, Rittberger and Schimmelfennig, 2013: 42; Moravcsik, 1998: 24-35). In addition, it becomes apparent from the analysis above that Germany has influenced the supranational agenda in accord with its national preferences, *inter alia*, during its Presidency of the Council of the European Union in the first half of 2007 (EU2007.de, 2007a: 1). This impression is supported by an analysis from Duffield and Westphal (2011: 175), who conclude that *'[i]n some respects, Germany has made EU energy policy a top priority, especially where doing so has been seen as a means of achieving Germany's goals of fighting climate change and, to a lesser extent, energy security. Indeed, energy policy was a special focus during the German presidency of the EU during the first half of 2007, which saw the adoption of a set of ambitious EU energy policy goals at the spring meeting of the Council (Silberberg 2006)'.* Taking into account both the indicated EU-level developments in light of RES-(E), and the legislative history of German renewable energy policies (see Tab. 17), gives rise to the impression that both levels are inextricably bound up to each other, a finding also deduced in the analysis by Bernd Hirschl (2008: 578). Finally, the analysis of the German legislative RES-E history justifies the conclusion that the domestic national preferences of Germany reflect an independent intrinsic approach to RES-E promotion, whose outcome is in accord with EU implications, not as a result, but as a influential originator, of these implications. Indeed, though applying a fairly different theoretical framework, this result is confirmed by other researches such as Hirschl (2008: 590), who concludes that German RES-E policy is independent from supranational implications in principle. On the contrary, German RES-E policies have been implemented as a pioneer policy and advocated on the supranational level. What does this mean for subordinate hypotheses (i) and (ii)? With respect to the overall renewable energy policy of Germany, it has been shown that the bulk of the overall policy tendencies have been derived within the domestic policy arena. Therefore it can be stated that the RES-E policy of Germany originates as a result to the domestic policy process. The German RES-E policy is in accordance with EU level implications, because of the absence of a degree of misfit in general.

What follows from the analysis above? The conclusion of the analysis in the 'EU

Level Energy Policy and the Context for RES-E' chapter was that subordinate hypothesis (ii) cannot be falsified in its entirety. Whilst, on the one hand, pre–Lisbon Treaty realties and Directive 2009/28/EC implementation-effectiveness evidence quite clearly illustrate a measurable impact of European legislation, Article 194(2) TFEU, on the other hand, defines absolute competence limits. German national intrinsic preferences, with respect to its polity design reflected in FCC case law, further strengthen the scope of these absolute competence limits. As confirmed in its 'Lisbon' decision (BVerfGE, 2009: 123 (267) at 331), the FCC claims ultimate authority to protect fundamental rights in the national system, but refrains *from striking down secondary European legislation per se'* (Tatham, 2013: 212). However, the German case in particular illustrates how national and supranational policies are inextricably bound up to each other. Taking this into account, the wording of hypothesis (ii), speaking of *exclusive results of the domestic policy process,* should be revised. Based on the evidence presented in this chapter, it can be concluded that there is a dominant or pioneer position to the national intrinsic logic with respect to German RES-E preferences, with consequences also for the supranational policy arena. In principal, based on, *inter alia,* FCC case law, and in accordance with, for exapmle, Article 194(2) TFEU, Germany considers itself to be one of the *'Masters of the Treaties'* (BVerfGE, 1993: 89 (155)), which grants it an ultimate veto position.

With respect to the German promotion of RES-E, this could occur in so far as the future EU implications of RES-E promotion would be considered a transgression of EU competencies, and if the policies themselves would not be in accordance with national preferences. It has become clear from the analysis above that neither aspect was not the case with respect to Directive 2009/28/EC and to German RES-E policy. Neither were the EU-level implications as a whole considered a transgression of EU competencies – on the contrary, they were dominantly shaped in accordance with German national preferences – nor was there a misfit of supranational and national RES-E preferences in principle. It follows from the above that subordinate hypothesis (ii) cannot be verified in its entirety. It becomes apparent, though, as stated above, that the RES-E policy of Germany is predominantly a result of the domestic policy process that reflects energy-specific interests of the dominant domestic group.

Tab. 18: Evaluation Matrix for Chapter 3 and 4

	European Union	GER	PL	GER and PL
Subordinate Hypothesis (I)	☺	☺	X	X
Subordinate Hypotehsis (II)	☺	☺	X	X
Subordinate Hypothesis (III)	X	X	X	X

With regard to subordinate hypothesis (i), and based on the analysis in the 'EU Level Energy Policy and the Context for RES-E' chapter, empirical evidence for Directive 2009/28/EC suggests that European renewable energy policies and measures excite change in Member States policies, but the change excited is not in its entirety in accordance with the change intended. The first aspect of subordinate hypothesis (i) (the supranational RES-(E) measures and policies of the European Union excite change in Member States RES-(E) policies) has thus been verified under conditions of usual reserve. With regard to the second aspect (if there is a degree of misfit between EU and domestic RES-(E) policies that causes adaptational pressure leading to the national promotion of RES-(E) in accordance with EU level implications) no conclusions could be drawn based on Chapter 3 . As has been stated above, additional insights from this chapter show that the overview of all policies and measures implemented in Germany before supranational implications based on Directive 2009/28/EC (see Tab. 16; NREAP Germany, 2010: 20-54) indicate an autonomous accordance of German renewable energy policy outcomes with European policy implications. In terms of these, arguably rather qualitative, indicators, and in light of subordinate hypothesis (i), supranational RES-(E) measures and policies of the European Union seem to have excited change in Germany's RES-(E) policies because there is a limited degree of misfit between EU and German domestic RES-(E) policies, and because adaptational needs *only* include rather detailed modifications in the overall national legislative context. These rather detailed modifications, however, are an exclusive result of the supranational policy process. By defining these indicators as qualitative rather than quantitative, emphasis should be given to the very narrow applicability of these findings. For an overview of the results, including Chapter 4 visualized in a basic evaluation matrix, see Tab. 18.

5. Poland

5.1 Polish Constitutional Context

What was true for the European and the German constitutional contexts is equally important for the Polish constitutional context. In order to understand the likelihood that renewable energy policies originating at the European level have an impact on the Member State Poland, it is necessary to understand the domestic judicial framework for the adaptation of secondary Union provisions. The set of problems in question, as was the case with Germany, should be linked to the overall framework of this study, namely Liberal Intergovernmentalism (Moravcsik, 1998) and Europeanization (Börzel and Risse, 2000; Radaelli, 2003). It furthermore should be linked to Grimm's (2001: 13-22) definitions about the relation of law and politics (see, *inter alia*, 2.2 Trans-, Supra-, and International Bound Policy Analysis Frameworks). The argument of the following chapter follows the descriptive design of an analysis by Alan F. Tatham (2013) of Central European constitutional courts in the face of EU Membership.

In light of the transfer of the exercise of sovereignty based on the 1997 Constitution of the Republic of Poland, five provisions play an important role. Article 90 allows for the transfer of competences to international organisations in some matters. Articles 9 and 91 allow for the direct effect of international agreements and supremacy over national laws and statutes. Articles 8 and 87 define the sources of the supremacy of the Polish Constitution (Dabrowska, 2009: 108).

Unlike the German Constitution – which, in its Preamble affirms the will of the people to be *'an equal partner in a united Europe'* (BMJ, 2012: 1), and provides for a special European integration clause (Article 23, see: BMJ, 2012: 8-9) – the Polish Constitution does not make use of the word "Europe" once (Sejm, 2014). Indeed, where, and how precisely, to put provisions regarding the possibility of Poland's membership of the European Union were the most hotly debated topics before the enactment of the 1997 Constitution and the constitutional referendum (Biernat, 1998: 339).

Text Box 6: Text of the Constitution of the Republic of Poland of 2nd April, 1997 as published in the Dziennik Ustaw No. 78, item 483 (Sejm, 2014)

Article 8

1. The Constitution shall be the supreme law of the Republic of Poland.

2. The provisions of the Constitution shall apply directly, unless the Constitution provides otherwise.

Article 9

1. The Republic of Poland shall respect international law binding upon it.

Article 87

1. The sources of universally binding law of the Republic of Poland shall be: the Constitution, statutes, ratified international agreements, and regulations.

2. Enactments of local law issued by the operation of organs shall be a source of universally binding law of the Republic of Poland in the territory of the organ issuing such enactments

Article 90

1.The Republic of Poland may, by virtue of international agreements, delegate to an international organization or international institution the competence of organs of State authority in relation to certain matters.

2. A statute, granting consent for ratification of an international agreement referred to in para.1, shall be passed by the Sejm by a two-thirds majority vote in the presence of at least half of the statutory number of Deputies, and by the Senate by a two-thirds majority vote in the presence of at least half of the statutory number of Senators.

3. Granting of consent for ratification of such agreement may also be passed by a nationwide referendum in accordance with the provisions of Article 125.

4. Any resolution in respect of the choice of procedure for granting consent to ratification shall be taken by the Sejm by an absolute majority vote taken in the presence of at least half of the statutory number of Deputies.

Article 91

1. After promulgation thereof in the Journal of Laws of the Republic of Poland (Dziennik Ustaw), a ratified international agreement shall constitute part of the domestic legal order and shall be applied directly, unless its application depends on the enactment of a statute.

2. An international agreement ratified upon prior consent granted by statute shall have precedence over statutes if such an agreement cannot be reconciled with the provisions of such statutes.

3. If an agreement, ratified by the Republic of Poland, establishing an international organization so provides, the laws established by it shall be applied directly and have precedence in the event of a conflict of laws.

Amongst those opposed to including special provisions to regulate this issue, Biernat (1998: 339) identifies two groups. The first, generally in favour of EU membership, believed this potential membership lie in the distant future. Hence, placing the relevant provisions in the Constitution would be premature. The second group was made up of opponents of a Polish EU Membership in general. In the end, the issue was addressed in Article 90 of the Constitution, stating that *'[t]he Republic of Poland may, by virtue of international agreements, delegate to an international organization or international institution the competence of organs of State authority in relation to certain matters'* (Sejm, 2014). Whilst there is no reference to European Integration, Biernat (1998: 400) concludes that *'in the light of the intentions of the framers of the Constitution [...] there can be no doubt that the provision in question is meant to refer primarily to future integration within the European Union'*. In fact, it has been noted that the more general nature of the formulation of the provision in question opens the 1997 Constitution in a broader manner to international and supranational laws, which has been shown to be a characteristic of the Polish Constitution (Biernat, 1998: 400; Tatham, 2013: 220). On the other hand, this means that, from the perspective of the Polish legal system, ratifying European Treaties is equal to ratifying international agreements (Dabrowska, 2009: 109).

Poland's accession to the EU was a complex act, but it did not need to pass specific amendments to its Constitution in order to accede (Tatham, 2013: 205 and 222). By signing the EU Accession Treaty on 16th April 2003 in Athens, Poland agreed that *'[f]rom the date of accession, the provisions of the original Treaties and the act adopted by the institutions and the European Central Bank before accession shall be binding on the new Member States and shall apply in those States under the conditions laid down in those Treaties and in this Act'* (OJoEU, 2003b: 33; Dabrowska, 2009: 109; Tatham, 2013: 222; CT, 2003: 1). The general European accession conditions and criteria as reflected, *inter alia,* in Article 2 of the *'Act concerning the conditions of the accession of [...] the Republic of Poland [...] and the adjustment to the Treaties on which the European Union is founded'* (OJoEU, 2003b: 33) establish that new Member States have to accept the entire *aquis communautaire,*[11] including the jurisprudence and decisions of the ECJ regarding the general principles of the Community, and the direct effect and supremacy of EU law (Dabrowska, 2009: 109; Biernat, undated: 2).

Based on Polish Constitution Art. 90(2) and Art. 90(3), *'[a] statute, granting consent for ratification of an international agreement referred to in para.1, shall be passed by the Sejm [...]. (G)ranting of consent for ratification of such agreement may also be passed by a nationwide referendum in accordance with the provisions of Article 125'* (Sejm, 2014). After the Accession Treaty was signed in Athens, the Sejm (lower house of the Polish parliament) adopted a resolution calling a referendum. Accordingly, a referendum on Polish Membership in the European Union was held on 7th and 8th June 2003. At the time, 77,45% of the electorate voted in favour of a EU Membership. Consequently – and after the Polish Constitutional Tribunal (CT) ruled that the nation-wide referendum conformed to the Polish Constitution – the President of the Republic of Poland ratified the Accession Treaty. Poland became a Member of the European Union on 1st May 2004 (Tatham, 2013: 222; CT, 2003: 1 and 3).

---

[11] The acquis communautaire is the accumulated body of European Union (EU) law and obligations from 1958 to the present day. It comprises all the EU's treaties and laws (directives, regulations, decisions), declarations and resolutions, international agreements, and the judgments of the Court of Justice. It also includes action that EU governments take together in the Area of Freedom, Security and Justice, and under the Common Foreign and Security Policy (Miller, 2011).

Unlike the German Constitution, which pursuant to Article 23(1) provides that the Federation *'transfer sovereign powers by a law with the consent of the Bundesrat'* (BMJ, 2012: 8), whilst simultaneously defining limits to that transfer (Biernat, 1998: 405; BMJ, 2012: 8-9), the Polish Constitution (here Art. 90(1)) only makes mention of the delegation of *'the competence of organs of State authority in relation to certain matters'* (Sejm, 2014). Ultimately, the Polish Constitution does not permit the transfer of sovereignty, but the exercise of specific competencies as opposed to the power itself (Tatham, 2013: 223). However, the Constitution neither explicitly specifies the limits up to which the competencies of organs of state authority can be transferred, nor does it uphold a positive indication as to what these *'certain matters'* involve (Biernat, 1998: 405). However, the lack of explicit limitations in the referred to provision, it must be concluded, does exclude the transfer of sovereignty in all cases – that is, it does not indicate the absence of constitutional limitations at all (Tatham, 2013: 223, Biernat, 1998: 405). On the other hand, the Polish Constitution recognises no unalterable provisions, as is the case pursuant to Article 79(3) of the German Constitution (BMJ, 2012: 25; Tatham, 2013: 223). *'It should be accepted that*

limits on transferring powers and "matters" that are not suitable for transfer pursuant to the provision in question may be imposed by other constitutional provisions' (Biernat, 1998: 405). These other constitutional provisions refer to the highest principles of government, essential to the identity of the Polish state (reflected in Chapter I of the Polish Constitution), and (reflected in Chapter II of the Polish Constitution) provisions proclaiming legal rights, and human and citizen freedoms (Biernat, 1998: 406; Sejm, 2014).

It becomes apparent that the rather general language of, inter alia, Article 90 of the Polish Constitution, leaves room for interpretation, which in terms of the Polish Constitution falls within the scope of the Polish Constitutional Tribunal (CT). What is true for the ECJ, its case law, and its effect on European constitutionalism – based on Article 19 TEU and the 17th Declaration of the Final Act of the Treaty of Lisbon (OJoEU, 2010: 83; OJoEU, 2007a: 256) – and what is true for the FCC, based on Article 93 of the German Constitution (BMJ, 2012: 33), is also true for the CT pursuant to Article 188 of the Polish Constitution (Sejm, 2014; Banaszkiewicz, 2005: 5). Tatham (2013: 226) emphasises that only after EU accession did the CT have the chance to mould the constitutional landscape in response to the implications of European law in the national system.

In its Accession Treaty Case, the CT defines the principal that *'[t]he accession of Poland to the European Union did not undermine the supremacy of the Constitution over the whole legal order within the field of sovereignty of the Republic of Poland'* (CT, 2005a : 7). Indeed, and in accordance with the Polish Constitution, the CT concludes that *'it is insufficiently justified to assert that the Communities and the European Union are "supranational organisations" [...]. The expression "supranational organisation" is not mentioned in the Accession Treaty, nor in the Acts constituting an integral part thereof or any provision of secondary Community law'* (CT, 2005a: 7-8). Furthermore, *'neither Article 90(1) nor Article 91(3) authorise delegation to an international organisation of the competence to issue legal acts or take decisions contrary to the Constitution, being the "supreme law of the Republic of Poland" (Article 8(1))'* (CT, 2005a: 8). In light of these principles, Tatham (2013: 240) argues that the recognition of national sovereignty, which remains inviolate, and the relation of powers transferred, shared or retained between the domestic and supranational level, form the very essence of the Accession Treaty Case. Whilst the CT emphasises the ultimate supremacy of the Polish Constitution, it also acknowledges the fact that *'[t]he concept and model of European law created a new situation, wherein, within each Member State, autonomous legal orders co-exist and are simultaneously operative. Their interaction may not be completely described by the traditional concepts of monism and dualism regarding the relationship between domestic law and international law. The existence of the relative autonomy of both, national and Community, legal orders in no way signifies an absence of interaction between them. Furthermore, it does not exclude the possibility of a collision between regulations of Community law and the Constitution'* (CT, 2005a: 8).

It furthermore concludes – and here a parallel to the FCC's conclusion in the 'Milk-powder' case can be drawn (BVerfGE, 1971: 31 (145) at 97) – that *'the direct review of the conformity with the Constitution of particular decisions of the ECJ [...] does not fall within the Constitutional Tribunal's scope of jurisdiction'* (CT, 2005a: 10).

The CT is, with respect to the case law cited here, and as some argue, in general heavily influenced by the FCC. This is especially true for the implications of the 'Solange I' and 'Solange II' judgements (Tatham, 2013: 242; Pernice, 2006: 28-29; BVerfGE, 1974: 37 (271)). The CT argues, in light of the basic rights guaranteed in the Constitution, that norms concerning the rights and freedoms of individuals are the minimum and unsurpassable threshold that cannot be questioned or lowered because of European provisions (Tatham, 2013: 243). This also follows the FCC's judgment that the Member States remain the *'Masters of the Treaties'*, who have conferred, as an example of international organisations, the exercise of some of their powers to the EU, which may in principal also be terminated (Bogdandy, 1999: 41; BVerfGE, 1993: 89 (155) at 112; Tatham, 2013: 83; Franzius, 2010: 25 and 34 and 43; Pernice, 2006: 30-31; Grimm, 2001: 288). With respect to the *'possibility of a collision between regulations of Community law and the Constitution'* the CT concludes, that *'[...]][i]n such an event the Nation as the sovereign, or a State authority organ authorised by the Constitution to represent the Nation, would need to decide on: amending the Constitution; or causing modifications within the Community provisions; or, ultimately, on Poland's withdrawal from the European Union'* (CT, 2005a: 9).
In terms of reviewing harmonised secondary EU legislation, the CT goes even further than the FCC (Tatham, 2013: 265). In its *SK 45/09* judgment, the

CT points out 'that it is necessary to draw a distinction between examining the conformity of the acts of EU secondary legislation to the Treaties, i.e. EU primary law, on the one hand, and examining their conformity to the Constitution, on the other. The institution that ultimately determines the conformity of EU regulations to the Treaties is the Court of Justice of the European Union, and as regards the conformity to the Constitution – the Constitutional Tribunal' (CT, 2011: 21). It goes on by stating: 'What needs to be considered is the effects of a judgment of the Constitutional Tribunal in the case of adjudication that the norms of EU secondary legislation are inconsistent with the Constitution. In the context of the acts of Polish law, the said non-conformity results in declaring the unconstitutional norms to be no longer legally binding (Article 190(1) and (3) of the Constitution). [...] The consequence of the ruling of the Constitutional Tribunal would be to rule out the possibility that the acts of EU secondary legislation would be applied by the organs of the Polish state and would have any legal effects in Poland. [...] What should be noted is that such a consequence of the Tribunal's ruling would be difficult to reconcile with the obligations of a Member State and the aforementioned principle of sincere cooperation (Article 4(3) of the TEU). [...] Undoubtedly, the ruling declaring the non-conformity of EU law to the Constitution should have the character of ultima ratio, and ought to appear only when all other ways of resolving a conflict between Polish norms and the norms of the EU legal order have failed' (CT, 2011: 25). The point of view that there is a possible indirect review jurisdiction of the CT has also been stressed in the European Arrest Warrant Case (CT, 2005b: 11-12). 'The obligation to implement the Framework Decisions is a constitutional requirement stemming from Article 9 of the Constitution, but its enactment does not assure automatically and in every case the material conformity of the provision of derivative EU law and of legislative acts implementing them to the national law with the norms of the Constitution. The basic function of the Constitutional Tribunal in the political system consists of reviewing the conformity of normative acts with the Constitution, and the same task applies also to situations, where the claim of unconstitutionality concerns that part of the scope regulated by a legislative act, which serves the purposes of implementation of EU law' (CT, 2005a: 11-12). With both judgements, Tatham (2013: 265) concludes, the CT fired a warning shot across the bows of the ECJ. Generally, however, the CT tries, as can also be seen in the wording of the judgements above (e.g. *ultima ratio* implications), to assure a nuanced approach to constitutional difficulties concerning EU law *vis-à-vis* national law (Tatham, 2013: 265; CT, 2011: 25).

Like the FCC (BVerfGE, 2009: 123 (267)), the CT had the chance to judge the constitutionality of the Lisbon Treaty (CT, 2010: 63). The *Lisbon Treaty Case* summarises the principles of the case law reflected so far in the judgments of the CT, and re-empathises both the CT's stance on the Constitution as the supreme law of the Republic of Poland – including the primacy of the binding force of the Constitution – and the understanding of mutual dependencies between Member States and the EU, conferring part of the competencies of state organs onto the Union (CT, 2010: 21 and 33-34). Explicitly specifying the limits up to which the competencies of organs of state authority can be transferred, including a positive indication of what the *'certain matters'* of Constitution Article 90(1) imply (Biernat, 1998: 405), the CT states: *'Regardless of the difficulties related to setting a detailed catalogue of inalienable competences, the following should be included among the matters under the complete prohibition of conferral: decisions specifying the fundamental principles of the Constitution and decisions concerning the rights of the individual which determine the identity of the state, including, in particular, the requirement of protection of human dignity and constitutional rights, the principle of social justice, the principle of subsidiarity, as well as the requirement of ensuring better implementation of constitutional values and the prohibition to confer the power to amend the Constitution and the competence to determine competence'* (CT, 2010: 22-23). But the CT also highlights that *'[t]he constitution-maker has decided that the system of law which is binding in the territory of the Republic of Poland will have a multi-faceted character, and will encompass, apart from legal acts constituted by Polish legislative bodies, acts of international and Community law. [...] It is the task of the Polish constitution-maker and legislator to resolve the problem of democratic legitimacy of the measures provided for in the Treaty, applied by the competent bodies of the Union'* (CT, 2010: 30-31 and 36). The latter statement quite clearly indicates that it is the responsibility of the legislator to decide on changes to the Constitution so as to avoid conflict between EU and domestic norms (Tatham 2013: 266).

Although concerning the issue of reviewing harmonised secondary law (see European Arrest Warrant Case and SK 45/09), the effective annulment of European Directives and Regulations remains an option, the CT's position can be characterised as trying to balance *'the principle of the protection of national sovereignty in European integration and the principle of a favourable predisposition towards European integration and co-operation between States'* (Tatham, 2013: 265 and 267). This, however, seems unlikely with respect to a secondary law such as Directive 2009/28/EC, because the constitution of Poland does not, like Article 194 TFEU, explicitly include provisions with respect to energy (Sejm, 2014; OJoEU, 2008: 134-135).

5.1.1 Summary

After the transfer of sovereignty based on the 1997 Constitution of the Republic of Poland, five provisions play an important role. Article 90 allows for the transfer of competencies to international organisations in some matters. Articles 9 and 91 allow for the direct supremacy of international agreements over national laws and statutes. Articles 8 and 87 define the sources of the supremacy of the Polish Constitution (Dabrowska, 2009: 108).

Unlike the German Constitution, which pursuant to Article 23(1) allows the Federation to *'transfer sovereign powers by a law with the consent of the Bundesrat'* (BMJ, 2012: 8), whilst simultaneously defining limits to that transfer (Biernat, 1998: 405; BMJ, 2012: 8-9), the Polish Constitution (here Art. 90(1)) only mentions the delegation of 'the competence of organs of State authority in relation to certain matters' (Sejm, 2014). The Polish Constitution therefore does not permit the transfer of sovereignty, but the exercise of specific competencies, as opposed to the power itself (Tatham, 2013: 223). However, the Constitution neither explicitly specifies the limits up to which the competencies of organs of state authority can be transferred, nor does it give a positive indication of what the 'certain matters' mentioned involve (Biernat, 1998: 405). However, the lack of explicit limitations in the relevant provision, it must be concluded, does exclude the transfer of sovereignty in all cases – that is, it does not indicate the absence of any constitutional limitations at all (Tatham, 2013: 223, Biernat, 1998: 405). On the other hand, the Polish Constitution recognises no unalterable provisions, as is the case pursuant to Article 79(3) of the German Constitution (BMJ, 2012: 25; Tatham, 2013: 223): 'It should be accepted that limits on transferring powers and "matters" that are not suitable for transfer pursuant to the provision in question may be imposed by other constitutional provisions' (Biernat, 1998: 405). In its Accession Treaty Case, the CT defines the principal that '[t]he accession of Poland to the European Union did not undermine the supremacy of the Constitution over the whole legal order within the field of sovereignty of the Republic of Poland' (CT, 2005a : 7). Indeed, and in accordance with the Polish Constitution, the CT concludes that 'it is insufficiently justified to assert that the Communities and the European Union are "supranational organisations" [...]. The expression "supranational organisation" is not mentioned in the Accession Treaty, nor in the Acts constituting an integral part thereof or any provision of secondary Community law' (CT, 2005a: 7-8).

The CT is, with respect to the case law cited here, and as some argue, in general heavily influenced by the FCC. This is especially true for the implications of the 'Solange I' and 'Solange II' judgements (Tatham, 2013: 242; Pernice, 2006: 28-29; BVerfGE, 1974: 37 (271)). The CT argues, in light of the basic rights guaranteed in the Constitution, that norms concerning the rights and freedoms of individuals are the minimum and unsurpassable threshold that cannot be questioned or lowered because of European provisions (Tatham, 2013: 243). It also follows the FCC's judgment that the Member States remain the 'Masters of the Treaties', who have conferred, as an example of international organisations, the exercise of some of their powers to the EU, which may in principal also be terminated (Bogdandy, 1999: 41; BVerfGE, 1993: 89 (155) at 112; Tatham, 2013: 83; Franzius, 2010: 25 and 34 and 43; Pernice, 2006: 30-31; Grimm, 2001: 288). In terms of reviewing harmonised secondary EU legislation, the CT goes even further than the FCC, and indicates the possibility of an effective annulment of European Directives (Tatham, 2013: 265). However, the CT's position can be characterised as trying to balance 'the principle of the protection of national sovereignty in European integration and the principle of a favourable predisposition towards European integration and co-operation between States' (Tatham, 2013: 265 and 267). This seems unlikely, however, with respect to secondary law such as Directive 2009/28/EC, because the constitution of Poland does not, like Article 194 TFEU, explicitly include provisions with respect to energy (Sejm, 2014; OJoEU, 2008: 134-135).

5.2 Polish Transposition of Directive 2009/28/EC and NREAP

Directive 2009/28/EC, as part of the Climate and Energy Package, was signed by the President of the European Parliament and the President of the Council on 23rd April 2009 and published in the Union's Official Journal on 5th June 2009 (see also Tab. 1) (OJoEU, 2009a-f: 11-145). Pursuant to Article 27 of Directive 2009/28/EC, Member States have to transpose the laws, regulations and administrative provisions necessary for complying with the Directive by 5th December 2010 (OJoEU, 2009b: 44). On 21st March 2013, the European Commission notified the public in a press release that it referred Poland to the ECJ for failing to transpose EU rules (EC, 2013e). The form of order sought from the ECJ was to *'declare that, by failing to adopt the laws, regulations and administrative provisions necessary to ensure compliance with Directive 2009/28/EC of the European Parliament and of the Council of 23 April 2009 on the promotion of the use of energy from renewable sources and amending and subsequently repealing Directive 2001/77/EC and 2003/30/EC, and in any event by not notifying the Commission of such provisions, the Republic of Poland has failed to fulfil its obligation under Article 27 of that directive'* (OJoEU, 2013b: 8). The Commission furthermore proposed to impose a penalty payment of EUR 133.228,80 from the day on which the judgment was delivered, and to issue an order to the Republic of Poland to pay the costs of the proceedings (OJoEU, 2013b: 8). The suggested penalty, the Commission argued, is based on the duration and the gravity of the infringement. A formal notice of non-transposition of Directive 2009/28/EC was sent to Poland as early as January 2011, followed by a Reasoned Opinion in March 2012 (EC, 2013e). Given the on-going infringement procedure mentioned above, it remains unclear whether, and to what extent, Poland transposed Directive 2009/28/EC. To approach this matter, it is necessary to take a look at the framework indicators of the Directive – that is, the development of the Polish NREAP. Pursuant to Article 4(3) of Directive 2009/28/EC (OJoEU, 2009b: 28-29), the Polish Minister of Economic Affairs published a forecast document indicating: (a) the estimated excess production of energy from renewable sources compared to the indicative trajectory, and (b) the estimated demand for energy from renewable sources to be satisfied by means other than domestic production in January 2010 (Minister of Economic Affairs, 2010a: 2). According to the forecast document, Poland in 2020 will have an excess of energy from renewable sources of 0,48% with respect to indicative trajectories defined in the Directive (see Tab. 3). In addition, the document states, there will be no need to use renewable energy from abroad in order to meet the 2020 targets (Minister of Economic Affairs, 2010a: 2-3). According to the same Article 4 of Directive 2009/28/EC, each Member State is obliged to adopt a National Renewable Energy Action Plan (NREAP) in order to communicate Member States' national targets for transport, electricity, heating and cooling (OJoEU, 2009b: 28). The document was due on 30th June 2010 (OJoEU, 2009b: 28). According to Ancygier (2013: 339), it was not before 25th May 2010 that a draft version of the Polish NREAP was presented to the Polish public for consultation (Minister for Economic Affairs, 2010b). Consequently,

Poland was amongst the last Member States to send the official version of its NREAP to Brussels. An *'Addendum to the National Renewable Energy Action Plan'* was sent to the Commission at the end of 2011 (NREAP Poland, 2010; Addendum NREAP Poland, 2011; Ancygier, 2013: 339-341). Comparing the three documents raises questions concerning the integrity of the calculations of the RES trajectories, and of the explanatory power of measures described for achieving those targets.

Tab. 19: Polish overall RES targets as defined in Annex I of Directive 2009/28/EC (Minister for Economy Affairs, 2010b: 16; NREAP Poland, 2010: 18; Addendum NREAP Poland, 2011)

National over all target based on Annex I of Directive 2009/28/EC	Draft	NREAP	Addendum
Share of energy from renewable sources in gross final consumption of energy in 2005 (%)	7,2%	7,2%	n.a.
Target for share of energy from renewable sources in gross final consumption of energy 2005 (%)	15%	15%	n.a.
Expected total consumption in 2020 after adjustment (ktoe)	69.203	69.200	n.a.
Expected amount of energy from renewable sources in accordance with the target 2020 (ktoe)	10.380	10.380,50	n.a.

Based on Annex I of Directive 2009/28/EC (see Tab. 3 and Tab. 19), the share of energy from RES in gross final energy consumption was 7,2% in 2005. The legally binding target for Poland as to be reached by 2020 is at least 15%.

Tab. 20: National 2020 target and estimated trajectory of energy from renewable sources in heating and cooling, electricity and transport according to NREAP Draft (Minister of Economic Affairs, 2010b: 18)

	2010	2011	2012	2013	2014	2015	2016	2017	2018	2019	2020
Renewable sources of energy - heating and cooling (%)	12	12,3	12,6	12,9	13,2	13,6	14,30	15	15,7	16,2	17
Renewable energy sources - electricity (%)	6,2	7,22	8,22	9,25	10,4	11,52	12,97	14,35	15,68	17,73	19,43
Renewable energy sources - transport (%)	5,3	5,8	6,6	7,2	7,8	8,3	8,7	9,1	9,6	10	10,2
Renewable energy sources, total (%)	9,1	9,6	10,2	10,7	11,2	11,7	12,5	13,2	13,9	14,7	15,5
Of which through cooperation mechanism (%)	n.a.	0	0	0	0	0	0	0	0	n.a.	0
Surplus for cooperation mechanism (%)	n.a.	0,8	1,4	1,1	1,6	1	1,8	0,9	1,7	n.a.	0,5

Tab. 21: National 2020 target and estimated trajectory of energy from renewable sources in heating and cooling, electricity and transport according to NREAP (NREAP Poland, 2010: 21)

	2010	2011	2012	2013	2014	2015	2016	2017	2018	2019	2020
Renewable sources of energy - heating and cooling (%)	12,29	12,54	12,78	13,05	13,29	13,71	14,39	15,02	15,68	16,5	17,05
Renewable energy sources - electricity (%)	7,53	8,85	10,19	11,13	12,19	13	13,85	14,68	15,64	16,78	19,13
Renewable energy sources - transport (%)	5,84	6,3	6,67	7,21	7,48	7,73	7,99	8,49	9,05	9,59	10,14
Renewable energy sources, total (%)	9,58	10,09	10,6	11,05	11,45	11,9	12,49	13,11	13,79	14,58	15,5
Of which through cooperation mechanism (%)	n.a.	0	0	0	0	0	0	0	0	n.a.	0
Surplus for cooperation mechanism (%)	n.a.	1,31	1,82	1,48	1,88	1,16	1,75	0,81	1,48	n.a.	0,5

Tab. 22: National 2020 target and estimated trajectory of energy from renewable sources in heating and cooling, electricity and transport according to NREAP (Addendum NREAP Poland, 2011: 6)

	2010	2011	2012	2013	2014	2015	2016	2017	2018	2019	2020
Renewable sources of energy - heating and cooling (%)	12,29	12,54	12,78	13,05	13,29	13,71	14,39	15,02	15,68	16,5	17,05
Renewable energy sources - electricity (%)	7,53	8,85	10,19	11,13	12,19	13	13,85	14,68	15,64	16,78	19,13
Renewable energy sources - transport (%)	5,84	6,56	7,27	7,79	8,05	8,37	8,62	9,34	10,09	10,83	11,36
Renewable energy sources, total (%)	9,58	11,16	10,74	11,21	11,61	12,08	12,66	13,35	14,09	14,94	15,85
Of which through cooperation mechanism (%)	n.a.	0	0	0	0	0	0	0	0	n.a.	0
Surplus for cooperation mechanism (%)	n.a.	1,4	1,98	1,67	2,07	1,37	1,95	1,08	1,82	n.a.	0,85

Tab. 23: Estimate of the total contribution (installed capacity, gross electricity consumption) anticipated in Poland of each technology using renewable energy sources with regard to the binding targets for 2020 and the indicative trajectories for the share of energy from renewable sources in the electricity sector in the period 2010-2014 according to NREAP Draft including operating hours in h/a (Minister of Economic Affairs, 2010b: 131)

	2005			2010			2011		
	MW	GWh	h/a	MW	GWh	h/a	MW	GWh	h/a
Hydropower:	915	2.201	2.405	952	2.279	2.394	962	2.311	2.402
< 1 MW	72	358	4.972	102	357	3.500	106	371	3.500
1 MW - 10 MW	174	504	2.897	178	534	3.000	184	552	3.000
> 10 MW	669	1.339	2.001	672	1.388	2.065	672	1.388	2.065
from pumped storage power plant	0	0	0	0	0	0	0	0	0
Geothermal power:	0	0	0	0	0	0	0	0	0
Solar energy	0	0	0	1	1	1.000	1	2	2.000
photovoltaics	0	0	0	1	1	1.000	1	2	2.000
concentrated solar energy	0	0	0	0	0	0	0	0	0
Tides, waves, other ocean energy:	0	0	0	0	0	0	0	0	0
Wind energy:	121	136	1.124	1.100	2.310	2.100	1.550	3.255	2.100
land-based	121	136	1.124	1.100	2.310	2.100	1.550	3.255	2.100
offshore	0	0	0	0	0	0	0	0	0
small installations	0	0	0	0	0	0	0	0	0
Biomass:	286	1.451	5.073	380	6.028	15.863	450	7.110	15.800
solid	268	1.340	5.000	300	5.700	19.000	350	6.700	19.143
biogas	18	111	6.167	80	328	4.100	100	410	4.100
liquid biofuels	0	0	0	0	0	0	0	0	0
Overall:	1.091	3.787	3.471	2.433	10.618	4.364	2.963	12.678	4.279
from combined heat and power	55	1.451	26.382	130	1.874	14.415	155	2.215	14.290

2012			2013			2014		
MW	GWh	h/a	MW	GWh	h/a	MW	GWh	h/a
972	2.343	2.410	982	2.375	2.419	992	2.407	2.426
110	385	3.500	114	399	3.500	118	413	3.500
190	570	3.000	196	588	3.000	202	606	3.000
672	1.388	2.065	672	1.388	2.065	672	1.388	2.065
0	0	0	0	0	0	0	0	0
0	0	0	0	0	0	0	0	0
2	2	1.000	2	2	1.000	2	2	1.000
2	2	1.000	2	2	1.000	2	2	1.000
0	0	0	0	0	0	0	0	0
0	0	0	0	0	0	0	0	0
2.010	4.308	2.143	2.520	5.327	2.114	3.030	6.491	2.142
2.000	4.300	2.150	2.450	5.267	2.150	2.900	6.380	2.200
0	0	0	0	0	0	0	0	0
10	8	800	70	60	857	130	111	854
720	8.192	11.378	940	8.774	9.334	1.180	9.438	7.998
600	7.700	12.833	800	8.200	10.250	1.000	8.700	8.700
120	492	4.100	140	574	4.100	180	738	4.100
0	0	0	0	0	0	0	0	0
3.704	14.845	4.008	4.444	16.478	3.708	5.204	18.338	3.524
240	2.556	10.650	310	2.747	8.861	390	2.979	7.638

Tab. 24: Estimate of the total contribution (installed capacity, gross electricity consumption) anticipated in Poland of each technology using renewable energy sources with regard to the binding targets for 2020 and the indicative trajectories for the share of energy from renewable sources in the electricity sector in the period 2015-2020 according to NREAP Draft including operating hours in h/a (Minister of Economic Affairs, 2010b: 132)

	2015			2016			2017		
	MW	GWh	h/a	MW	GWh	h/a	MW	GWh	h/a
Hydropower:	1.002	2.439	2.434	1.012	2.471	2.442	1.002	2.503	2.498
< 1 MW	122	427	3.500	126	441	3.500	130	455	3.500
1 MW - 10 MW	208	624	3.000	214	642	3.000	220	660	3.000
> 10 MW	672	1.388	2.065	672	1.388	2.065	672	1.388	2.065
from pumped storage power plant	0	0	0	0	0	0	0	0	0
Geothermal power:	0	0	0	0	0	0	0	0	0
Solar energy	2	2	1.000	2	2	1.000	3	3	1.000
photovoltaics	2	2	1.000	2	2	1.000	3	3	1.000
concentrated solar energy	0	0	0	0	0	0	0	0	0
Tides, waves,other ocean energy:	0	0	0	0	0	0	0	0	0
Wind energy:	3.540	7.541	2.130	4.060	8.784	2.164	4.580	9.860	2.153
land-based	3.350	7.370	2.200	3.800	8.550	2.250	4.250	9.563	2.250
offshore	0	0	0	0	0	0	0	0	0
small installations	190	171	900	260	234	900	330	297	900
Biomass:	1.530	9.893	6.466	1.630	10.348	6.348	1.780	11.008	6.184
solid	1.300	8.950	6.885	1.350	9.200	6.815	1.400	9.450	6.750
biogas	230	943	4.100	280	1.148	4.100	380	1.558	4.100
liquid biofuels	0	0	0	0	0	0	0	0	0
Overall:	6.074	19.875	3.272	6.704	21.605	3.223	7.385	23.374	3.165
from combined heat and power	505	3.157	6.250	545	3.334	6.117	610	3.614	5.925

2018			2019			2020		
MW	GWh	h/a	MW	GWh	h/a	MW	GWh	h/a
1.032	2.535	2.456	1.042	2.567	2.464	1.152	2.969	2.577
134	469	3.500	138	483	3.500	142	497	3.500
226	678	3.000	232	696	3.000	238	714	3.000
672	1.388	2.065	672	1.388	2.065	772	1.758	2.277
0	0	0	0	0	0	0	0	0
0	0	0	0	0	0	0	0	0
3	3	1.000	3	3	1.000	3	3	1.000
3	3	1.000	3	3	1.000	3	3	1.000
0	0	0	0	0	0	0	0	0
0	0	0	0	0	0	0	0	0
5.100	11.210	2.198	5.620	12.315	2.191	6.650	15.210	2.287
4.700	10.810	2.300	5.150	11.845	2.300	5.600	13.160	2.350
0	0	0	0	0	0	500	1.500	3.000
400	400	1.000	470	470	1.000	550	550	1.000
1.930	11.668	6.046	2.230	12.943	5.804	2.530	14.218	5.620
1.450	9.700	6.690	1.500	9.950	6.633	1.550	10.200	6.581
480	1.968	4.100	730	2.993	4.100	980	4.018	4.100
0	0	0	0	0	0	0	0	0
8.065	25.416	3.151	8.895	27.828	3.128	10.335	32.400	3.135
675	3.894	5.769	815	4.482	5.499	955	5.069	5.308

Tab. 25: Estimate of the total contribution (installed capacity, gross electricity consumption) anticipated in Poland of each technology using renewable energy sources with regard to the binding targets for 2020 and the indicative trajectories for the share of energy from renewable sources in the electricity sector in the period 2010-2014 according to NREAP including operating hours in h/a (NREAP Poland: 134)

	2005			2010			2011		
	MW	GWh	h/a	MW	GWh	h/a	MW	GWh	h/a
Hydropower:	915	2.201	2.405	952	2.279	2.394	962	2.311	2.402
< 1 MW	72	358	4.972	102	357	3.500	106	371	3.500
1 MW - 10 MW	174	504	2.897	178	534	3.000	184	552	3.000
> 10 MW	669	1.339	2.001	672	1.388	2.065	672	1.388	2.065
from pumped storage power plant	0	0	0	0	0	0	0	0	0
Geothermal power:	0	0	0	0	0	0	0	0	0
Solar energy	0	0	0	1	1	1.000	2	1	500
photovoltaics	0	0	0	1	1	1.000	2	1	500
concentrated solar energy	0	0	0	0	0	0	0	0	0
Tides, waves,other ocean energy:	0	0	0	0	0	0	0	0	0
Wind energy:	121	136	1.124	910	1.911	2.100	1.260	2.646	2.100
land-based	121	136	1.124	910	1.911	2.100	1.260	2.646	2.100
offshore	0	0	0	0	0	0	0	0	0
Biomass:	55	1.451	26.382	114	3.838	33.667	192	4.568	23.792
solid	25	1.340	53.600	40	3.472	86.800	71	3.948	55.606
biogas	30	111	3.700	74	366	4.946	121	620	5.124
liquid biofuels	0	0	0	0	0	0	0	0	0
Overall:	1.091	3.787	3.471	1.977	8.029	4.061	2.416	9.526	3.943
from combined heat and power	55	1451	26.382	114	3.838		192	4.568	

2012			2013			2014		
MW	GWh	h/a	MW	GWh	h/a	MW	GWh	h/a
972	2.343	2.410	982	2.375	2.419	992	2.407	2.426
110	385	3.500	114	399	3.500	118	413	3.500
190	570	3.000	196	588	3.000	202	606	3.000
672	1.388	2.065	672	1.388	2.065	672	1.388	2.065
0	0	0	0	0	0	0	0	0
0	0	0	0	0	0	0	0	0
2	2	1.000	2	2	1.000	2	2	1.000
2	2	1.000	2	2	1.000	2	2	1.000
0	0	0	0	0	0	0	0	0
0	0	0	0	0	0	0	0	0
1.610	3.381	2.100	2.010	4.221	2.100	2.510	5.271	2.100
1.610	3.381	2.100	2.010	4.221	2.100	2.510	5.271	2.100
0	0	0	0	0	0	0	0	0
270	5.298	19.622	348	6.029	17.325	426	6.759	15.866
102	4.424	43.373	134	4.900	36.567	165	5.376	32.582
168	874	5.202	214	1.129	5.276	261	1.383	5.299
0	0	0	0	0	0	0	0	0
2.854	11.024	3.863	3.342	12.626	3.778	3.930	14.438	3.674
270	5.298	19.622	348	6.029	17.325	426	6.759	15.866

Tab. 26: Estimate of the total contribution (installed capacity, gross electricity consumption) anticipated in Poland of each technology using renewable energy sources with regard to the binding targets for 2020 and the indicative trajectories for the share of energy from renewable sources in the electricity sector in the period 2015-2020 according to NREAP including operating hours in h/a (NREAP Poland: 135)

	2015			2016			2017		
	MW	GWh	h/a	MW	GWh	h/a	MW	GWh	h/a
Hydropower:	1.002	2.439	2.434	1.012	2.471	2.442	1.022	2.503	2.449
< 1 MW	122	427	3.500	126	441	3.500	130	455	3.500
1 MW - 10 MW	208	624	3.000	214	642	3.000	220	660	3.000
> 10 MW	672	1.388	2.065	672	1.388	2.065	672	1.388	2.065
from pumped storage power plant	0	0	0	0	0	0	0	0	0
Geothermal power:	0	0	0	0	0	0	0	0	0
Solar energy	2	2	1.000	3	2	667	3	2	667
photovoltaics	2	2	1.000	3	2	667	3	2	667
concentrated solar energy	0	0	0	0	0	0	0	0	0
Tides, waves, other ocean energy:	0	0	0	0	0	0	0	0	0
Wind energy:	3.010	6.321	2.100	3.510	7.371	2.100	4.010	8.421	2.100
land-based	3.010	6.321	2.100	3.510	7.371	2.100	4.010	8.421	2.100
offshore	0	0	0	0	0	0	0	0	0
Biomass:	504	7.489	14.859	688	8.868	12.890	872	10.247	11.751
solid	196	5.852	29.857	281	6.757	24.046	367	7.662	20.877
biogas	308	1.637	5.315	407	2.111	5.187	506	2.585	5.109
liquid biofuels	0	0	0	0	0	0	0	0	0
Overall:	4.518	16.251	3.597	5.213	18.712	3.589	5.907	21.173	3.584
from combined heat and power	504	7.489,00	14.859	688	8.868	12.890	872	10.247	11.751

2018			2019			2020		
MW	GWh	h/a	MW	GWh	h/a	MW	GWh	h/a
1.032	2.535	2.456	1.042	2.567	2.464	1.052	2.599	2.471
134	469	3.500	138	483	3.500	142	497	3.500
226	678	3.000	232	696	3.000	238	714	3.000
672	1.388	2.065	672	1.388	2.065	672	1.388	2.065
0	0	0	0	0	0	0	0	0
0	0	0	0	0	0	0	0	0
3	3	1.000	3	3	1.000	3	3	1.000
3	3	1.000	3	3	1.000	3	3	1.000
0	0	0	0	0	0	0	0	0
0	0	0	0	0	0	0	0	0
4.510	9.471	2.100	5.260	11.716	2.227	6.110	13.541	2.216
4.510	9.471	2.100	5.260	11.716	2.227	6.110	13.541	2.216
0	0	0	0	0	0	0	0	0
1.057	11.625	10.998	1.241	13.004	10.479	1.425	14.383	10.093
452	8.567	18.954	538	9.472	17.606	623	10.377	16.657
604	3.058	5.063	703	3.532	5.024	802	4.006	4.995
0	0	0	0	0	0	0	0	0
6.602	23.634	3.580	7.546	27.290	3.616	8.590	30.526	3.554
1.057	11.625	10.998	1.241	13.004	10.479	1.425	14.383	10.093

As Ancygier (2013: 339) points out, although the total contribution (installed capacity, gross electricity consumption) anticipated in Poland for each technology using renewable energy sources is presented in these documents (see Tab. 23, Tab. 24, Tab. 25, and Tab. 26), it remains unclear how they were calculated. Based on the draft NREAP (Minister of Economic Affairs, 2010b: 131-132) in 2020, biomass, for example, would account for an installed capacity of 1.425 MW, producing 14.383 GWh of gross electricity. This would mean producing electricity for 10.093 h/a. Unfortunately, the year only consists of 8.760 h (24 x 365 = 8.760). Indeed, the numbers are questionable for the entire decade leading up to the 2020 trajectory. Right at the beginning, in 2005, biomass was supposed to have had an installed capacity of 55 MW leading to 1.451 GWh of produced electricity. This would account for 26.382 h of electricity production from biomass in 2005.

Overall, based on the draft document, by 2020 the installed capacity of all renewable energy sources is anticipated to be 8.590 MW leading to 30.526 GWh. This equals 3.557 h/a (40% of the year's total hours) of electricity production from all facilities using renewable energy technologies. In comparison, the German NREAP indicates that all its renewable energy facilities produce energy for 1.956 h/a (22% of the year's total hours) by 2020 (see Tab. 15). Unfortunately, neither the Polish nor the German NREAP provides mathematical details on, for example, the technology-specific assumptions for full power when calculating the estimates provided for in the tables in question (Tab. 14, Tab. 15, Tab. 23, Tab. 24, Tab. 25, and Tab. 26).

Furthermore, the document fails to explain what kind of future policies and measures the Polish legislator envisions in order to set framework conditions for the anticipated growth rates of the different renewable energy sources and renewable energy technologies (see Tab. 20, Tab. 23, Tab. 24) (Ancygier, 2013: 339). This can be illustrated with respect to wind energy. According to the draft document, the installed capacity of wind facilities is to grow from 121 MW in 2005 to 6.110 MW in 2020 (see Tab. 23, and Tab. 24). This would mean the role of wind energy with regard to the overall installed capacities of renewable facilities increases from 11% in 2005 to 71% in 2020. However, policy implications of the draft document refer to the policy conditions implemented before Directive 2009/28/EC, which in principle reflect the implementation necessities of Directive 2001/77/EC and 2003/30/EC designed for differing policy goals (Ancygier, 2013: 339-334; PWEA, 2010: 7).

The inconsistencies of the draft version of the NREAP did not go unnoticed by the interested public groups. Whilst it is not surprising in general that representatives of the renewable energy interest groups would criticise aspects of the action plan, the degree of criticism is astonishing. In an opinion of the Polish Wind Energy Association developed in cooperation with the Polish Institute for Renewable Energy, the authors conclude that *'[t]he numerical data presented in the draft NREAP indicate the authors' aspirations to formally and mechanically fulfil the requirements of Directive 2009/28/EC, [...] rather than to make an actual attempt to reach a final outcome in accordance with the*

principles of science and to discuss its determinants and consequences. [...]
The presented data is not substantiated and highly uncertain as well as difficult
or impossible to verify' (PWEA, 2010: 2). *'Having regard to the above, PWEA*
applies for rejecting the draft NREAP in full and developing it again' (PWEA,
2010: 9).

In terms of the integrity of the calculations of the RES trajectories, and the explanatory power of policy measures for achieving those targets, the official NREAP of Poland sent to the Commission in late 2010 does not fundamentally change the impression gained from the draft version. As a response to the mathematical inadequacies of the draft NREAP regarding the estimate of the total contribution (installed capacity, gross electricity consumption) anticipated in Poland, it becomes apparent that the numbers have been subject to revision. However, with respect to biomass, for example, it seems that the values for installed capacities were changed in order to be more plausible, rather than to reflect empirical indications (NREAP Poland, 2010: 134-135). Following the new document, biomass would account for an installed capacity of 2.530 MW (as opposed to 1.425 MW according to the draft version), producing 14.218 GWh of electricity in 2020 (see Tab. 26). These numbers would lead to 5.619 operating hours of all biomass facilities. However, the document still fails to justify the numbers. It also remains unclear how the numbers for the base year 2005 could be changed from 55 MW installed biomass capacity (draft version) to 286 MW (NREAP). Besides these corrections, biomass data given for the years 2010-2013 still indicate higher yearly operating hours for the facilities in question than the year consists of (see Tab. 25) (NREAP Poland, 2010: 134-135).

As for policy legislation to promote the use of energy from renewable sources, the NREAP of Poland provides 15 respective pieces of legislation, quoted below: (NREAP Poland, 2010: 25-26):

1. The Energy Act of April 1997.

2. Act of 2nd July 2004 on freedom of economic activity.

3. Act of 6th December 2008 on excise duty.

4. Environmental Act of 27th April 2001.

5. Regulation of the Minister of Economy of 14th August 2007 on detailed scope of obligations in respect to obtaining certificates of origin and submitting them for cancellation, payment of a substitution fee, purchase of electricity and heat from renewable sources, as well as the obligation to confirm the data on the amount of electricity produced from a renewable energy source.

6. Regulation of the Minister of Economy of 4th May 2007 on detailed conditions for the electricity system functioning.

7. Regulation of the Minister of Economy of 2nd July 2007 on detailed principles for defining and calculating tariffs and principles for settlements in electric energy trading.

8. Regulation of the Minister of Economy of 3rd February 2009 on granting public aid for investment involving construction or extension of units producing electricity or heat from renewable sources.

9. Energy Policy of Poland until 2030 (with appendices) adopted by the Council of Ministers on 10th November 2009.

10. Act of 26th October 2000 on commodity exchanges, together with Act of 29th July on public offers and the conditions for introducing financial instruments to the organised trading system, and on public companies, as well as the Regulation of the Council of Ministers of 22nd December 2009 on special procedure and conditions for introducing property rights to trading on the exchange

11. Regulation of the Minister of the Environment of 2nd June 2010 on detailed technical conditions for qualifying a part of energy recovered from heat treatment of municipal waste.

12. Regulations of the Commodity Energy Exchange, and the related document Conditions of trading of property rights to certificates of origin for energy produced from renewable energy sources [sic.

13. Regulation of the Register of Certificates of Origin kept by the Commodity Energy Exchange S.A.

14. Operating and Maintenance Instructions of the Transmission Grid drawn up and published by PSE Operator S.A.

15. Financial priority programme of the National Fund for Environment Protection and Water Management.

Given the sheer number of legislative acts relevant to the implementation of the national policy to promote the use of energy from renewable sources, Poland acknowledges the need for legislative transparency in the area of RES to clarify its support possibilities. The NREAP therefore announces the preparation of an Act on energy from renewables sources in total gross demand for energy as a main legislative project (NREAP Poland, 2010: 30). *'Consequently, the new planned legislation will be aimed at directing stronger, systematic and multi-dimensional support for sustainable development of the renewable sector. Planned activities for the development of RES would require several amendments to the legislation with respect to: definitions, overall objectives and measures necessary to achieve the objectives, principles of calculation of the share of energy from RES, administrative procedures, legislation and codes, installer information and certification, guarantee of electricity RES origin, access to grids and their operations, as well as reporting'* (NREAP Poland, 2010: 30). This quotation clearly indicates that the inadequacies of the NREAP are identified, and that Poland desires to correct them. In its 2011 *'Addendum to the National Renewable Energy Action Plan'*, Poland reiterates its position on working on a draft RES Act that will:

- establish a transparent and cost-effective system promoting the usage of RES energy;

- specify the principles of supporting energy generation from RES, based on national objective and international obligations;

- establish and systematise the support mechanisms for energy from RES included in the Energy Law Act;

- improve the implementation of the provision of Directive 2009/28/EC (Addendum NREAP, 2011: 9-10).

The Addendum, published on 2nd December 2011, also illustrates that, one year after publishing the NREAP, the Polish legislator did not agree on the adoption of one of its main legislative projects.[12] Even by the end of 2011, the drafted regulation *'has not been subject to neither inter-department arrangements nor social consultations as of yet'* (Addendum NREAP Poland, 2011: 11). Instead, the authors of the Addendum emphasise that the bulk of issues addressed by Directive 2009/28/EC were already implemented by statutory matters such as the Energy Law (Addendum NREAP Poland, 2011: 11). On 22nd March 2012, the European Commission sent a Reasoned Opinion to Poland because it has not informed the Commission of the full transposition of Directive 2009/28/EC into its national legislation (EC, 2012b: 1). In a press release, the Commission stated: *'If the Member States do not comply with their legal obligation within two months, the Commission may decide to refer them to the Court of Justice'* (EC, 2012b: 1). As mentioned above, on 21st March 2013 the European Commission notified the public in a press release that it referred Poland to the ECJ for failing to transpose EU rules (EC, 2013e).

According to an analysis of the implementation status of the RES Directive published by Rybski and Stoczkiewicz (2013: 5), as of June 2013 it is impossible to identify regulations for electricity in the Polish legal system that would bring Polish law in line with the supranational requirements. In the same month, however, the Sejm started working on an amendment act to, *inter alia,* the Energy Law, in order to avoid a potential financial penalty as a result of

---

[12] Pursuant to Article 27 of Directive 2009/28/EC, Member States had to transpose the laws, regulations and administrative provisions necessary to comply with the Directive by 5th December 2010 (OJoEU, 2009b: 44)

the infringement procedure initiated by the Commission. What is known as the Small Tri-Pack, adopted by the Sejm on 21st June 2013, is meant to nullify complaints for non-transposition and accelerate work on the Large Tri-Pack, including a new Gas Law, an amendment to the Energy Law, and the adoption of the envisioned RES Act (Reuters, 2013; ADVOC, 2013). The Small Tri-Pack became effective on 11th September 2013 (Lexology, 2013). Following Rybski and Stoczkiewicz (2013: 16), it is unlikely whether the Small Tri-Pack bill will result in the Commission's withdrawing its action against Poland. In addition, according to a press release of the Minister of Economic Affairs (2014), the Council of Ministers adopted an RES Act on 8th April 2014. Overall, and strictly based on official documents, it remains true that until there is a judgment by the ECJ, the Republic of Poland has failed to notify the European Commission of provisions transposing Directive 2009/28/EC on time. Based on the indicators cited above, it seems highly likely the Republic of Poland failed to fulfil its obligation under Article 27 of Directive 2009/28/EC by September 2013 (OJoEU, 2013b: 8; OJoEU, 2009b: 44).

In order to guarantee compatibility with the '4.2 German Transposition of Directive 2009/28/EC and NREAP' section, implications for Poland deriving from Article 13, 16 and 22(3) of Directive 2009/28/EC will be incorporated in the following section (OJoEU, 2009b: 32, 35 and 42).

5.2.2 Summary

On 21st March 2013 the European Commission notified the public in a press release that it referred Poland to the ECJ for failing to transpose EU rules (EC, 2013e). The form of order sought from the ECJ was to *'declare that, by failing to adopt the laws, regulations and administrative provisions necessary to ensure compliance with Directive 2009/28/EC of the European Parliament and of the Council of 23 April 2009 on the promotion of the use of energy from renewable sources and amending and subsequently repealing Directive 2001/77/EC and 2003/30/EC, and in any event by not notifying the Commission of such provisions, the Republic of Poland has failed to fulfil its obligation under Article 27 of that directive'* (OJoEU, 2013b: 8). The Commission furthermore proposed to impose a penalty payment of EUR 133.228,80 from the day on which the judgment was delivered and to ussue an order to the Republic of Poland to pay the costs of the proceedings (OJoEU, 2013b: 8). The penalty suggested, the Commission argued, was based on the duration and the gravity of the infringement. A formal notice of non-transposition of Directive 2009/28/EC was sent to Poland as early as January 2011, followed by a Reasoned Opinion in March 2012.

According to the same Article 4 of Directive 2009/28/EC, each Member State is obliged to adopt a National Renewable Energy Action Plan (NREAP) in order to communicate Member States' national targets for transport, electricity, heating and cooling (OJoEU, 2009b: 28). The document was due on 30th June 2010 (OJoEU, 2009b: 28). According to Ancygier (2013: 339), it was not before 25th

May 2010 that a draft version of the Polish NREAP was presented to the Polish public for consultation (Minister for Economic Affairs, 2010b). Accordingly, Poland was amongst the last Member States to send the official version of its NREAP to Brussels. An *Addendum to the National Renewable Energy Action Plan* was send to the Commission by the end of 2011 (NREAP Poland, 2010; Addendum NREAP Poland, 2011; Ancygier, 2013: 339-341). Comparing the three documents raises questions concerning the integrity of the calculations of the RES trajectories, and of the explanatory power of measures described for achieving those targets.

The inconsistencies of the draft version of the NREAP did not go unnoticed by the interested public groups. In an opinion of the Polish Wind Energy Association, developed in cooperation with the Polish Institute for Renewable Energy, the authors conclude that *'[t]he numerical data presented in the draft NREAP indicate the authors' aspirations to formally and mechanically fulfil the requirements of Directive 2009/28/EC, [...] rather than to make an actual attempt to reach a final outcome in accordance with the principles of science and to discuss its determinants and consequences. [...] The presented data is not substantiated and highly uncertain as well as difficult or impossible to verify'* (PWEA, 2010: 2). *'Having regard to the above, PWEA applies for rejecting the draft NREAP in full and developing it again'* (PWEA, 2010: 9). In terms of the integrity of the calculations of the RES trajectories, and the explanatory power of policy measures described for achieving those targets, the official NREAP of Poland send to the Commission in late 2010 does not fundamentally change the impression gained from the draft version.

Given the sheer number of legislative acts relevant to the implementation of the national policy to promote the use of energy from renewable sources, Poland acknowledges the need for legislative transparency in the area of RES to clarify its support possibilities. The NREAP therefore announces the preparation of an Act on energy from renewables sources in total gross demand for energy as a main legislative project (NREAP Poland, 2010: 30). *'Consequently, the new planned legislation will be aimed at directing stronger, systematic and multi-dimensional support for sustainable development of the renewable sector. Planned activities for the development of RES would require several amendments to the legislation with respect to: definitions, overall objectives and measures necessary to achieve the objectives, principles of calculation of the share of energy from RES, administrative procedures, legislation and codes, installer information and certification, guarantee of electricity RES origin, access to grids and their operations, as well as reporting'* (NREAP Poland, 2010: 30). This quotation clearly indicates that the inadequacies of the NREAP are identified, and that Poland desires to correct them.

On 22nd March 2012 the European Commission sent a Reasoned Opinion to Poland because it has not informed the Commission of the full transposition of Directive 2009/28/EC into its national legislation (EC, 2012b: 1). As mentioned above, on 21st March 2013 the European Commission notified the public in a press release that is referred Poland to the ECJ for failing to transpose EU rules (EC, 2013e).

In 2013, the Sejm started working on an amendment act to, *inter alia,* the Energy Law, in order to avoid a potential financial penalty as a result of the infringement procedure initiated by the Commission. What is known as the Small Tri-Pack, adopted by the Sejm on 21st June 2013, is meant to nullify complaints for non-transposition, and accelerate work on the Large Tri-Pack, including a new Gas Law, an amendment to the Energy Law, and the adoption of the envisioned RES Act (Reuters, 2013; ADVOC, 2013). The Small Tri-Pack became effective on 11th September 2013 (Lexology, 2013). In addition, according to a press release of the Minister of Economic Affairs (2014), the Council of Ministers adopted an RES Act on 8th April 2014. Overall, strictly based on official documents, it remains true, until there is a judgment by the ECJ, the Republic of Poland has failed to notify the European Commission of provisions transposing Directive 2009/28/EC on time. Based on the indicators cited here, it seems highly likely the Republic of Poland failed to fulfil its obligation under Article 27 of Directive 2009/28/EC by September 2013 (OJoEU, 2013b: 8; OJoEU, 2009b: 44).

5.3 Implementation Evaluation of Directive 2009/28/EC in Poland

In accordance with the section '4.3 Implementation Evaluation of Directive 2009/28/EC in Germany', evaluation of the implementation of Directive 2009/28/EC in Poland will be predominantly based on a 'renewable energy progress and biofuels sustainability' report produced for the European Commission by Ecofys, Fraunhofer, Becker Bütter Held (BBH), Energy Economics Group, and Windrock International (Ecofys, 2012). Whilst this might cause criticism in terms of a lack of counterfactual reasoning – given there is only one major source – it is deemed adequate for the following reasons: (i) Empirical information about the implementation effectiveness serves as one aspect in the overall thesis design. It lies beyond the scope of this thesis to conduct a new comprehensive implementation analysis for all European Member States to counterpoint the study in question. (ii) Whilst additional information on certain aspects reflected in the evaluation report might be available from different sources (e.g. statistical agencies from the Member States in question), they do not apply the same evaluation methodology, which makes a data synopsis illegitimate. (iii) The research in question has been carried out for the Commission and informs its renewable energy progress report (EC, 2013b), and the 'commission staff working' document (EC, 2013c). It therefore not only reflects the basis for official evaluation statements in the political process, but also guarantees a common analysis basis for inter-state comparability. It furthermore provides the elements of evidence for further decisions in the context of infringement procedure for non-communication of national transposition measures and/or non-transposition of supranational provisions.

5.3.1 Past Progress Evaluation

As can be seen from 'Tab. 4: Actual and planned RES Shares 2009-2011/2012 (Ecofys, 2012: 18)' in conjunction with 'Tab. 21: National 2020 target and estimated trajectory of energy from renewable sources in heating and cooling, electricity and transport according to NREAP (NREAP Poland, 2010: 21)', Poland missed its NREAP indicative target RES share in 2010 by 0,09 percentage points, which equals a negative deviation of -0,99% (Ecofys, 2012: 18; NREAP Poland, 2010: 17). Notwithstanding this minor negative deviation, according to Ecofys (2012: 20), Poland is among the sixteen Member States that had a growth rate in 2009/2010 above the required average, and is therefore in a good situation to reach its overall RES target in 2020.

With respect to the non–legally binding RES-E target for 2010, Poland undershot its target with a negative deviation of its actual 2010 share from the planned 2010 share of -11,02% (see 'Tab. 5: Actual and planned RES-E shares (NREAP_ EU27, 2010; Eurostat, 2013b)' and 'Tab. 21: National 2020 target and estimated trajectory of energy from renewable sources in heating and cooling, electricity and transport according to NREAP (NREAP Poland, 2010: 21)') (Ecofys, 2012: 21-22 and 27; NREAP Poland, 2010: 21; EC, 2013c: 4). Poland is among the six Member States that missed their respective targets by more than 10%. With respect to its RES-E growth between 2009/2010, Poland, whilst showing good growth, would need higher than average growth rates to move from the NREAP 2010 RES-E target to the 2020 RES-E target (Ecofys, 2012: 23).

Following below is a differentiated description of the growth of RES-E technologies from 2009 to 2010. Please regard the data in conjunction with 'Tab. 25: Estimate of the total contribution (installed capacity, gross electricity consumption) anticipated in Poland of each technology using renewable energy sources with regard to the binding targets for 2020 and the indicative trajectories for the share of energy from renewable sources in the electricity sector in the period 2010-2014 according to NREAP including operating hours in h/a (NREAP Poland: 134)':

- Hydropower: According to Ecofys (2012: 25), Poland achieved a growth rate of electricity produced from hydropower of > 10 MW (2,3%) in 2009/2010. In the category small hydro, which combines installations of < 1 MW and 1-10, Poland had a negative growth rate of -0,71% (Ecofys, 2012: 25). Note: The latter category is in distinction to the categories indicated in the NREAP, since it combines two subcategories used in the NREAP (see Tab. 25) (NREAP Poland, 2010: 134, Ecofys, 2012: 33).

- Photovoltaic: Ecofys (2012: 25 and 26) indicates, based on 1,67 GWh electricity produced by the RES-E photovoltaic-technology in 2010, a growth rate of 20,57% in between 2009 and 2010. According to Eurostat (2014a), however, photovoltaic electricity generation accounted for 0 GWh in 2010

- Wind: Ecofys (2012: 25 and 26) indicates, based on 1.700 GWh electricity produced by the RES-E onshore wind-technology in 2010, a growth of 31,53% in between 2009 and 2010. According to Eurostat (2014a), Poland had 1.077 GWh of wind in 2009 and 1.664 GWh in 2010, which would account for a growth rate of roughly 55%.

- Biomass: In 2010, electricity generation from solid biomass accounted for 5.905 GWh, leading to a growth rate of 16,95% in from 2009 to 2010. Electricity produced from biogas accounted for 398 GWh, leading to a growth rate of 19,87% in the same period (Ecofys, 2012: 25 and 26). Eurostat (2014a) does not provide data on biomass or biogas for this period.

The assessment of overall policies and measures for the promotion of renewable energies in Poland as indicted in the NREAP leads the authors of the Ecofys report (2012: 101) to conclude that Poland only partially fulfilled its overall NREAP policy commitments, and did not fulfil its NREAP policy commitments for the promotion of RES-E. With respect to the latter, support levels for each technology are *'poor'*, whilst the long-term security of support for the RES sectors receives a *'fair'* rating (Ecofys, 2012: 101). This judgment does seem to confirm the analysis implications of the circumstances described, concerning the integrity of Poland's NREAP with regard to the calculations of the RES trajectories, and with regard to the explanatory power of measures described for achieving those targets (see '5.2 Polish Transposition of Directive 2009/28/EC and NREAP'). It explicitly results from the fact that, at the time evaluation of the Ecofys report was conducted, Poland *'did not adopt the Act on Renewable Energy Sources that it announced in its NREAP'* (Ecofys, 2012: 118).

With respect to the overall assessment of administrative procedures (see Tab. 6: Assessment of administrative procedures in the Member States (Ecofys, 2012: 186-187; EC, 2013c: 9)'), Poland received a *'needs improvement'* rating. Following the Ecofys report (2012: 197), building permits, locations permits, and grid connection for renewable energy facilities have to be approved in three different decision procedures. Whilst this was already identified as a potential obstaclein the NREAP, Poland's progress report does not mention any changes in this area.

With respect to Polish progress in electricity grid integration (see Tab. 7: Member State rating on progress in electricity grid integration (Ecofys, 2012: 177, 179 and 181)'), it receives a *'fair'* rating for overall RES grid integration. According to the same source, it receives a *'fair'* rating for addressing grid capacity limitations, whilst for rules for bearing and sharing the costs of grid development, it is considered *'advanced'* (Ecofys, 2012: 179 and 181).

The quantitative assessment of the future deployment of RES in Poland, based on the Green-X model (see Text Box 4: Short characterisation of the model Green-X by EEG (EEG, 2007: 3-4)') leads the authors of the Ecofys report (2012: 66) to predict an expected share of RES in Germany under the CPI-Scenario of max. 9,6%,, and max. 10,7% under the CPI+PPI-Scenario in 2020 (see also section '3.6.4 Expected RES shares in 2020'). With respect to the latter evaluation, the authors of the Ecofys report (2012: 66) do not expect Poland to reach its overall RES target of 15% by 2020 based on measures and policies referred to in its NREAP.

5.3.2 Summary

As can be seen from 'Tab. 4: Actual and planned RES Shares 2009-2011/2012 (Ecofys, 2012: 18)' in conjunction with 'Tab. 21: National 2020 target and estimated trajectory of energy from renewable sources in heating and cooling, electricity and transport according to NREAP (NREAP Poland, 2010: 21)', Poland missed its NREAP indicative target RES share in 2010 by 0,09 percentage points, which equals a negative deviation of -0,99% (Ecofys, 2012: 18; NREAP Poland, 2010: 17).

With respect to the non–legally binding RES-E target for 2010, Poland undershot its target with a negative deviation of its actual 2010 share from the planned 2010 share of -11,02% (see 'Tab. 5: Actual and planned RES-E shares (NREAP_ EU27, 2010; Eurostat, 2013b)' and 'Tab. 21: National 2020 target and estimated trajectory of energy from renewable sources in heating and cooling, electricity and transport according to NREAP (NREAP Poland, 2010: 21)') (Ecofys, 2012: 21-22 and 27; NREAP Poland, 2010: 21; EC, 2013c: 4). Poland is among the six Member States that missed their respective targets by more than 10%. With respect to its RES-E growth in 2009/2010, Poland, whilst showing good growth, would need higher than average growth rates to move from the NREAP 2010 RES-E target to the 2020 RES-E target (Ecofys, 2012: 23).

The assessment of overall policies and measures for the promotion of renewable energies in Poland as indicted in the NREAP leads the authors of the Ecofys report (2012: 101) to conclude that Poland only partially fulfilled its overall NREAP policy commitments, and did not fulfil its NREAP policy commitments for the promotion of RES-E. With respect to the latter, support levels for each technology are *'poor'*, whilst the long-term security of support for the RES-Sectors received a *'fair'* rating (Ecofys, 2012: 101). This judgment seems to confirm the implications of the circumstances concerning the integrity of Poland's NREAP with regard to the calculations of the RES trajectories, and with regard to the explanatory power of measures described for achieving those targets (see '5.2 Polish Transposition of Directive 2009/28/EC and NREAP'). This is a result of the fact that, at the time evaluation of the Ecofys report was conducted, Poland *'did not adopt the Act on Renewable Energy Sources that it announced in its NREAP'* (Ecofys, 2012: 118).

The quantitative assessment of the future deployment of RES in Poland, based on the Green-X model (see Text Box 4: Short characterisation of the model Green-X by EEG (EEG, 2007: 3-4)') leads the authors of the Ecofys report (2012: 66) to predict an expected share of RES in Poland under the CPI-Scenario of max. 9,6%, and max. 10,7% under the CPI+PPI-Scenario in 2020 (see also section '3.6.4 Expected RES shares in 2020'). With respect to the latter evaluation, the authors of the Ecofys report (2012: 66) do not expect Poland to reach its overall RES target of 15% based on measures and policies referred to in its NREAP by 2020. The results of the future deployment assessment are contradictory to the predictions based on the RES growth rates in 2009/2010, where Poland is among the sixteen Member States that had a growth rate above the required average, and therefore in a good situation to reach its overall RES target in 2020 (Ecofys, 2012: 20).

5.4 Polish National Preferences

So far, the supranational energy policy context for the promotion of renewable energies in the Union has been the focal point of the analysis. Emphasis has been given to the broader legislative context leading, in particular, to the adoption of Directive 2009/28/EC. Additional emphasis has been given to the European polity framework based on the respective European treaties, the interpretation of the treaties in light of national primary law (based on case law of the ECJ), and the treaty provisions that enable the adoption of renewable energy policies. The aforementioned aspects have been described, interlinked and analysed. Furthermore, a first evaluation of the implementation effectiveness of Directive 2009/28/EC in the European Union has been sketched.

In a second step, the German domestic response to the European stimuli was described. This discussed the German polity with respect to the European polity, the interpretation of the German constitution in light of European primary law (based on case law by the CT) and the implementation of Directive 2009/28/EC both in legal and empirical terms. With respect to explanatory factor (ii) (see '2.3.2 Applied Theory'), focus was placed on the domestic policy process, which reflects issue-specific interests determined by geopolitical and economic factors through national preferences. In a third step, the Polish domestic response to the European stimuli has been described. Again, this discussed the Polish polity with respect to the European polity, the interpretation of the Polish constitution in light of European primary law (based on case law by the CT) and the implementation of Directive 2009/28/EC both in legal and empirical terms. With respect to explanatory factor (ii) (see '2.3.2 Applied Theory'), it is now time to focus on the domestic policy process, which reflects issue-specific interests determined by geopolitical and economic factors through national preferences. The overall theory framework of this thesis, reflected in the model illustrated in Fig. 2, includes theories derived from Liberal Intergovernmentalism (See: '2.3 Applied Framework, Theory and Model'). The following aspect should be re-emphasized: the nation state in Liberal Intergovernmentalism is considered to be a unitary and rational actor. This assumption is not to be mistaken with the idea that states are unitary in their internal politics. The unitary actor assumption, which maintains the idea that each nation state acts in international negotiations 'as if' with a single voice, subsumes the idea that 'once particular objectives arise out of [...] domestic competition, states strategize as unitary actors vis-à-vis other states in an effort to realize them' (Moravcsik, 1998: 22). The same holds for the assumption that nation states act rationally. The idea is that states make internal decisions 'as if' efficiently conducting a weighted choice based on a stable set of underlying principles. Both assumptions, however, should not be over emphasised (Moravcsik, 1998: 23). Fundamental goals of states are neither uniform nor fixed. State preferences vary over time and in response to changes in, for example, economic, ideological, or geopolitical environments (Moravcsik, 1998: 22-23). It is the goal of this section to summarise the particular objectives that arise out of domestic competition (Moravcsik, 1998: 22). Thus, the following analysis is not to be mistaken for a nation-bound policy analysis (see '2.1 Nation-Bound Policy Analysis Frameworks'). For an extensive nation-bound, multi-level policy analysis with respect to German RES-E promotion, see Bernd Hirschl (2008). However, features described in the nation-bound policy analysis theory approaches inform the *'metatheoretical language'* (Ostrom, 2007: 25) reflected in the overall theory framework, and is to be understood in conjunction with, in particular, the applied Liberal Intergorvenmentalism approaches underlying the on-going discussion.

5.4.1 Polish Energy Mix

In 2012, the production of primary energy in Poland amounted to 3.035,6 PJ (Paiz.gov.pl, 2013: 4). For energy production, coal remains the most important raw material in the republic. Poland is the largest coal producer in the European Union, making it account for 80% of its energy production (Paiz.gov.pl, 2013: 4; Cwiek-Karpowicz, Gawlikowska-Fyk and Westphal, 2013: 3; EC, 2011d: 125). With regard to primary energy consumption in 2012, hard coal accounted for 41%, oil for 24%, gas for 13% and lignite for 12%. Peat and wood accounted for 5%, with the same amount being consumed from other sources (see Fig.10) (Paiz.gov.pl, 2013: 4). In 2012, Poland had to import 94,7% of its demand for oil and 73,8% of its gas demand (Eurostat, 2014). However, with the rather large occurrence of indigenous coal resources – out of which the country exports large quantities to other EU Member States – the Polish score with regard to energy dependence (30,7%) is well below the EU average (53,3%) (see Fig. 11) (Eurostat, 2014; Cwiek-Karpowicz, Gawlikowska-Fyk and Westphal, 2013: 3; EC, 2007c: 2)

Fig. 10: Primary Energy Consumption Poland in 2012 (Paiz.gov.pl, 2013: 4)

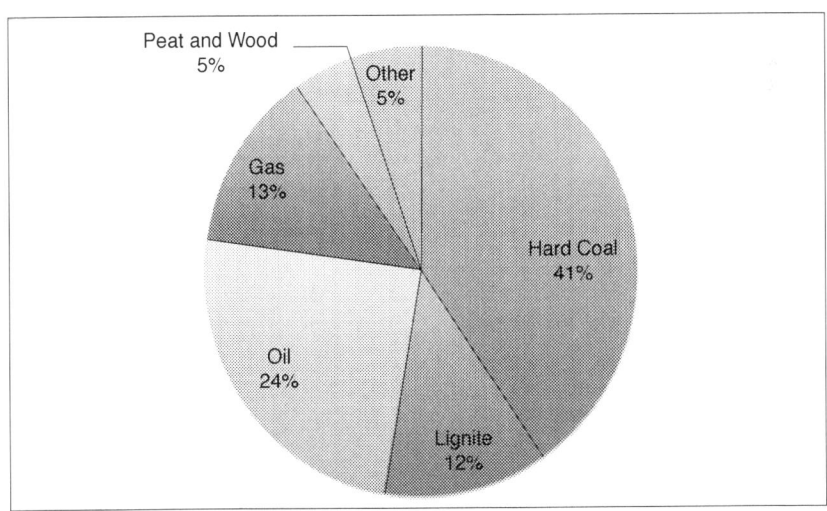

Poland imports 90% of its Gas from Russia, and 10% from Germany (EC, 2011d: 126). According to the U.S. Energy Information Administration (EIA, 2013), Poland imports oil primarily from Russia. Overall, with regard to its external energy supplies, the country is heavily dependent on Russian sources (Cwiek-Karpowicz, Gawlikowska-Fyk and Westphal, 2013: 4).

Fig. 11: Net Energy Imports Poland in 2012 (Eurostat, 2014)

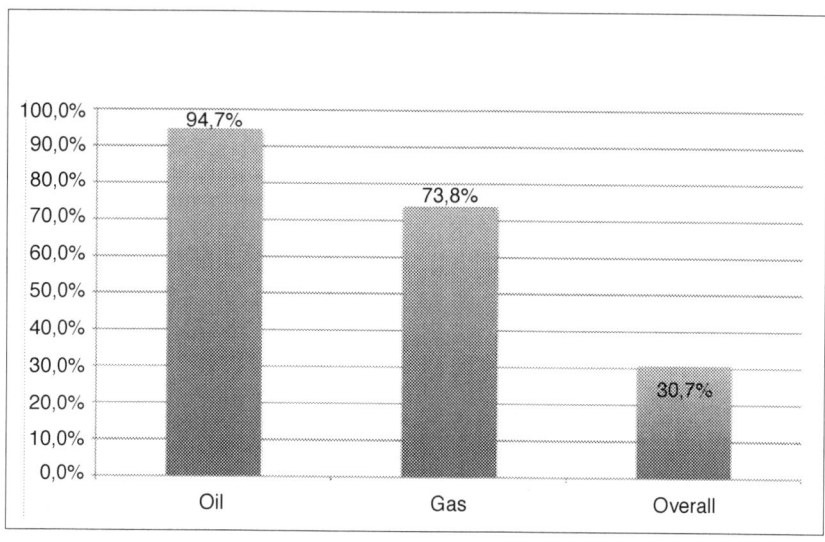

In 2012, 82% of the electric power produced in Poland, was delivered by Combined Heat and Power Plants using traditional fuels (hard coal: 54,5%, and lignite: 25,2%) (Paiz.gov.pl, 2013: 5). Wind power plants accounted for 6,3%, and hydro power plants for 5,7%. So-called Autoproducer Combined Heat and Power Plants – that is, plants producing electricity for their own industrial facilities – accounted for 5% (Paiz.gov.pl, 2013: 5).

Fig. 12: Gross Electricity Generation in 2012 (Paiz.gov.pl, 2013: 5)

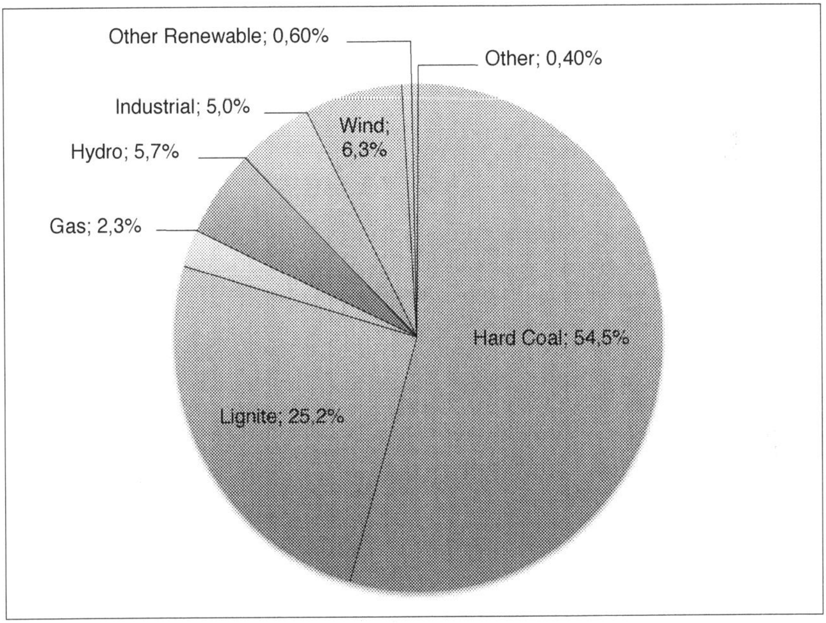

5.4.2 Domestic Energy Policies as Indicator of National Objectives

On 10th November 2009, Polish Council of Ministers adopted the 'Energy Policy of Poland until 2030' (Ministry of Economy, 2009a). In the same year, the European Union adopted the Climate and Energy Package, part of which is the Renewable Energy Directive 2009/28/EC (EC, 2007a: 3). The Polish Council of Ministers states in its adopted policy document that *'[a]s a Member State of the European Union, Poland actively participates in devising the Community energy policy, it also implements its main objectives under the specific domestic conditions taking into account the protection of interests of customers, the energy resources and technological conditions of energy generation and transmission'* (Ministry of Economy, 2009a: 4).

The primary direction of the energy policy reflected in the 2030 strategy can be summarised as follows (Ministry of Economy, 2009a: 4-5):

- Improve energy efficiency;

- Enhance security of fuel and energy supplies;

- Diversify the electricity generation structure by introducing nuclear energy;

- Develop the use of renewable energy sources (including biofuels);

- Develop competitive fuel and energy markets;

- Reduce the environmental impact of the power industry.

Energy efficiency and security of energy supplies, *inter alia,* by introducing nuclear power and relying on coal as the main fuel for the power industry, are the main foci of the Polish energy strategy reflected in the cited document (Ministry of Economy, 2009a: 4-9).

In order to accomplish the above objectives, in particular the optimal use of domestic coal deposits, measures envisaged include, *inter alia:* (a) the abolition of legal barriers for the exploitation of new hard coal and lignite deposits; (b) the identification and protection of national resources of hard coal and lignite; (c) intensified geological research to extend the coal resource base; and (d) the safeguard of access to coal resources (Ministry of Economy, 2009a: 9-10).

With respect to the diversification of the electricity generation *' the Council of Ministers of 13 January 2009 imposed an obligation on all process participants to take intensive actions aimed at setting the stage for implementing the nuclear energy production programme in Poland'* (Ministry of Economy, 2009a: 15). One of the specific objectives in this context is to adapt the Polish legal system so that the process of developing a nuclear energy sector in Poland is efficient (Ministry of Economy, 2009a: 16).

Developing the use of renewable energy sources is not explicitly classified as a measure for diversification of the electricity system in the 'Energy Policy of Poland until 2030' (Ministry of Economy, 2009a: 16 and 17). Indeed, whilst measures with respect to coal are considered a 'main objective' of the energy policy, and objectives with respect to the introduction of nuclear energy are considered a *'primary objective'* of the energy policy, the development of the use of energy from renewable sources is prioritised, being of *'considerable importance'* (Ministry of Economy, 2009a: 9, 15 and 17). The main policy objective in the field of RES as expressed in the policy document generally reflects the Polish implications of Directive 2009/28/EC – that is, increasing the use of renewable energy sources within final energy use to 15% by 2020, and the share of biofuels for transport to 10% by 2020 (Ministry of Economy, 2009a: 18; OJoEU, 2009b: 17).

To gain an overview of how Polish energy policy indicates national preferences in the last one and a half decades, 1997 marks a reasonable starting point. In this year, the Polish legislator adopted an Energy Law that, for the first time, would regulate all issues concerning the energy sector in a comprehensive way. It marks the year a new centre-right government, under Prime Minister Jerzy Buzek, took power (Ancygier, 2013: 240). Most importantly, it is the year of the constitutional referendum, and the enactment of the 1997 Constitution (Biernat, 1998: 229). Owing to the impeding Membership of Poland in the European Union, preparations to adapt the *aquis communautaire,* including Directive 2001/77/EC on the promotion of electricity produced from renewable

energy sources, also started to influence the Polish energy and climate policy (Ancygier, 2013: 240; OJoEC, 2001: 33; Dabrowska, 2009: 109).
The research institution EC BREC (EC Baltic Renewable Energy Centre) organised a first conference to discuss a strategy for the development of renewable sources of energy in November 1997, including representatives of the relevant Polish ministries and of the European Commission (Podrygala, 2008: 72). Results of a second conference organised by EC BREC in collaboration with the Polish parliament one year later laid the basis for a parliament resolution on behalf of the Parliament Committee for the Environment, including four key demands (Podrygala, 2008: 73):

- The Polish government must commit to a defined share of the use of energy from renewable sources;

- Development of a strategy for the promotion of renewable sources of energy by the end of 1999;

- Integration of this strategy into the overall Polish energy and climate policy;

- Creation of a legal and financial framework to enable interested actors to develop a renewable energy sector.

In July 1999, the Sjem accepted the resolution in question with a majority of 395 to 5 (Ancygier, 2013: 241). Furthermore, the parliament declared its willingness to engage actively in the creation of a Renewable Energy Act (Podrygala, 2008: 74). In response to the resolution, the Prime Minister initially assigned the task to develop such a strategy to the Ministry of Economy. Since the Ministry in question did not show much enthusiasm dealing with the subject, it was later assigned to the Ministry of Environment (Podrygala, 2008: 74). On 23rd August 2001, the strategy was adopted by parliament in a second version based on a draft adopted by Council of Ministers and prepared by the Ministry of Environment with participation of EC BREC a year earlier. The document sets targets for the share of energy from renewable sources defined at 7,5 % for 2010, and 14% for 2020 (Ancygier, 2013: 241; Podrygala, 2008: 76). Since the strategy also envisaged technology-specific renewable energy development programs, EC BREC – on behalf of the Ministry for Environment – started to prepare an executive program for wind energy in 2002. The Minister of the Economy in the Council of Ministers, however, blocked the adoption of the programme after a new centre-left government, led by Prime Minister Miller, took over from the government led by Prime Minister Buzek in October 2001 (Ancygier, 2013: 242; Podrygala, 2008: 79; Princeton.edu (interlinked with Wikipedia), 2014). Within the same legislative period, on 29th November 2000, the Sejm adopted the Act on an Atomic Law (National Atomic Energy Agency, 2014; OECD, 2008: 3). The general provision of the law enshrines, *inter alia, 'activities related to peaceful uses of atomic energy, involving actual and potential exposures to ionizing radiation emitted by artificial radioactive sources, nuclear materials, ionizing radiation generating devices, radioactive waste and spent nuclear fuel'* (Act of Parliament of 29 November 2000 Atomic Law, in: National Atomic Energy Agency, 2014). In the same year, the Sjem also adopted an amendment to the 1997 Energy Law. Subject of the amendment act was, *inter alia,* the introduction of a quota scheme – as opposed to feed-in tariffs as in the case of Germany (see above) – for the promotion of renewable energies. The introduction of a quota scheme, aiming at guaranteeing the purchase of a defined physical volume of electricity produced from RES by the electricity supplier, was prepared by the Ministry of Economy parallel to the preparation of the strategy for the renewable energy sector by the Ministry of the Environment (Podrygala, 2008: 80-81, Ancygier, 2013: 281). Whilst the amendment to the Energy Law was rapidly processed, and since the introduction of a quota schemes was a by-product of the overall amendment act, this particular proposal did not cause any major opposition during the parliamentary readings (Podrygala, 2008: 81).

Under the same government, in 2003, the Ministry of the Environment adopted a climate policy strategy for Poland for up to 2020 (Ministry of the Environment, 2003; Ancygier, 2013: 243). Whilst the document explicitly refers to the above-mentioned strategy for the development of renewable energy sources as part of the legal framework for the implementation of a Polish climate policy, developing RES is not considered a short-term objective until 2006 (Ministry of the Environment, 2003: 13, 15, and 21-22). The medium and long-term objectives and measures (2007-2012, and 2013-2020) of the climate policy, however, include *'the promotion, development and growth of the use of new and renewable energy sources, the technologies of CO2 sequestration, as well as advanced and innovative, environmentally friendly technologies and the identification and elimination of barriers to their use'* (Ministry of the Environment, 2003: 22). The wording regardings this latter aspect is interesting, because it includes nuclear energy without explicitly mentioning it. Nuclear power could be considered both a new energy source for the Polish energy system and – contentiously – an environmentally friendly technology. As a result of the 1999 resolution on the development of a strategy for the promotion of renewable energies, and the willingness of the Sjem to work on a Renewable Energy Act, in 2003/2004, EC BREC presented a first comprehensive proposal for a draft Renewable Energy Act. The Ministry of Environment rejected the act proposal without much explanation (Ancygier, 2013: 245, Podrygala, 2008: 86-87). Instead, the Ministry of Economy came forward in 2005 with an energy policy strategy for up to 2025, adopted under the new minority left-wing government lead by Prime Minister Belka (Ministry of Economy, 2005; Princeton.edu (interlinked with Wikipedia), 2014). The document summarises long-term plans for up to 2025 with regard to, inter alia, generation capacities of domestic fuels and energy sources (which marks the first headline of the document), and with regard to the development of the use of renewable energy sources (which marks the sixth headline of the document) (Ministry of Economy, 2005: 16). With respect to the first aspect, the document states that *'[i]n view of the need to diversify the primary energy carriers and the need to limit greenhouse gases emissions to atmosphere, the introduction of nuclear energy to the domestic system becomes substantiated'* (Ministry of Economy, 2005: 16). With respect to the latter argument, the document states that *'[t]he rational use of renewable energy sources is one of the important elements of the country's sustainable development. [...] The share of electricity from RES in the total gross consumption of electricity in the country should obtain 7,5% in 2010. [...] This is in line with the indicative quantitative objective stipulated for Poland in Directive 2001/77/EC [...]'* (Ministry of Economy, 2005: 26). Whilst overall, the document extends options for the development of renewable energies, it ignores the target of a 14% share of RES-E by 2020 mentioned in the 2001 renewable energy strategy. Furthermore, it explicitly addresses for the first time the need for the introduction of nuclear power to the Polish energy sector (Ancygier, 2013: 243-244).

Whilst, with the adoption of the amendment act to the Energy Law in 2000, the legislator provided a legal basis for the introduction of a quota scheme for electricity from renewable sources, it took another five years for the introduction of tradable green certificates (Podrygala, 2008: 64, Ancygier, 2013: 281). The absence of tradable green certificates is problematic, because it adds network access charges for the transmission of electricity to the overall costs of renewables, owing to the lack of the division of physical and virtual trading. If renewable energy producers receive certificates for each megawatt hour of electricity produced that can be purchased by the electricity suppliers to fulfil their obligations under the quota scheme, then transmission costs become obsolete, for it is not important where the megawatt hour of renewable energy has been produced, merely that it has been produced (Podrygala, 2008: 64, Ancygier, 2013: 281-282). The introduction of tradable green certificates originally intended as part of the rejected draft Renewable Energy Act prepared by EC BREC in 2003/2004 was incorporated in a 2005 amendment to the Energy Law adopted by the Sejm in March 2005 (Podrygala, 2008: 65 and 89).

The year 2005 also marks the beginning of a period of right-wing governments in Poland, starting with a two-year government coalition led by the Law and Justice Party under Prime Minister Marcinkiewicz (Ancygier, 2013: 246). In 2007, two years after he took office, Jaroslaw Kaczynski replaced Marcinkiewicz, also a member of his party, as Prime Minister (Princeton.edu (interlinked with Wikipedia), 2014). In 2007, the Kaczynski cabinet was replaced by a government coalition lead by the Civic Platform under Prime Minister Donald Tusk, who in 2014 still serves at Prime Minister (Princeton.edu (interlinked with Wikipedia), 2014). The period in question also comprehends the first half of 2007, when Germany held the Presidency of the Council of the European Union, under which the Council adopted a comprehensive energy action plan and the European institutions started their work on the Climate and Energy Package (see Tab. 1) (EU2007.de, 2007a: 1; Council of the European Union, 2007: 20). According to Riedl (2008: 23), until the end of the negotiations over the Climate and Energy Package in the Council of the European Union, the threat of vetoing the package was the strongest from Poland. After Donald Tusk took office, he immediately pointed out that, with the Climate and Energy Package, Poland accepted unfavourable conditions. By adding, that *'[i]t was possible to veto the climate and energy pact in March 2007, when President Kaczyński and Prime Minister Kaczyński decided on this'*, he implicitly expressed the government's opposition to the preferences of the Climate and Energy Package (Premier.gov. pl, 2013). Back to the national arena. Ancygier (2013: 247) points out: *'Despite significant differences between the Law and Justice Party and Civic Platform, both parties followed similar energy and climate policies. Two important processes in the energy sector started by the Law and Justice-led government were continued by the government dominated by the Civic Platform.'* The first process referred to here regards the adoption of a programme for the electricity sector by the Ministry of Economy in 2006 concerning the consolidation of the energy sector by creating four vertically integrated energy groups. The second process concerns the implementation of the Polish nuclear strategy by appointing a government representative for Alternative Sources of Energy (Ancygier, 2013: 247). It is interesting to note that the very title of this newly created office refers to the medium and long-term objectives of the 2003 climate policy strategy, including *'the promotion, development and growth of the use of new and renewable energy sources'* (Ministry of the Environment, 2003: 22). Krzysztof Zareba, who was appointed to the position, not only pointed out the necessity of building 2-3 nuclear power plants, but was also considered an advocate for renewable energies. At the time Zarbea took office, members of the Law and Justice Party also focused on a new draft proposal for a Renewable Resource Law, which in the end was rejected by the government (Ancygier, 2013: 248).

After the inauguration of Donald Tusk as Prime Minister in October 2007, the position of Representative for Alternative Sources of Energy was cancelled (Ancygier, 2013: 248). Instead, the government started working on the 'Energy Policy of Poland until 2030', described above, and a 'Polish Nuclear Power Program' (Ministry of Economy, 2009a; Ministry of Economy, 2011).

The nuclear power programme prepared by the Ministry for Economy aims at enabling the Polish Republic to *'keep its obligations in the areas of sustainable development and provisions of electricity at reasonable prices and with consideration of environment protection requirements'* (Ministry of Economy, 2011: 7). Major goals of the programme are to implement nuclear power in Poland, *inter alia*, by establishing a framework for the development and operation of nuclear power (Ministry of Economy, 2011: 11). The document envisages the amendment of the Atomic Energy Law, *'the purpose of which is to specify security requirements for nuclear power facilities, including the construction and operation of nuclear power facilities at the highest attainable level according to international requirements and recommendations'* (Ministry of Economy, 2011: 45). The Sejm confirmed the amendments foreseen in the Nuclear Power Programme by adopting a respective act on amending the Act on Atomic Law and other Laws on 13th May 2011 (National Atomic Energy Agency, 2014: 1; Ancygier, 2013: 250). The Amendment of the Atomic Law in 2011 led to interesting situation, that *'a non-existing source of energy was regulated by a separate legal act, whereas the regulations concerning renewable sources of energy, which should cover 14% of Poland's energy consumption by 2020, were spread among a number of ordinances and patched into the Energy Law, which after over 50 amendments was becoming increasingly illegible'* (Ancygier, 2013: 250).

Tab. 27: Legislative history Polish renewable energy policy (for sources see references text)

EU	1996	Directive 96/92/EG — Common rules for internal market electricity
	2001	Directive 2001/77/EC — RES-E promotion in internal market
	2004	Aquis communautaire — Polish EU accession
	2006	Green Paper — A European Strategy for Sustainable, Competitive and Secure Energy
	2013	Commission Communication — An Energy Policy for Europe
	2001	Directive 2009/28/EC — National RE targets for final energy in 2020
	2013	Green Paper — New climate and energy framework up to 2030
Government Poland under:		Prime Minister J. Buzek (1997-2001)
		Prime Minister L. Miller (2001-2004)
		PM M. Belka (2004-2005)
		PM K. Marcinkiewicz (2005-06)
		PM J. Kaczynski (2006-2007)
		Prime Minister D. Tusk (2007-incumbent)
Non-legislative Documents	2000	Government Strategy for Renewable Energy Sector: Objectives: 1.) 7,5% share of RES-E by 2010 2.) 14% share of RES-E by 2020
	2002	Ministry of Environment: Proposal for first executive program for wind energy (not adopted)
	2003	Ministry of Environment: Polish Climate Policy until 2020 (adopted) Objectives: 1.) Diversification energy sources 2.) Enhancement
	2005	Ministry of Economy: Polish Energy Policy until 2025 (adopted) Objectives: 1.) Diversification energy sources (with nuclear) 2.) 7,5% share of RES-E by 2010
	2006	Ministry of Economy: Program for the electricity sector Objectives: 1.) Diversification of energy sector 2.) Creation of four vertically integrated energy groups
	2009	Ministry of Economy: Polish Energy Policy until 2030 (adopted) Objectives: 1.) Diversification energy sources (with nuclear) 2.) 7,5% share of RES-E by 2010 3.) 15% share of RES-E by 2020
RES-E Legislation	1997	Energy Law — Subject: Comprehensive regulation of energy sector
	1999	Parliament Resolution — Subject: Development of renewable sources of energy
	2000	Amendment Energy Law — Introducing, inter alia, quota scheme for electricity from renewable sources
	2003	EC BREC: Draft proposal Renewable Energy Act (rejected)
	2005	Amendment Energy Law — Introducing, inter alia, tradable green certificates
	2007	Law and Justice Party: Draft proposal Renewable Resource Law (rejected)
	2011	Ministry of Economy: 1st Draft proposal Renewable Energy Act (rejected)
	2012	Ministry of Economy: 2nd Draft proposal Renewable Energy Act (on-going)
Nuclear-Power Legislation	2000	Atomic Law: Subject: Activities related to the peaceful uses of atomic energy
	2010	Ministry of Economy: Polish Nuclear Power Program (adopted)
	2011	Amendment Atomic Law: Subject: Specify security requirements for nuclear facilities

As mentioned various times, European Directive 2009/28/EC, as part of the Climate and Energy Package, was signed by the President of the European Parliament and the President of the Council on 23rd April 2009 and published in the Union's Official Journal on 5th June 2009 (see also Tab. 1) (OJoEU, 2009a-f: 11-145). With regard to the Polish transposition of the respective Directive, see '5.3 Implementation Evaluation of Directive 2009/28/EC in Poland'. It should be mentioned at this stage however, that, after the rejected draft proposals for an RES act in 2003 and 2007, neither a first nor second draft of the Renewable Energy Act proposed by the Ministry of Economy in 2011 and 2012 was adopted by the Sejm. Even the creation of a Renewable Energy Department within the Ministry of Economy, to manage the implementation of Directive 2009/28/EC, did not fundamentally accelerate, let alone finalise, the legislative process for the adaption of an RES Act to transpose fully the renewables Directive in time (Ancygier, 2013: 250-253). However, according to a press release of the Minister of Economic Affairs (2014), the Council of Ministers adopted an RES Act on 8th April 2014. Overall, and strictly based on official documents available to date, it remains true, until there is a judgment by the ECJ, the Republic of Poland has failed to notify the European Commission of provisions transposing Directive 2009/28/EC on time. Based on the cited indicators, it seems highly likely that the Republic of Poland, at least by September 2013, failed to fulfil its obligation under Article 27 of Directive 2009/28/EC (OJoEU, 2013b: 8; OJoEU, 2009b: 44).

The overall legislative history of renewable energy promotion as part of the Polish energy policy is summarised in Tab. 27. Note that the selection of non-legislative documents and legislative acts in this section is substantial, but should not be misunderstood as an exhaustive description of the Polish (renewable) energy policy process. As a concluding remark, it can be claimed that energy efficiency and energy security encapsulate the main focus of the Polish energy strategy (Venjakob, 2012: 167). Whilst measures with respect to coal are considered a *'main objective'* of the energy policy, and objectives with respect to the introduction of nuclear energy are considered a *'primary objective'*, the development of the use of energy from renewable sources is prioritised as being of *'considerable importance'* (Ministry of Economy, 2009a: 9, 15 and 17). Following this hierarchy, the promotion of renewable energies seems to hold a subordinate position in the overlying energy policy as defined in the 2030 energy policy paper (Venjakob, 2011: 169). A statement made by Polish Prime Minister, Donald Tusk, seems to confirm this impression. In March 2013 he declared that Poland *'would fulfil its 15 per cent renewable energy target by 2020 but "nothing more"'* (Ancygier and Szulecki, 2013a: 1). As Podrygala (2008: 95) points out, it can be concluded (also from the indications given in the selection of non-legislative and legislative documents at hand), that the Ministry of the Economy is the main opponent of the promotion of renewable energies, and is a dominant actor in shaping the national preferences of the Polish energy policy. It is furthermore worth mentioning that, in the period 2004-2008 alone, Poland had five governments, each of which was dedicated to shaping the country's overall energy policy (Riedl, 2008: 10).

5.4.3 Summary

On 10th November 2009, the Polish Council of Ministers adopted the 'Energy Policy of Poland until 2030' (Ministry of Economy, 2009a). In the same year, the European Union adopted the Climate and Energy Package, part of which is the Renewable Energy Directive 2009/28/EC (EC, 2007a: 3).
Energy efficiency and security of energy supplies, *inter alia,* by introducing nuclear power and by relying on coal as the main fuel for the power industry, are the main foci of the Polish energy strategy as reflected in the cited document (Ministry of Economy, 2009a: 4-9).
In order to accomplish the above objectives, in particular the optimal use of domestic coal deposits, measures envisaged include, *inter alia:* (a) the abolition of legal barriers for the exploitation of new hard coal and lignite deposits; (b) the identification and protection of national resources of hard coal and lignite; (c) an intensified geological research to extend the coal resource base; and (d) the safeguard of access to coal resources (Ministry of Economy, 2009a: 9-10).
The research institution EC BREC (EC Baltic Renewable Energy Centre) organised in November 1997 a first conference to discuss a strategy for the development of renewable sources of energy, including representatives from the relevant Polish ministries and of the European Commission (Podrygala, 2008: 72). Results of a second conference organised by EC BREC in collaboration with the Polish parliament one year later laid the basis for a parliament resolution on behalf of the Parliament Committee for the Environment. In July 1999, the Sjem accepted the resolution in question with a majority of 395 to 5 (Ancygier, 2013: 241). Furthermore, the parliament declared its willingness to engaged actively in the creation of a Renewable Energy Act (Podrygala, 2008: 74).
After the inauguration of Donald Tusk as Prime Minister in October 2007, the government started working on the 'Energy Policy of Poland until 2030' and a 'Polish Nuclear Power Program' (Ministry of Economy, 2009a; Ministry of Economy, 2011; Ancygier, 2013: 248). The Sejm confirmed the amendments foreseen in the Nuclear Power Programme by adopting an act amending the Act on Atomic Law and other Laws on 13th May 2011 (National Atomic Energy Agency, 2014: 1; Ancygier, 2013: 250). The Amendment of the Atomic Law in 2011 lead to interesting situation, where *'a non-existing source of energy was regulated by a separate legal act, whereas the regulations concerning renewable sources of energy, which should cover 14% of Poland's energy consumption by 2020, were spread among a number of ordinances and patched into the Energy Law, which after over 50 amendments was becoming increasingly illegible'* (Ancygier, 2013: 250).
With regard to the Polish transposition of the respective Directive, see '5.3 Implementation Evaluation of Directive 2009/28/EC in Poland'. It should be mentioned at this stage that neither a first nor second draft for a Renewable Energy Act proposed by the Ministry of Economy in 2011 and 2012 was adopted by the Sejm. Even the creation of a Renewable Energy Department within the Ministry of Economy, to manage the implementation of Directive 2009/28/EC, did not fundamentally accelerate, let alone finalise, the legislative process

for the adaption of an RES Act to transpose fully the renewables Directive in time (Ancygier, 2013: 250-253). However, according to a press release of the Minister of Economic Affairs (2014), the Council of Ministers adopted an RES Act on 8th April 2014. Overall, and strictly based on official documents available to, it remains true that, until there is a judgment by the ECJ, the Republic of Poland has failed to notify the European Commission of provisions transposing Directive 2009/28/EC on time. Based on the cited indicators it seems highly likely the Republic of Poland, at least by September 2013, failed to fulfil its obligation under Article 27 of Directive 2009/28/EC (OJoEU, 2013b: 8; OJoEU, 2009b: 44).

The overall legislative history of renewable energy promotion as part of the Polish energy policy is summarised in Tab. 27. Note that the selection of non-legislative documents and legislative acts in this section is substantial, but should not be misunderstood as an exhaustive description of the Polish (renewable) energy policy process. As a concluding remark, it can be claimed that energy efficiency and energy security encapsualte the main focus of the Polish energy strategy (Venjakob, 2012: 167). Whilst measures with respect to coal are considered a *'main objective'* of the energy policy, and objectives with respect to the introduction of nuclear energy are considered a *'primary objective'*, the development of the use of energy from renewable sources is prioritised as being of *'considerable importance'* (Ministry of Economy, 2009a: 9, 15 and 17). Following this hierarchy, the promotion of renewable energies seems to hold a subordinate position in the overlying energy policy as defined in the 2030 energy policy paper (Venjakob, 2011: 169). A statement made by Polish Prime Minister Donald Tusk seems to confirm this impression. In March 2013, he declared that Poland *'would fulfil its 15 per cent renewable energy target by 2020 but "nothing more"'* (Ancygier and Szulecki, 2013a: 1). As Podrygala (2008: 95) points out, it can be concluded (also from the indications given in the selection of non-legislative and legislative documents at hand) that the Ministry of the Economy is the main opponent of the promotion of renewable energies, and is a dominant actor in shaping the national preferences of Polish energy policy. It is furthermore worth mentioning that, in the period 2004-2008 alone, Poland had five governments, each of which was dedicated to shaping the country's overall energy policy (Riedl, 2008: 10).

5.5 Analysis

Based on the overall hypothesis in light of the multi-level policy dependence theory applied here, three subordinate hypotheses have been identified in the theory section. To identify constraints and dependencies of the European Integration of RES-(E) promotion, the explanatory factors implicitly describe three different influencing factors that follow a top-down, vertical logic, in turn leading to a horizontal logic. The first subordinate hypothesis (i) claims supranational renewable energy measures and policies produce compliance at the level of Member States (Radaelli, 2003: 19). Subordinate hypothesis (ii) counterpoints the first by claiming the European Union lacks the executive authority to expect compliance at the Member States level. Instead, it is national preferences that dominate the Member States' policy outcomes. Subordinate hypothesis (iii) is not relevant for the analysis based on Chapter 5.

As has been stressed, the overall framework includes theories derived from Liberal Intergovernmentalism and Europeanization. They are reflected in subordinate hypothesis (i) (Europeanization) and subordinate hypothesis (ii) (Liberal Intergovernmentalism) and read as follows:

(i) The supranational RES-(E) measures and policies of the European Union, when expressed in legislation, excite change in Member States' RES-(E) policies only when there is a degree of misfit between EU and domestic RES-(E) policies that causes adaptational pressure, leading to the national promotion of RES-(E) in accordance with EU level implications.

(ii) The RES-E policy of a Member State is an exclusive result of the domestic policy process that reflects energy-specific economic interests of the dominant domestic group, and is determined by geopolitical and economic interest.

Since any progress in the promotion of RES-(E) can only be measured at the Member States level, a model has been derived where the EU Member States are the focal point. In the applied model, the dependent variable are the Member State RES-E policies of Poland and Germany, with and ultimate focus being on the vertical and horizontal impacts on the RES-E policy of Poland. It has been stressed that a mono-causal approach cannot explain the impact on the dependent variable 'Member State RES-E'. Therefore there cannot be a single independent variable (e.g. defined at EU level) (Radaelli, 2003: 27-28). Instead, there is a set of three decisive variables, occurring at the EU level, the national level and the inter-state level. Fig. 2 illustrates their interrelations. Since all three influencing factors for the promotion of RES-E in Member States (supranational extrinsic, national intrinsic, inter-state interdependencies) will be considered, the analysis aims at identifying general insights about constraints and dependencies of European RES-E promotion mechanisms.

The Polish reaction to 'supranational extrinsic adaptational pressure' and its 'intrinsic national logic' were the analytical focus of this chapter. Polish RES-E policies, as well as the underlying Polish polity with respect to supranational and in particular national implications, reflect it. Note that at this point the dependent variable is 'Polish RES-E policy'. Fig. 13 illustrates the analytical focus of this chapter.

If subordinate hypothesis (i) should be accepted as a valid explanation, the questions are: On which judicial basis does Poland implement secondary provisions adopted by the European Union? Did the secondary provision in question excite change in Poland's RES-E policy? And: If change has been excited, was it because of a misfit between EU and Polish RES-E policies?

If, in order to consider alternative explanations for the empirical evidence presented in this chapter (Ladrech, 2010: 40-41), subordinate hypothesis

(ii) should serve as a valid explanation, then the question is: Can the RES-E policy of the EU Member State Poland be considered an exclusive result of the domestic policy process? What kinds of indicators verify or falsify subordinate hypothesis (ii)?

Fig. 13: Applied model of adaptational pressure for Polish RES-E policy

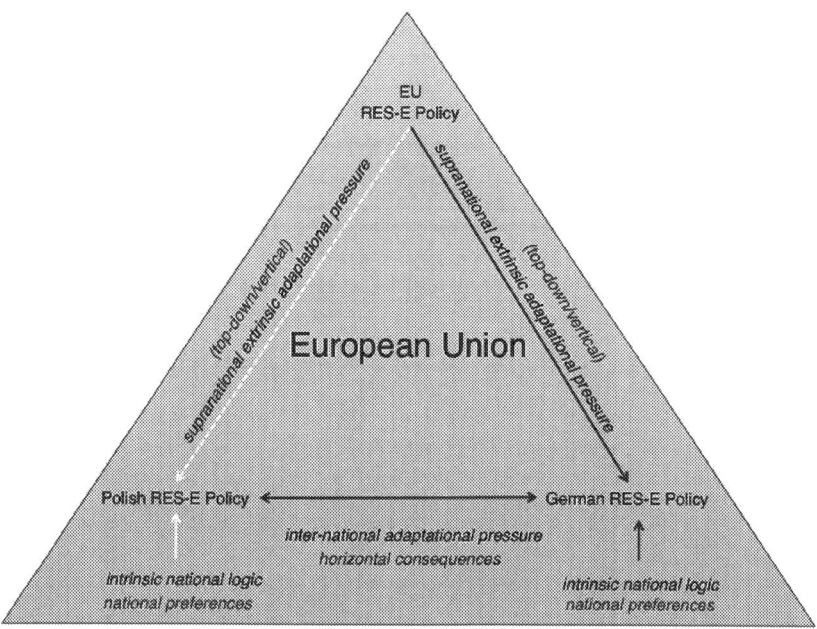

With respect to subordinate hypothesis (i), the Polish judicial framework for the adaptation of secondary provisions at the supranational level has been described. As a reminder it should be reiterated that, based on the arguments in light of the EU constitutional context, it has been concluded that there is a recognised constitutional principle of the European Union reflected in the *sui generis* nature of EU law, its supremacy, direct effect, pre-emption and protection of fundamental rights (Tatham, 2013: 15). But can the coherent logic of the supranational polity system of the EU be integrated into the pre-existing sovereign polity logic of the Member State Poland without constraints?

In light of the transfer of the exercise of sovereignty based on the 1997 Constitution of the Republic of Poland, five provisions play an important role (see Text Box 6). Article 90 allows for the transfer of competences to international organisations in some matters. Articles 9 and 91 allow for the direct effect of international agreements and supremacy over national laws and statutes. Articles 8 and 87 define the sources of the supremacy of the Polish constitution (Dabrowska, 2009: 108). Poland's accession to the EU was a complex act, but it did not need to pass specific amendments to its Constitution in order to accede (Tatham, 2013: 205 and 222). By signing the EU Accession Treaty on 16th April 2003 in Athens, Poland agreed that *'[f]rom the date of accession, the provisions of the original Treaties and the act adopted by the institutions and the European Central Bank before accession shall be binding on the new Member States and shall apply in those States under the conditions laid down in those Treaties and in this Act'* (OJoEU, 2003b: 33; Dabrowska, 2009: 109; Tatham, 2013: 222; CT, 2003: 1).

> Unlike the German constitution, which pursuant to Article 23(1) allows the Federation to *'transfer sovereign powers by a law with the consent of the Bundesrat'* (BMJ, 2012: 8), whilst simultaneously defining limits to that transfer (Biernat, 1998: 405; BMJ, 2012: 8-9), the Polish constitution (here Art. 90 (1)) only makes mention of the delegation of *'the competence of organs of State authority in relation to certain matters'* (Sejm, 2014). Ultimately, the Polish constitution therefore does not permit the transfer of sovereignty, but the exercise of specific competencies, as opposed to the power itself (Tatham, 2013: 223). However, the Constitution neither explicitly specifies the limits up to which the competencies of organs of state authority can be transferred, nor does it uphold a positive indication as to what the 'certain matters' referred to involve (Biernat, 1998: 405). The lack of explicit limitations in the provision, it must be concluded, does, however, exclude the transfer of sovereignty in all cases, and it does not indicate the absence of constitutional limitations at all (Tatham, 2013: 223, Biernat, 1998: 405). On the other hand, the Polish constitution recognises no unalterable provisions, as is the case pursuant to Article 79(3) of the German Constitution (BMJ, 2012: 25; Tatham, 2013: 223). *'It should be accepted that limits on transferring powers and "matters" that are not suitable for transfer pursuant to the provision in question may be imposed by other constitutional provisions'* (Biernat, 1998: 405). These other constitutional provisions refer to the highest principles of government, essential to the identity of Polish statehood (reflected in Chapter I of the Polish Constitution) and (reflected in Chapter II of the Polish Constitution) provisions proclaiming legal rights, and human and citizen freedoms (Biernat, 1998: 406; Sejm, 2014). It becomes apparent that the rather general language of, inter alia, Article 90 of the Polish constitution, leaves room for interpretation, which in terms of the Polish constitution falls within the scope of the Polish Constitutional Tribunal (CT).

In its Accession Treaty Case, the CT defines the principal that *'[t]he accession of Poland to the European Union did not undermine the supremacy of the Constitution over the whole legal order within the field of sovereignty of the Republic of Poland'* (CT, 2005a : 7). Indeed, and in accordance with the Polish Constitution, the CT concludes that *'it is insufficiently justified to assert that the Communities and the European Union are "supranational organisations" [...]. The expression "supranational organisation" is not mentioned in the Accession*

Treaty, nor in the Acts constituting an integral part thereof or any provision of secondary Community law' (CT, 2005a: 7-8). Furthermore, 'neither Article 90(1) nor Article 91(3) authorise delegation to an international organisation of the competence to issue legal acts or take decisions contrary to the Constitution, being the "supreme law of the Republic of Poland" (Article 8(1))' (CT, 2005a: 8). Like the FCC (BVerfGE, 2009: 123 (267)) the CT had the chance to judge on the constitutionality of the Lisbon Treaty (CT, 2010: 63). The Lisbon Treaty Case summarises the principles of the case law so far reflected in the judgments of the CT, and re-empathises both the CT's stance on the constitution as being the supreme law of the Republic of Poland – including the primacy of the binding force of the constitution – and the understanding of mutual dependencies among Member States and the EU, relating to conferring part of the competencies of state organs to the Union (CT, 2010: 21 and 33-34).

> The CT is, with respect to the case law cited here, and as some argue,in general heavily influenced by the FCC. This is especially true for the im plications of the 'Solange I' and 'Solange II' judgements (Tatham, 2013: 242; Pernice, 2006: 28-29; BVerfGE, 1974: 37 (271)). The CT argues, in light of the basic rights guaranteed in the Constitution, that norms concern- ing the rights and freedoms of individuals are the minimum and unsupass- able threshold that cannot be questioned or lowered because of European provisions (Tatham, 2013: 243). It also follows the FCC's judgment that the Member States remain the 'Masters of the Treaties', who have conferred, as an example of international organisations, the exercise of some of their powers to the EU, which may in principal also be terminated (Bogdandy, 1999: 41; BVerfGE, 1993: 89 (155) at 112; Tatham, 2013: 83; Franzius, 2010: 25 and 34 and 43; Pernice, 2006: 30-31; Grimm, 2001: 288). With respect to the 'possibility of a collision between regulations of Com- munity law and the Constitution' the CT concludes, that '[...] [i]n such an event the Nation as the sovereign, or a State authority organ authorised by the Constitution to represent the Nation, would need to decide on: amend- ing the Constitution; or causing modifications within the Community provi- sions; or, ultimately, on Poland's withdrawal from the European Union' (CT, 2005a: 9). In terms of reviewing harmonised secondary EU legislation, the CT points out in its SK 45/09 judgment 'that it is necessary to draw a distinction between examining the conformity of the acts of EU secondary legislation to the Treaties, i.e. EU primary law, on the one hand, and examining their conformity to the Constitution, on the other. The institution that ultimately determines the conformity of EU regulations to the Treaties is the Court of Justice of the European Union, and as regards the conformity to the Constitution – the Constitutional Tribunal' (CT, 2011: 21). It goes on by stating: 'What needs to be considered is the effects of a judgment of the Constitutional Tribunal in the case of adjudication that the norms of EU secondary legislation are inconsistent with the Constitution. In the context of the acts of Polish law, the said non-conformity results in declaring the

unconstitutional norms to be no longer legally binding (Article 190(1) and (3) of the Constitution)' (CT, 2011: 25).

The point of view that there is a possible indirect review jurisdiction of the CT has also been stressed in the *European Arrest Warrant Case* (CT, 2005b: 11-12). *'The obligation to implement the Framework Decisions is a constitutional requirement stemming from Article 9 of the Constitution, but its enactment does not assure automatically and in every case the material conformity of the provision of derivative EU law and of legislative acts implementing them to the national law with the norms of the Constitution. The basic function of the Constitutional Tribunal in the political system consists of reviewing the conformity of normative acts with the Constitution, and the same task applies also to situations, where the claim of unconstitutionality concerns that part of the scope regulated by a legislative act, which serves the purposes of implementation of EU law'* (CT, 2005a: 11-12). With both judgements, Tatham (2013: 265) concludes, the CT fired a warning shot across the bows of the ECJ. Generally, however, the CT tries, as can also be seen in the wording of the judgements above (e.g. *ultima ratio* implications), to assure a nuanced approach to constitutional difficulties concerning EU law vis-à-vis national law (Tatham, 2013: 265; CT, 2011: 25).

What does this mean for the national implementation of the Climate and Energy Package, and Directive 2009/28/EC in particular? As has been stated, the particular case of federal court judgements in light of European implications is similar in the Polish and German cases. Therefore, in the case of Poland, the relation of the domestic and the supranational polity design, including interpretations of the CT – see '5.1 Polish Constitutional Context – and the ECJ – see '3.4 EU Constitutional Context in light of EU Renewable Energy Policy' – has not led to polity constraints regarding the transposition necessities of Directive 2009/28/EC (see also '5.2 Polish Transposition of Directive 2009/28/EC and NREAP'). It is important to note that, after all, the CT's position can be characterised as trying to balance *'the principle of the protection of national sovereignty in European integration and the principle of a favourable predisposition towards European integration and co-operation between States'* (Tatham, 2013: 265 and 267). However, the uncertainties caused by Article 194 TFEU, prohibiting any measures that affect a *'Member State's right to determine the conditions for exploiting its energy resources, its choice between different energy sources and the general structure of its energy supply'* (OJoEU, 2008: 134-135), might not only be subject to ECJ decisions, but might also be subject to the ultimate authority of the CT to protect fundamental rights in the national system. This is even more so, since the CT, unlike the FCC, does not explicitly refrain from striking down harmonised secondary EU legislation per se (Tatham, 2013: 265). With respect to secondary law such as Directive 2009/28/EC, this could prove to be of less importance, because the constitution of Poland does not, unlike Article 194 TFEU, explicitly include provisions with respect to energy (Sejm, 2014; OJoEU, 2008: 134-135). Furthermore, pre–

Lisbon Treaty, Directive 2009/28/EC was not been considered a measure *'significantly affecting a Member State's choice between different energy sources and the general structure of its energy supply'* (OJoEU, 2008: 133; German Advisory Council on the Environment, 2011: 175). The legislative packages in question, and particularly Directive 2009/28/EC, were adopted through the ordinary legislative procedure, and therefore under non-exclusive participation of the Member States to excite change pursuant to each provisions content. But for the future adoption of renewable energy policies, the Polish polity – by way of CT decisions based on the Polish Constitution in light of an intrinsic national logic – might constitute a potential veto point. Additionally, it is remarkable – and a strong indicator in favour of subordinate hypothesis (ii) – that the CT does not acknowledge the European Union to be a supranational organisation, since *'it is insufficiently justified to assert that the Communities and the European Union are "supranational organisations" [...].The expression "supranational organisation" is not mentioned in the Accession Treaty, nor in the Acts constituting an integral part thereof or any provision of secondary Community law'* (CT, 2005a: 7-8).

On 21st March 2013, the European Commission notified the public in a press release that it referred Poland to the ECJ for failing to transpose EU rules (EC, 2013e). The form of order sought from the ECJ was to *'declare that, by failing to adopt the laws, regulations and administrative provisions necessary to ensure compliance with Directive 2009/28/EC of the European Parliament and of the Council of 23 April 2009 on the promotion of the use of energy from renewable sources and amending and subsequently repealing Directive 2001/77/EC and 2003/30/EC, and in any event by not notifying the Commission of such provisions, the Republic of Poland has failed to fulfil its obligation under Article 27 of that directive'* (OJoEU, 2013b: 8). The Commission furthermore proposed to impose a penalty payment of EUR 133.228,80 from the day on which the judgment was delivered, and to issue an order to the Republic of Poland to pay the costs of the proceedings (OJoEU, 2013b: 8). The suggested penalty, the Commission argued, is based on the duration and the gravity of the infringement. A formal notice of non-transposition of Directive 2009/28/EC was sent to Poland as early as January 2011, followed by a Reasoned Opinion in March 2012 (EC, 2013e). Given the on-going infringement procedure mentioned above, it remains unclear whether, and to what extent, Poland transposed Directive 2009/28/EC. To approach this matter, it is necessary to take a look at the framework indicators of the Directive – that is, the development of the Polish NREAP. The document was due on 30th June 2010 (OJoEU, 2009b: 28). According to Ancygier (2013: 339), it was not before 25th May 2010 that a draft version of the Polish NREAP was presented to the Polish public for consultation (Minister for Economic Affairs, 2010b). Consequently, Poland was amongst the last Member States to send the official version of its NREAP to Brussels. An *'Addendum to the National Renewable Energy Action Plan'* was sent to the Commission at the end of 2011 (NREAP Poland, 2010; Addendum NREAP Poland, 2011; Ancygier, 2013: 339-341). Comparing the three documents raises questions concerning the integrity of the calculations of the RES trajectories, and of the explanatory power of measures described for achieving those targets. After the Commission referred Poland to the Court for failing to transpose EU rules, the Sejm started working on an amendment act to, *inter alia,* the Energy Law, in order to avoid a potential financial penalty as a result of the initiated infringement procedure (EC, 2013e). What is known as the Small Tri Pack, adopted by the Sejm on 21st June 2013, is meant to nullify complaints for non-transposition and accelerate work on the Large Tri-Pack, including a new Gas Law, an amendment to the Energy Law, and the adoption of the envisioned RES Act (Reuters, 2013; ADVOC, 2013). The Small Tri-Pack became effective on 11th September 2013 (Lexology, 2013). Following Rybski and Stoczkiewicz (2013: 16), it is unlikely whether the Small Tri-Pack bill will result in the Commission's withdrawing its action against Poland. In addition, according to a press release of the Minister of Economic Affairs (2014), the Council of Ministers adopted an RES Act on 8th April 2014.

> Given an analysis of the implementation status of the RES Directive published by Rybski and Stoczkiewicz (2013: 5), as of June 2013 it is impossible to identify regulations for electricity in the Polish legal system that would bring Polish law in line with the supranational requirements. Whilst, according to a press release of the Minister of Economic Affairs (2014), the Council of Ministers adopted an RES Act on 8th April 2014, it remains true that, strictly based on official documents, until there is a judgment by the ECJ, the Republic of Poland has failed to notify the European Commission of provisions transposing Directive 2009/28/EC in time. In any case, pursuant to Article 27 of Directive 2009/28/EC, Poland was obliged to bring into force the laws, regulations and administrative provisions necessary to comply with the Directive 2009/28/EC by 5th December 2010 (OJoEU, 2009b: 44). If the Act of 1st April 2014 is designed to bring into full accord national legislation with supranational requirements resulting from the Directive in question, it would still be four years too late. Poland did produce a NREAP. However, the documents, including a draft version and an addendum, only formally fulfil the requirements pursuant to Article 4 of Directive 2009/28/EC (OJoEU, 2009b: 28; PWEA, 2010: 2). As Ancygier (2013: 339) points out, although the total contribution (installed capacity, gross electricity consumption) anticipated in Poland for each technology using renewable energy sources is presented in the referred to documents (see Tab. 23, Tab. 24, Tab. 25, and Tab. 26), it remains unclear how they were calculated. Furthermore the documents fail to explain what kind of future policies and measures the Polish legislator envisions in order to provide correct framework conditions for the anticipated growth rates of the different renewable energy sources and renewable energy technologies (see Tab. 20, Tab. 23, Tab. 24) (Ancygier, 2013: 339). Indeed, the NREAP announces the preparation of an Act on energy from renewables sources as an important legislative project, but, as has been stated, little happened until at least the end of 2013 (NREAP Poland, 2010: 30; Minister of Economic Affairs: 2014). Overall, it seems that

Poland is very reluctant to transpose supranational policy implications. Unlike Germany, where an autonomous accordance of its renewable energy policy outcomes with respect to overall European policy implications could be detected, the opposite seems to apply to Poland. There seems to be a rather distinct misfit between the pre-existing national legal framework with regard to the promotion of RES and the demands arising out of the supranational stimulus. The adaptational necessities acknowledged in the Polish NREAP, touching upon almost all relevant aspects with respect to the promotion of renewable energies, are evidence of this. *'Planned activities for the development of RES would require several amendments to the legislation with respect to: definitions, overall objectives and measures necessary to achieve the objectives, principles of calculation of the share of energy from RES, administrative procedures, legislation and codes, installer information and certification, guarantee of electricity RES origin, access to grids and their operations, as well as reporting'* (NREAP Poland, 2010: 30). In terms of these rather qualitative indicators, and in light of subordinate hypothesis (i), it seems that supranational RES-(E) measures and policies of the European Union did not excite the required change in Poland's RES-(E) policies by the end of 2013, exactly because the degree of misfit between EU and Polish domestic RES-(E) policy preferences were and still are to severe. However, the latter statement needs to be put into perspective. The initial adaptational pressure arising out of the requirements stipulated in Directive 2009/28/2009 did lead to a formal transposition by way of adopting an NREAP and announcing the creation of an RES Act.

These conclusions are to a certain extinct supported by the results of the implementation evaluation of Directive 2009/28/EC. As can be seen from 'Tab. 4: Actual and planned RES Shares 2009-2011/2012 (Ecofys, 2012: 18)', in conjunction with 'Tab. 21: National 2020 target and estimated trajectory of energy from renewable sources in heating and cooling, electricity and transport according to NREAP (NREAP Poland, 2010: 21)' and Tab. 22, Poland missed its NREAP indicative target RES share in 2010 by just under 0,09 percentage points, which equals a negative deviation of -0,99% (Ecofys, 2012: 18; NREAP Poland, 2010: 17). According to Ecofys (2012: 20), Poland had a growth rate in 2009/2010 above the required average, and is in a good situation to reach its overall RES target in 2020. With respect to the non–legally binding RES-E target for 2010, Poland undershot its target with a negative deviation of its actual 2010 share from the planned 2010 share of -11,02% (Ecofys, 2012: 21-22 and 27; NREAP Poland, 2010: 21; EC, 2013c: 4). It is among the six Member States that missed their respective target by more than 10%. With respect to its RES-E growth between 2009/2010, Poland would need higher than average growth rates to move from the NREAP 2010 RES-E target to the 2020 RES-E target (Ecofys, 2012: 23). With respect to the overall assessment of administrative procedures – see Tab. 6: Assessment of administrative procedures in the Member States (Ecofys, 2012: 186-187; EC, 2013c: 9)' – Poland receives a *'needs improvement'* rating. With respect to Polish progress in electricity grid integration – see 'Tab. 7: Member State rating on progress in electricity grid inte-gration (Ecofys, 2012: 177, 179 and 181)' – it receives a *'fair'* rating for overall RES grid integration. According to the same source, it receives a *'fair'* rating for addressing grid capacity limitations, whilst for rules for bearing and sharing the costs of grid development it is considered *'advanced'* (Ecofys, 2012: 179 and 181).

> Given the empirical evaluation of the implementation realities of Directive 2009/28/EC, the analysis shows ambivalent results for Poland. On the one hand, it is *'in a good situation'* to reach its overall RES targets in 2020. On the other hand, it would need *'higher than average growth rates'* to move from its 2010 RES-E target to the 2020 RES-E target (Ecofys, 2012: 20 and 23). According to a quantitative assessment of the future deployment of RES in Poland (based on the Green-X model), the authors of the Ecofys report (2012: 66) do not expect Poland to reach its overall 2020 RES target of 15 % based on the measures and policies referred to in its NREAP. The assessment of overall policies and measures for the promotion of renewable energies in Poland as indicated in the NREAP leads the authors of the Ecofys report (2012: 101) to conclude that Poland did only partially fulfil its overall NREAP policy commitments, and did not fulfil its NREAP policy commitments for the promotion of RES-E. With respect to the latter, support levels for each technology are *'poor'*, whilst the long-term security of support for the RES-Sectors receives a *'fair'* rating (Ecofys, 2012: 101). This judgment does seem to confirm the analysis implications of the described circumstances concerning the integrity of Poland's NREAP with regard to the calculations of the RES trajectories, and with regard to the explanatory power of measures described for achieving those targets (see '5.2 Polish Transposition of Directive 2009/28/EC and NREAP'). This is a result of the fact that, at the time evaluation of the Ecofy report was conducted, Poland *'did not adopt the Act on Renewable Energy Sources that it announced in its NREAP'* (Ecofys, 2012: 118). In order to validate these findings, it is important to pay closer attention to the national legislative context and its policy dependencies. At this point, it is not possible to apply findings from the implementation evaluation with regard to subordinate hypothesis (i) or (ii).

With respect to the applied model of this thesis, and the need for the consideration of an 'intrinsic national logic' for Poland's RES-(E) policies, the respective overall legislative history has been shown to be an indicator. Seen from the perspective of Liberal Intergovernmentlism, it is the domestic policy outcome that subsumes the particular objectives that arise from domestic competition, and therefore reflect the 'intrinsic national logic' (Moravcsik, 1998: 22). The overall legislative history of renewable energy promotion as part of the Polish energy policy described in the previous section is summarised in 'Tab. 27'. Note that the selection of non-legislative documents and legislative acts in this section is substantial, but should not be misunderstood as an exhaustive description of the Polish (renewable) energy policy process. Based on the cited indicators, it can be concluded that energy efficiency and energy security encapsulate the main foci of the Polish energy strategy (Venjakob, 2012: 167). Whilst measures with respect to coal are considered a *'main objective'* of the energy policy, and objectives with respect to the introduction of nuclear energy are considered a *'primary objective'* of the energy policy, the development of the use of energy from renewable sources is prioritised, being of *'considerable importance'* (Ministry of Economy, 2009a: 9, 15 and 17). The fact that the Sejm did nont adopt either a first or second draft of the Renewable Energy Act proposed by the Ministry of Economy in 2011 and 2012, but did agree on adopting legislation to accelerate the introduction of nuclear power, as envisioned in the Nuclear Power Programme, seems to confirm this hierarchy of priorities (Ministry of Economy, 2011; National Atomic Energy Agency, 2014). Even the creation of a Renewable Energy Department within the Ministry of Economy, to manage the implementation of Directive 2009/28/EC, did not fundamentally accelerate, let alone finalise, the legislative process for the adaption of an RES Act (Ancygier, 2013: 250-253). However, according to a press release of the Minister of Economic Affairs (2014), the Council of Ministers adopted an RES Act on 8th April 2014. Overall, it can be concluded that the Republic of Poland failed to notify the European Commission of provisions transposing Directive 2009/28/EC on time. Based on the cited indicators, it seems highly likely the Republic of Poland, at least by September 2013, failed to fulfil its obligation under Article 27 of Directive 2009/28/EC (OJoEU, 2013b: 8; OJoEU, 2009b: 44).

> This analysis of Polish national energy policy preferences gives rise to the conclusion that the issue-specific policy intentions of the dominant domestic group do not include the promotion of renewable energies as priority in the overlying energy policy. This becomes evident, inter alia, in the 2030 energy policy paper (Venjakob, 2011: 169, Leuffen, Rittberger and Shimmelfennig, 2013: 42; Moravcsik, 1998: 24-35). A statement made by Polish Prime Minister, Donald Tusk, seems to confirm this impression. In March 2013, he declared that Poland *'would fulfil its 15 per cent renewable energy target by 2020 but "nothing more"'* (Ancygier and Szulecki, 2013a: 1). Additionally, according to Riedl (2008: 23), until the end of the negotiations of the Climate and Energy Package in the Council of the European Union – a result of which, inter alia, was Directive 2009/28/EC – the threat of vetoing the package was the strongest from Poland. As Podrygala (2008: 95) points out, it can be concluded that the Ministry of the Economy is the main opponent of the promotion of renewable energies, and is a dominant actor in shaping the national preferences of the Polish energy policy. The fact that, in the period 2004-2008 alone, Poland had five governments, each of which was dedicated to shaping the country's overall energy policy, reflects a lack of consent for a general policy direction in favour for RES promotion (Riedl, 2008:10). The zigzag course with respect to the promotion of renewable energy, which becomes apparent, *inter alia,* in the (at least) three rejected attempts to adopt an RES Act formally from 2003-2011 supports this conclusion (see

Tab. 27). The situation is fundamentally different when it comes to nuclear energy. Although the adoption of a renewable energy strategy from 1999 to 2001 is met with significant attention in the scientific literature referred to here, and the adoption of a first Act on an Atomic Law as early as 2000 seems to have happened without much recognition, neither from the public nor the scientific sphere, the goal of introducing nuclear energy has not been subject to fundamental opposition in the national policy process (Ancygier, 2013; Venjakob, 2011; Podrygala, 2008; Riedl, 2008). Ever since the introduction of the Act on an Atomic Law, nuclear energy as an option for *'new energy sources'* has been consequently promoted in the domestic policy arena, independent of the acting governments' desire to shape the country's overall energy policy in accordance with their own agendas. The 2003 climate policy strategy, the 2005 energy policy strategy, the 2003 creation of a government representative for Alternative Sources of Energy, the 2009 energy policy, the 2011 Nuclear Power Programme, and – ultimately – the 2011 Amendment Act to the Act on Atomic Law all indicate an overall issue-specific consent for Polish national energy policy preferences for the introduction of nuclear energy (Ministry of Environment, 2003; Ministry of Economy, 2005; Ministry of Economy, 2009a; Ministry of Economy, 2011; Ancygier, 2013: 247; National Atomic Energy Agency, 2014).

Of course, studies applying a nation-bound policy analysis, such as Ancygier (2013) or Podrygala (2008), by focusing exclusively on domestic Polish developments in multi-level realities, derive a more differentiated picture with regard to, for example, national advocacy coalitions in light of RES-E promotion. But, whilst the latter analytical approach has explicitly been disregarded in the applied overall theory design, the studies referred to confirm the impression that issue-specific policy intentions reflecting the 'intrinsic national logic' of Poland's RES-(E) policies almost certainly consider the promotion of renewable energies as subordinate in the overlying energy policy (Ancygier, 2013: Podrygala, 2008: 95-97). *'Having compared the interests of the actors in energy sector and their impact on renewable energy and climate policies at the European level and national levels it can clearly be seen that there exists a significant discrepancy between the interest that found expression in the European directives and the laws aimed at implementing these directives in Poland. This discrepancy results from three factors: the interest of the actors related to energy, the ability of these actors to influence the policy-making process at the respective level, and the particular interests of those who directly shape and formulate the policy [...] Finally, the interests of the decision-makers at the European and national levels differ significantly'* (Anygier, 2013: 267 and 258). Analysis of the Polish legislative energy policy justifies the conclusion that the domestic national preferences of Poland reflect an independent intrinsic approach, influenced only in its formal policy outcome by an extrinsic supranational adaptational pressure.

What follows from the analysis above? The conclusion of the analysis in the 'EU Level Energy Policy and the Context for RES-E'- chapter was that subordinate hypothesis (ii) cannot be falsified in its entirety. Whilst, on the one hand, pre-Lisbon Treaty realties and Directive 2009/28/EC implementation-effectiveness evidence quite clearly illustrate a measurable impact of European legislation, Article 194(2) TFEU, on the other hand, defines absolute competence limits. The conclusion of the analysis in the '4. German' chapter also resulted in the finding that subordinate hypothesis (ii) cannot be verified in its entirety. Neither were the EU-level implications as a whole considered a transgression of EU competencies – on the contrary, they were dominantly shaped in accordance with German national preferences – nor was there a misfit of supranational and national RES-E preferences in principle. It became apparent, however, that the RES-E policy of Germany is predominantly the result of the domestic policy process that reflects energy-specific interests of the dominant domestic group.

What follows from the Polish case? The Polish national intrinsic preferences with respect to its polity design, reflected in CT case law, further strengthen the scope of absolute competence limits with respect to both supranational polity and policy implications. Like the FCC (BVerfGE, 2009: 123 (267)), the CT had the chance to judge on the constitutionality of the Lisbon Treaty (CT, 2010: 63). The *Lisbon Treaty Case* summarised the principles of the case law reflected in the judgments of the CT and re-empathised both the CT's stance on the Constitution as being the supreme law of the Republic of Poland – including the primacy of the binding force of the Constitution – and the understanding of mutual dependencies among Member States and the EU, relating to conferring part of the competencies of state organs to the Union (CT, 2010: 21 and 33-34). Whilst, based on CT case law in accordance with, for example, Article 194(2) TFEU, the CT states that there is a possible indirect review jurisdiction of the CT (CT, 2005b: 11-12), the EU level implications – in particular those for the promotion of renewable energies – have not been considered a transgression of EU competencies. At least, the CT has not yet had the opportunity to judge on the matter. On the contrary, although the analysis of Polish national energy policy preferences gives rise to the conclusion that the 'intrinsic national logic' of Poland's RES-(E) policies does not take the promotion of renewable energies to be a priority in the overlying energy policy, the initial adaptational pressure arising from the requirements stipulated in the supranational arena (e.g. Directive 2009/28/EC) did lead to a formal transposition of supranational implications (e.g. by way of adopting a NREAP and announcing the creation of an RES Act). Furthermore, after the European Commission increased adaptational pressure by referring Poland to the Court, both adoption of the Small Tri-Pack, as well as adoption of an RES Act, were observed in the national arena (Reuters, 2013; ADVOC, 2013; Minister of Economy Affairs, 2014). That being said, Poland, as has been shown, is to date very reluctant to transpose its supranational policy implications. This analysis identified a distinct misfit between European requirements and the pre-existing national legal framework for the promotion of RES. Ultimately Poland will face a proceeding for failing to transpose Directive

2009/28/EC in front of the ECJ, because Poland did not (at least by the end of 2013) bring into force the laws, regulations and administrative provisions necessary to comply with the Directive 2009/28/EC by 5th December 2010 (OJoEU, 2009b: 44; EC, 2013e). It follows that the RES-(E) policy of Poland – in the observed timeframe – is a non-exclusive but distinctly dominant result of the domestic policy process. Therefore subordinate hypothesis (ii) cannot be verified in its entirety.

Tab. 28: Evaluation Matrix for Chapter 3, 4 and 5

	European Union	GER	PL	GER and PL
Subordinate Hypothesis (I)	☺	☺	☺	X
Subordinate Hypotehsis (II)	☺	☺	☺	X
Subordinate Hypothesis (III)	X	X	X	X

With regard to subordinate hypothesis (i), and based on the analysis in the 'EU Level Energy Policy and the Context for RES-E' chapter, empirical evidence for Directive 2009/28/EC suggests that European renewable energy policies and measures excite change in Member States' policies, but the change excited is not in its entirety in accordance with the change intended. Therefore, the first aspect of subordinate hypothesis (i) (the supranational RES-(E) measures and policies of the European Union excite change in Member States RES-(E) policies) has been verified under conditions of usual reserve. With regard to the second aspect (if there is a degree of misfit between EU and domestic RES-(E) policies that causes adaptational pressure leading to the national promotion of RES-(E) in accordance with EU level implications), no conclusions could be drawn based on Chapter 3. With respect to German renewable energy policy an autonomous accordance of German renewable energy policy outcomes as part of the overall energy policy with European implications has been identified. Therefore, in light of subordinate hypothesis (i), supranational RES-(E) measures and policies of the European Union were able to excite change in Germany's RES-(E) policies, because there was a limited degree of misfit between EU and German domestic RES-(E) policy preferences. As a result, subordinate hypothesis (ii) has not been verified in its entirety. What follows from the Polish case? Taking into account the more detailed analysis outlined above, and accepting the rather qualitative indicators as a basis for the on-going analysis, it seems reasonably likely that, in light of subordinate hypothesis (i), the supranational RES-(E) measures and policies of the European Union did not excite the required change in Poland's RES-(E) policies by the end of 2013 exactly because the degree of misfit between EU and Polish domestic RES-(E) policy preferences were and still are too severe. This impression is also confirmed in an analysis by Riedl (2008: 23), according to which, by the end of the negotiations of the Climate and Energy Package in the Council of the European Union – a result of which was Directive 2009/28/EC – the threat of vetoing the package was the strongest from Poland. Legitimately, the latter conclusion needs to be put into perspective. The initial adaptational pressure arising from the requirements stipulated in Directive 2009/28/2009 did lead to a formal transposition, by way of adopting a NREAP and announcing the creation of an RES Act. Furthermore, after the European Commission increased adaptational pressure, by referring Poland to the Court, both adoption of the Small Tri-Pack as well as adoption of an RES Act were observed in the national arena (Reuters, 2013; ADVOC, 2013; Minister of Economy Affairs, 2014). Timely interference of these actions might suggest a correlation, which at this point, however, cannot be subject to further inquiry. Accordingly, based on this chapter, subordinate hypothesis (ii) cannot be verified in its entirety. The conclusions of this chapter, and the results derived in this thesis so far, are visualised in Tab. 28.

6. Inter-National Interdependencies between Germany and Poland

As was stressed in the introduction, this is study foremost an empirical observation in the context of the promotion of RES-E in the European Union. The empirical analysis is based on the European Union (and the impact of its polity, and policy outcome on Member States' policies (Schimmelfennig and Sedelmeier, 2005: 5; Börzel and Risse, 2000: 1; Heritier, 2001: 3)), the Member State Germany (its reaction to EU adaptational pressure, its national preferences (Moravcsik, 1998: 24-35) and horizontal consequences (Radaelli, 2003: 17)) and the Member State Poland (its reaction to EU adaptational pressure, its national preferences and its reaction to German induced horizontal pressures). This last chapter deals with the horizontal consequences of renewable energy and in particular renewable energy electricity promotion with respect to inter-national interdependencies between the EU Member States Germany and Poland. In light of the subject matter and the theory applied, the following analysis processes subordinate hypothesis (iii), which reads as follows:

European inter-state connectedness, competition and economic pressure will foster the promotion of RES-E in Europe beyond pioneering Member States, if a pioneering State with sufficient power in the international system holds a dominant state and/or market position, reflected in its RES-E policy and/or RES-E market design.

With respect to the overall theory hypothesis the epistemological interest of this chapter is expressed in part (iii):

If (i) the European Union maintains an overall adaptational pressure by deciding on RES-E measures and policies for its Member States; (ii) pioneering Member States with sufficient power in the international system with regard to RES-E promotion (such as Germany) succeed in implementing an energy policy with a high share in electricity from renewable energy sources; and (iii) neighbouring Member States (such as Poland) are exposed to a pioneering Member State's RES-E policy and/or RES-E market design – **then,** non-pioneering Member States' RES-E measures and policies will adapt to the vertical (EU level) and horizontal (inter-state level) pressures.

In terms of the applied framework, subordinate hypothesis (iii) reflects both aspects from Europeanization and Liberal Intergovernmentalism. Whilst little Europeanization literature is available to specify the explanatory mechanisms of a horizontal approach, Radaelli (2003: 17) sums it up as follows: *'Horizontal Europeanization is a process of change triggered by the market and the choice of the consumer or by diffusion of ideas and discourses about the notion of good policy and best practice. More precisely [...] the horizontal mechanisms involve different forms of framing.'* He illustrates a vague idea of the process by claiming that, for example, regulatory competition is a *'mechanism starting with vertical prerequisites but that has horizontal consequences'* (Radaelli, 2003: 17).

Liberal Intergovernmentalism introduces a bargaining theory to describe the relevant negotiation processes concerning the distribution of gains from cooperation (Leuffen, Rittberger and Schimmelfennig, 2013: 48). In order to overcome suboptimal outcomes of coordination and cooperation, governments must negotiate how mutual gains are distributed amongst states. These negotiations, in the case of the European Community – as analysed by Moravcsik – are characterised by *'hard interstate bargaining, in which credible threats to veto proposals, to withhold financial side-payments, and to form alternative alliances excluding recalcitrant governments carried the day. The outcomes reflected the relative power of states – more precisely patterns of asymmetrical interdependence. Those who gained the most economically from integration compromised the most on the margin to realize it, whereas those who gained the least or for whom the costs of adaptation were highest imposed conditions'* (Moravcsik, 1998: 3; see also Leuffen, Rittberger and Schimmelfennig, 2013: 49). Again, the focus is on states as the dominant actors in the process. By combining both approaches, the assumption, as has been argued in the theory section, is that the relative power of states in the European Union, reflected in asymmetrical interdependence (Moravcsik, 1998:3) and economic interests triggered by the market and the choice of the consumer (Radaelli, 203: 17), leads to an establishment of the dominant state and/or market position, reflected in a particular policy and/or market design, over the position of the weaker state and/or market. With respect to the applied model, this this assumption leads to the postulate of an inter-national adaptational pressure in 'favour' of Germany and to the 'disadvantage' of Poland (see Fig. 4).

To take into account mechanisms *'starting with vertical prerequisites'* leading to *'horizontal consequences'*, a brief re-introduction to the regulatory framework conditions for a European gas and electricity market as a result of the Third Energy Package will be given. Both Polish and German renewable energy markets will then be compared, followed by brief introduction of energy market specifics, leading to an analysis of economic indicators potentially influencing Polish renewable energy and renewable energy electricity development (Radaelli, 2003: 17; OJoEU, 200b).

6.1 Supranational Framework Policies for Inter-State Interconnectedness

Whilst the key aspects of supranational framework policies with relevance to inter-state interconnectedness have been outlined in the 'EU Level Energy Policy and the Context for RES-E' chapter, some aspects should be re-emphasized at this stage. The overall EU-level energy policies and measures with relevance to, *inter alia,* RES-E promotion in the period up to 2020, are clustered in what is called the Climate Change Package and the Third Energy Package (EP, 2008: 1; EP, 2009: 1). They are comprised of eleven legislative acts, out of which six belong to the Climate Change Package, and five belong to the Third Energy Package (see 'Developing a European Energy Policy'). Directive 2009/28/EC, as part of the Climate Change Package, has been subject to extensive analysis in the previous chapters. In order to understand the supranational context for German–Polish interconnectedness with respect to renewable energies, and in order to maintain the top-down to horizontal logic of this thesis, attention shall now be given to the Third Energy Package.

The five pieces of legislation of the Third Energy Package aim at: (1) establishing an Agency for the Cooperation of Energy Regulators (Regulation (EC) No 713/2009; OJoEU, 2009g: 4); (2) setting fair rules for cross-border exchanges in electricity and facilitating the emergence of a wholesale electricity market (Regulation (EC) No 714/2009; OJoEU, 2009h: 17-18); (3) setting non-discriminatory rules for access conditions to natural gas transmission systems and to LNG facilities, as well as facilitating the emergence of a wholesale gas market (Regulation (EC) No 715/2009; OJoEU, 2009i: 39); (4) establishing common rules for the generation, transmission, distribution and supply of electricity (Directive 2009/72/EC; OJoEU, 2009j: 62); and (5) establishing common rules for the generation, transmission, distribution, supply and storage of natural gas as well as LNG, biogas, gas from biomass and other types of gas (Directive 2009/73/EC; OJoEU, 2009k: 101).

The European Parliament and the Council of the Union conclude that, despite Directive 2009/72/EC and Directive 2009/73/EC of the Third Energy Package, concerning common rules for the internal energy market, the market remains fragmented owing to insufficient interconnections between national networks (OJoEU, 2013a: 40). Furthermore, the ten-year network development plan published by ENTSO-E, including the trans-European energy networks for electricity framework (TEN-E) as defined in Decision No 1364/2006/EC (OJoEU, 2006a: 8-14), fails its intention by lacking 'vision, focus, and flexibility to fill identified infrastructure gaps' (OJoEU, 2013a: 39). Hence Regulation (EU) No 347/2013 of the European Parliament and the Council of 17 April 2013, on guidelines for trans-European energy infrastructure and repealing Decision No 1364/2006/EC, and amending Regulations (EC) No 713/2009, (EC) No 714/2009 and (EC) No 715/2009, were adopted (OJoEU, 2013a: 39).

The scope of this Regulation is to deliver guidelines for a timely development and interoperability of priority corridors and areas of trans-European energy infrastructure (OJoEU, 2013a: 44). The Regulation (a) addresses the identification of projects of common interest, (b) facilitates timely implementation of projects, (c) provides rules and guidance for cross-border allocation of costs, and (d) determines the conditions of eligibility for projects of common interest requiring financial assistance by the Union (OJoEU, 2013a: 45).

With respect to electricity, four priority electricity corridors are defined in Annex I of Regulation 347/2013 (OJoEU, 2013a: 62). These are:

(1) Northern Seas offshore grid (NSOG) concerning Member States: Belgium, Denmark, France, Germany, Ireland, Luxemburg, Netherlands, Sweden, United Kingdom.

(2) North–South electricity interconnections in Western Europe (NSI West Electricity) concerning: Austria, Belgium, France, Germany, Ireland, Italy, Luxembourg, Netherlands, Malta, Portugal, Spain, United Kingdom.

(3) North–South electricity interconnections in Central Eastern and South Eastern Europe (NSI East Electricity) concerning: Austria, Bulgaria, Croatia, Czech Republic, Cyprus, Germany, Greece, Hungary, Italy, Poland, Romania, Slovakia, Slovenia.

(4) Baltic Energy Market Interconnection Plan in electricity (BEMIP Electricity) concerning: Denmark, Estonia, Finland, Germany, Latvia, Lithuania, Poland, Sweden.

According to Article 3(4) of Regulation 347/2013, the *'Commission shall be empowered to adopt delegated acts in accordance with Article 16 that establish the Union list of projects of common interest ('Union List'), subject to the second paragraph of Article 172 of the TFEU. The Union list shall take the form of an Annex to this Regulation. [...] The first Union list shall be adopted by 30 September 2013'* (OJoEU, 2013a: 46). Accordingly, a *'COMMISSION DELEGATED REGULATION (EU) No 1391/2013 of 14 October 2013 amending Regulation (EU) No 347/2013 of the European Parliament and of the Council on guidelines for trans-European energy infrastructures as regards the Union list of projects of common interest'* was published in the Official Journal of the European Union on 21st December 2013 (OJoEU, 2013c: 28). Based on Article 3(4) of Regulation 347/2013, Article 1 of Regulation 1391/2013 announces that *'[a]n Annex VII is added to Regulation (EU) No 347/2013 [...]'*, establishing a *'Union list of projects of common interest ("union list") [...] '* (OJoEU, 2013a: 46; OJoEU, 2013c: 29 and 30). With respect to the priority corridor, North–South electricity interconnections in Central Eastern and South Eastern Europe (NSI East Electricity) as defined in Annex I of Regulation 347/2013 (see above), and Annex VII (as added to the Regulation in question through Regulation 1391/2013) identifies the *'Cluster Germany – Poland between Vierraden and Krajnik'* including the following PCIs (Projects of Common Interest) (OJoEU, 2013c: 35):

- Interconnection between Eisenhüttenstadt (DE) and Plewiska (PL), (PCI No. according to Regulation 347/2013: 3.14.1);

- Interconnection between Vierraden (DE) and Krajnik (PL), (PCI No. according to Regulation 347/2013: 3.15.1);

- Coordinated installation and operation of phase-shifting transformers on the interconnection lines between Krajnik (PL) – Vierraden (PL) and Mikulowa (PL) – Hagenwerder (DE), (PCI No. according to Regulation 347/2013: 3.15.2).

Further information on the development of the PCIs in question is available through an interactive map on the transparency platform of DG Energy (EC, 2014a). The PCI 3.14.1 involves German Transmission System Operator (TSO) 50Hertz Transmission GmbH and Polish TSO PSE S.A. (EC, 2104b: 1). The project, including a new line construction and substation on the route Polish border (Plewiska), is summarised as follows (EC, 2014b: 3): *'The new 400 kV double circuit OHL between Eisenhüttenstadt (DE) and Plewiska (PL) enhances the interoperability on the PL-DE profile. The Project contributes highly to the market integration by increasing the grid transfer capacity and lowering the system costs. Additionally, the new grid connection will positively influence the integration of RES installed mostly in the 50Hertz control area as well as in the PSE S.A. control area (mainly in northern Part of Poland)'* (EC, 2014b: 2). Furthermore, the project will increase Polish import capacity by 1.500 MW and German import capacity by 500 MW. The planned date of commissioning is the year 2022 (EC, 2014b: 2).

The project status of PCI 3.15.1, aiming at increasing the grid transfer capacity between Vierraden (DE) and Krajnik (PL), is summarised as follows (EC, 2014c: 2): *'A 380 kV switchyard in substation Vierraden has been built in 2012 and a 3 km new line substituting the 220 kV part of the line Krajnik – Vierraden has been erected in 2013. Further construction works on the existing line Krajnik – Vierraden (e.g. erection of five new pylons) are necessary, for which an additional permit has to be obtained. The documents for the permitting process are currently being prepared. Also the preparation for the extension of the 380 kV switchyard in substation Vierraden is being carried out'* (EC, 2014c: 2). Companies involved are TSO 50Hertz Transmission GmbH and TSO PSE S.A., and planned year of commissioning is 2017 (EC, 2014c: 2). The project will increase Polish import capacity by 500 MW and German import capacity by 1.500 MW (EC, 2014c: 2).

Impact and benefits of PCI 3.15.2, including the previously mentioned TSOs, is summarised as follows (EC, 2014d: 2): The installation of phase-shifting transformers (PSTs) *'will positively influence the integration of RES installed mostly in the 50Hertz control area as well as in the PSE S.A. control area (mainly in northern Poland). The project also increases system safety by providing means of control of unscheduled flows'* (EC, 2014d: 2). Negotiations between the German and Polish TSO are on-going. *'The location and commissioning dates depend on the agreement (2015/2017)'* (EC, 2014d: 2). The PCI increasing the capacity on the Polish Synchronous Profile – including the Polish–German border, Polish–Slovak border, and Polish–Czech border – will provide an additional 500 MW import and 1.500 MW export capacity for Poland (EC, 2014d: 2).

Overall, with respect to PCI 3.14.1 (due in 2022) and PCI 3.15.1 (due in 2017), both German and Polish import capacities will be increased by 2.000 MW (EC, 2014b: 2; EC, 2014c: 2). However ,in the short term (by 2017), the increase of capacity is 1.500 MW for Poland and 500 MW for Germany. The installation of phase-shifting transformers will add additional capacities of 500 MW in terms import and 1.500 MW in terms export with respect to the entire Polish Synchronous Profile (EC, 2014d: 2).

6.1.1 Summary

Key aspects of supranational framework policies with relevance to inter-state interconnectedness have been outlined in the 'EU Level Energy Policy and the Context for RES-E' chapter. The overall EU-level energy policies and measures with relevance to, *inter alia,* RES-E promotion in the period up to 2020 are clustered in what is called the Climate Change Package and the Third Energy Package (EP, 2008: 1; EP, 2009: 1). With respect to electricity, four priority electricity corridors are defined in Annex I of Regulation 347/2013 (OJoEU, 2013a: 62). According to Article 3(4) of Regulation 347/2013, the *'Commission shall be empowered to adopt delegated acts in accordance with Article 16 that establish the Union list of projects of common interest ('Union List'), subject to the second paragraph of Article 172 of the TFEU. The Union list shall take the form of an Annex to this Regulation. [...] The first Union list shall be adopted by 30 September 2013'* (OJoEU, 2013a: 46). With respect to the priority corridor, North–South electricity interconnections in Central Eastern and South Eastern Europe (NSI East Electricity) as defined in Annex I of Regulation 347/2013 (see above), and Annex VII (as added to the Regulation in question through Regulation 1391/2013) identifies the *'Cluster Germany – Poland between Vierraden and Krajnik'* including the following PCIs (Projects of Common Interest) (OJoEU, 2013c: 35):

- Interconnection between Eisenhüttenstadt (DE) and Plewiska (PL), (PCI No. according to Regulation 347/2013: 3.14.1);

- Interconnection between Vierraden (DE) and Krajnik (PL), (PCI No. according to Regulation 347/2013: 3.15.1);

- Coordinated installation and operation of phase-shifting transformers on the interconnection lines between Krajnik (PL) – Vierraden (PL) and Mikulowa (PL) – Hagenwerder (DE), (PCI No. according to Regulation 347/2013: 3.15.2).

Overall, with respect to PCI 3.14.1 (due in 2022) and PCI 3.15.1 (due in 2017), both German and Polish import capacities will be increased by 2.000 MW (EC, 2014b: 2; EC, 2014c: 2). However ,in the short term (by 2017), the increase of capacity is 1.500 MW for Poland and 500 MW for Germany. The installation of phase-shifting transformers will add additional capacities of 500 MW in terms import and 1.500 MW in terms export with respect to the entire Polish Synchronous Profile (EC, 2014d: 2).

6.2 Comparison RES-E Deployment in Germany and Poland

Before describing the German and Polish framework conditions with regard to European PCIs and the respective national electricity markets, a comparative look shall be taken at RES-E development in Germany and Poland, as indicated in their NREAPs. As becomes apparent from the analysis of national preferences for renewable energy, and in particular renewable energy electricity promotion, policy preferences differ substantially in Poland and Germany. This analysis is supported by Ancygier and Szulecki (2014) who conclude that '[i]t needs to be emphasised that whilst for Germany an ambitious RE policy fulfils the double goal of mitigating climate change and providing a new impulse to the German economy, from the perspective of the Polish government RE policy is seen primarily as a response to the (strongly German-inspired) European energy and climate policy' (Ancygier and Szulecki, 2013b: 3). Naturally, these differences are reflected in the national NREAPS. Whilst the trajectories for the share of energy from renewable sources in the electricity sector in the period 2015–2020 have been described independently for Poland and Germany (see Tab.14, Tab.15, Tab. 25, and Tab. 26), a direct comparison with regard to absolute numbers for the indicative trajectories in MW and GWh highlights the stark differences between the two countries (see Fig. 15 and Fig. 16). Overall, according to the indicative trajectories, by 2020 Germany will have installed ten times the capacity of Poland, whilst producing seven times as much electricity from renewable energies (see Tab. 29). The difference in efficiency when it comes to the utilisation of installed capacities can be explained by the differences in anticipated operating hours of renewable energy facilities in both countries. As has been described in the chapter on Poland, based on its NREAP by 2020, the installed capacity of all renewable energy sources is anticipated to be 10.335 MW, leading to 32.400 GWh. This equals 3.135 h/a (36% of the year's total hours) of electricity production from all facilities using renewable energy technologies. In comparison, the German NREAP indicates all its renewable energy facilities to produce energy for 1.956 h/a (22% of the year's total hours) by 2020 (see Tab. 15). Unfortunately, neither the Polish nor the German NREAP provides mathematical details on technology-specific assumptions for full power with regard to calculating the estimates provided in the tables in question (Tab. 14, Tab. 15, Tab. 23, and Tab. 26). The fact that, according to the CIA World Fact Book, both Polish and German climates are characterised as being temperate does not lead to insightful indications for explaining the difference in the anticipated operating hours either. (Of course, the latter argument in no way represents a legitimate scientific deduction by meteorological standards) (CIA, 2014a; CIA, 2014b)). The questions concerning the integrity of the physical and mathematical foundations for the calculations of the anticipated contribution of each technology, as reflected in the Polish indicative trajectories for the share of energy from renewables, might have some explanatory power, but this has been discussed already (see 'Polish Transposition of Directive 2009/28/EC and NREAP').

Finally, comparing anticipated technology development using renewable energy sources with regard to the binding targets for 2020, as indicated in the NREAPs, leads to the observation that both countries expect wind energy to play a significant role in the renewable energy technology mix (see Fig. 17 and Fig. 18). An estimated installed capacity of 6.650 MW wind energy in Poland by 2020 makes this technology reflect 64 % of the overall installed renewable energy capacities by 2020 (see Tab. 26). An estimated installed capacity of 45.750 MW wind energy in Germany by 2020 makes this technology reflect 41% of the overall installed renewable energy capacities by 2020 (see Tab.15).

Fig. 14: Overall Renewable Energy Installed Capacity [MW] in Germany and Poland (NREAP) (Zepeda, 2014: 4; NREAP Germany, 2010; NREAP Poland, 2010)

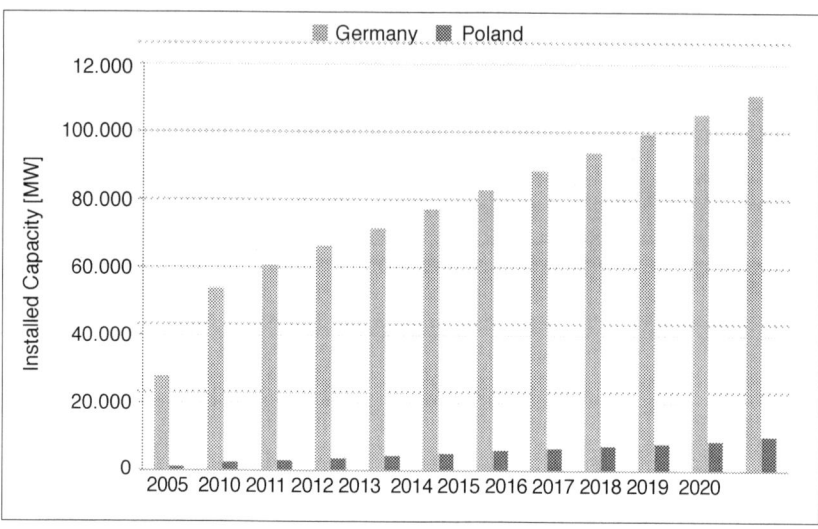

Fig. 15: Overall Gross Electricity Generation from Renewable Sources [GWh] in Germany and Poland (NREAP) (Zepeda, 2014: 3; NREAP Germany, 2010; NREAP Poland, 2010)

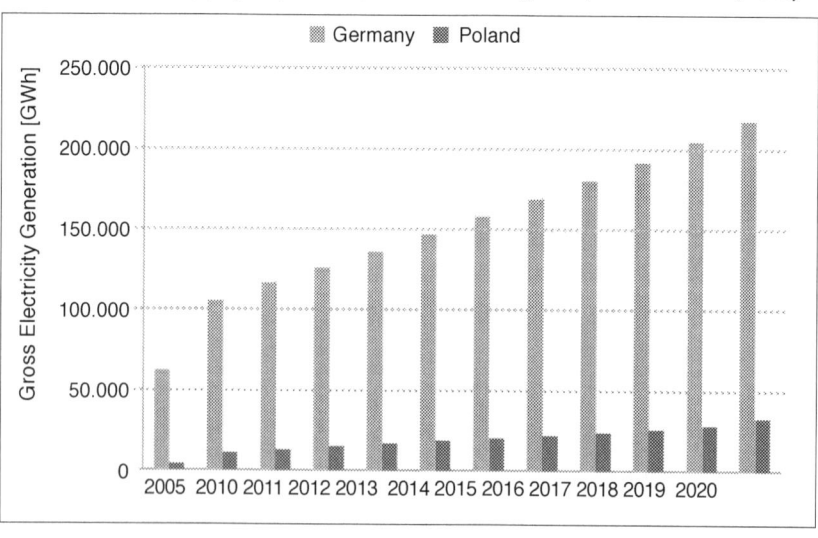

Fig. 16: Operation Hours of Renewable Energy [h] in Germany and Poland (NREAP) (Zepeda, 2014: 5; NREAP Germany, 2010; NREAP Poland, 2010)

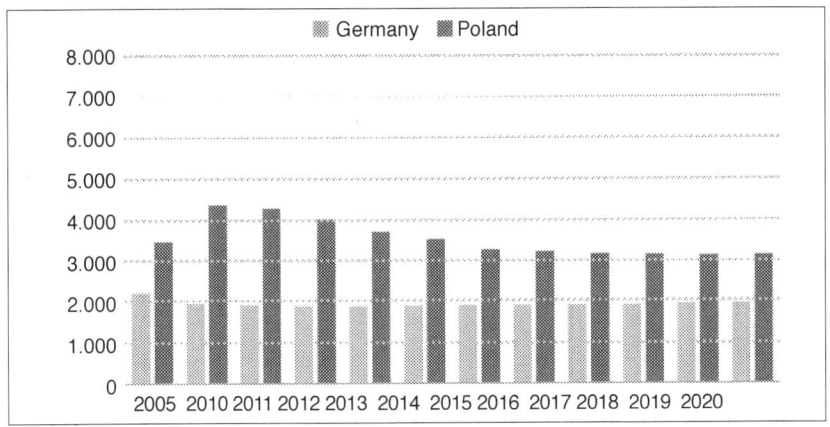

Tab. 29: Difference Generation and Installed Capacity Objectives for Germany and Poland in 2020 (NREAP) (Zepeda, 2014: 3; NREAP Germany, 2010; NREAP Poland, 2010)

	Objectives for Germany 2020		Objectives for Poland 2020	
	MW	GWh	MW	GWh
Hydropower:	4.309	20.000	1.152	2.969
< 1 MW	564	2.550	142	497
1 MW - 10 MW	1.043	4.500	238	714
> 10 MW	2.702	12.950	772	1.758
from pumped storage power plant	7.900	8.395	-	-
Geothermal power:	**289**	**1.654**	**-**	**-**
Solar energy	**51.753**	**41.389**	**3**	**3**
photovoltaics	51.753	41.389	3	3
concentrated solar energy	-	-	-	-
Tides, waves, other ocean energy:	**-**	**-**	**-**	**-**
Wind energy:	**45.750**	**104.435**	**6.650**	**15.210**
land-based	35.750	72.664	5.600	13.160
offshore	10.000	31.771	500	1.500
small installations	-	-	550	550
Biomass:	**8.825**	**49.457**	**2.530**	**14.218**
solid	4.792	24.569	1.550	10.200
biogas	3.796	23.438	980	4.018
liquid biofuels	237	1.450	-	-
Overall:	**110.934**	**216.935**	**10.335**	**32.400**
from combined heat and power	3.765	20.791	955	5.069
	Installed Capacity GE/PL ~ 10x		Generated RE GE/PL ~ 7x	

Fig. 17: Renewable Energy Installed Capacity [MW] in Germany 2005-2020 (Zepeda, 2014: 9; NREAP Germany, 2010; NREAP Poland, 2010)

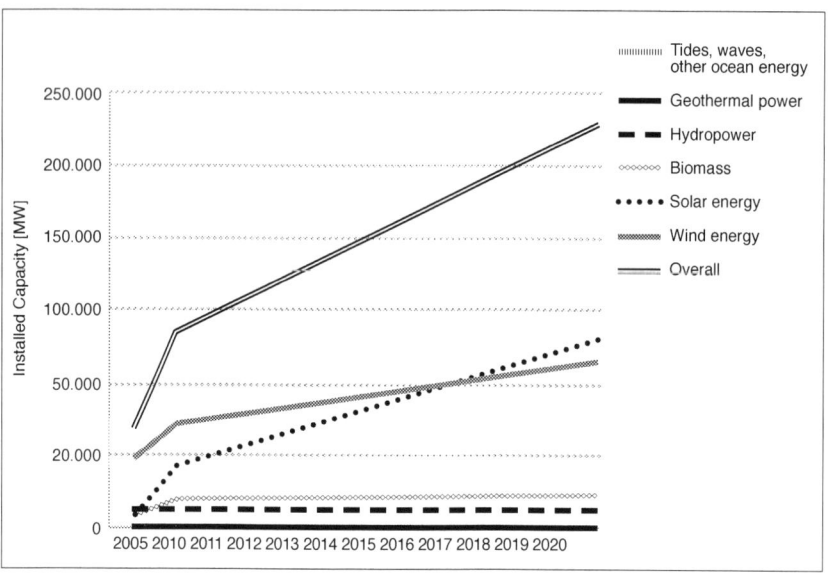

Fig. 18: Renewable Energy Installed Capacity [MW] in Poland 2005-2020 (Zepeda, 2014: 13; NREAP Germany, 2010; NREAP Poland, 2010)

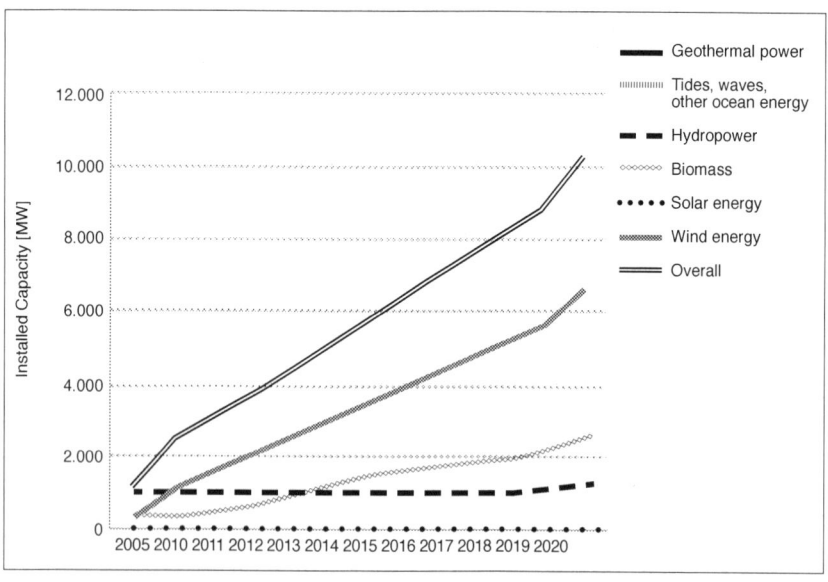

6.2.1 Summary

Whilst the trajectories for the share of energy from renewable sources in the electricity sector in the period 2015–2020 have been described independently for Poland and Germany (see Tab. 14, Tab. 15, Tab. 25, and Tab. 26), a direct comparison with regard to absolute numbers for the indicative trajectories in MW and GWh highlights the stark differences between the two countries (see Fig. 15 and Fig. 16). Overall, according to the indicative trajectories, by 2020 Germany will have installed ten times the capacity of Poland, whilst producing seven times as much electricity from renewable energies (see Tab. 29).

Comparing anticipated technology development using renewable energy sources with regard to the binding targets for 2020, as indicated in the NREAPs, leads to the observation that both countries expect wind energy to play a significant role in the renewable energy technology mix (see Fig.17 and Fig. 18). An estimated installed capacity of 6.650 MW wind energy in Poland by 2020 makes this technology reflect 64 % of the overall installed renewable energy capacities by 2020 (see Tab. 26). An estimated installed capacity of 45.750 MW wind energy in Germany by 2020 makes this technology reflect 41% of the overall installed renewable energy capacities by 2020 (see Tab. 15).

6.3 Consequences of German RES Deployment for Germany and Poland: Loop Flows and Spot Market Prices

The rapid growth of RES deployment in Germany has an impact on the adequacy of the German power system. Since a comprehensive assessment of this impact lies well beyond the scope of this thesis, a differentiated analysis can be found, inter alia, in a scenario outlook and adequacy report by the European Network of Transmission System Operators for Electricity, or a grid study on the integration of RES in Germany by the German Energy Agency (ENTSO-E, 2013a; Dena, 2013). Overall, with respect to the adequacy of the German power system the rapid RES deployment does not so much cause a problem of capacity, but of regional distribution of power plants in Germany (ENTSO-E, 2014: 47). *'The shutdown of the nuclear power plants causes a shortage of available reactive power. [...] German TSOs have identified the risk of high voltages for scenarios of very low load combined with high PV feed-in in Southern Germany. Thus, in the summer period the German TSOs may be faced with problems to meet (n-1)-security[13] rules affecting the violation of permitted voltage limits'* (ENTSO-E, 2014: 47). In the winter period, on the other hand, high wind feed-in in the North could cause problems to meet the (n-1)-security rules (ENTSO-E, 2013b: 19). German RES deployment also heavily affects the Polish power system. *'Extremely severe balancing conditions in the summer period may*

---

[13] n-1: Outage condition; specifies that one transmission system element is out of service (DG Energy, 2006: 7)

take place in case of long lasting heat spells leading to significant deterioration of Polish power balance (increase of load with simultaneous decrease of generating capacities due to higher forced outage rate of generators, worse cooling conditions and increase in network constraints). [...] [T]he risk of high unscheduled transit flows from through the Polish system (from the west to the south) during such weather conditions cannot be considered low any more (as a result of development of solar generation in Germany)' (ENTSO-E, 2014: 16). In the winter period, the risk of unscheduled flows remains, owing to the impact of wind feed-in from Germany (ENTSO-E, 2013b: 19-20).

Why do, in particular, solar and wind power feed-in in Germany affect the Polish power system? In interconnected grids – as is the case between Germany and Poland (Krajnik (PL) – Vierraden (PL) and Mikulowa (PL) – Hagenwerder (DE) (50Hertz, 2014: 1) – electricity flows independently of national borders from the generator to the consumer on paths of lowest electrical resistance (PSE and 50Hertz, 2014: 6). In a zonal market approach, as used in Europe, the difference between a physical flow and a schedule (commercial transactions scheduled on a given border) is called an unplanned flow (PSE and 50Hertz, 2014: 6). Three terms shall be defined (CEPS+MAVIR+PSE+SEPS, 2013: 9):

- realised schedules: commercial flows through a given cross-border profile comprising long-term nominations, day-ahead nominations and intraday nominations;

- measured load flows/physical flows: measured physical cross-border flows;

- unplanned power flows: the difference of measured load flows/physical flows and realised schedules.

In the case of Germany and Poland, unplanned flows can reach significant volumes, with schedules going from Poland to Germany, whilst physical flows go in the opposite direction (PSE and 50Hertz, 2014: 7). Between January 2010 and December 2012, measured physical flows on the border between the two countries in question had the opposite direction to the commercial schedules . During this period there was a permanent and high level of unplanned flows from Germany to Poland in between 500 and 1.500 MW (CEPS+MAVIR+PSE+SEPS, 2013: 13). Fig.19 illustrates the development of monthly average values of realised schedules, measured load flows and unplanned flows between Germany and Poland.

Fig. 19: Realised Schedules, Measured Load Flows and Unplanned Flows between Germany and Poland (CEPS+MAVIR+PSE+SEPS, 2013: 13)

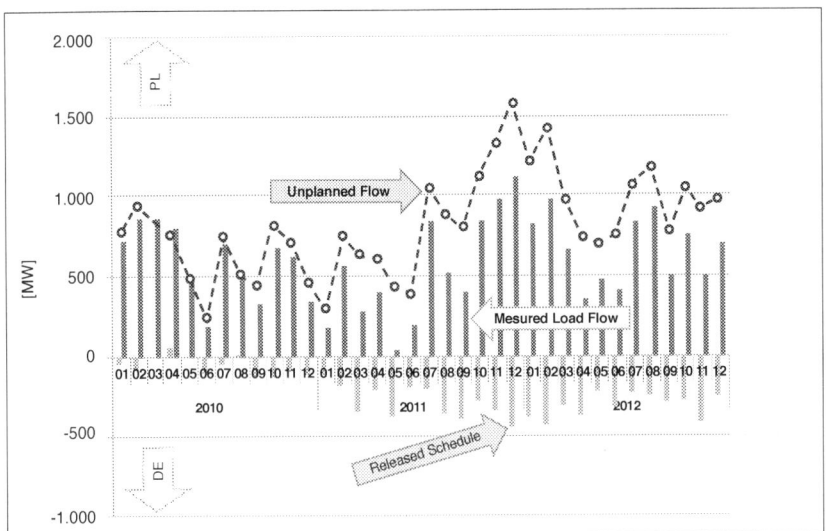

A significant influence of physical power flows – and in consequence unplanned power flows with regard to the example in question – can be observed in between the control area or bidding zone of the TSO 50Hertz (operating in the northern and eastern part of Germany) and PSE, the Polish system operator. The reason for these unplanned physical flows is the highly fluctuating power production of wind and solar plants in the 50Hertz area (PSE and 50Hertz, 2014: 14). In Fig. 20, the green graph represents the sum of solar and wind power production in Germany, whilst the blue graph represents the physical power flows or measured load flows between the 50Hertz and PSE area. During the period covered in Fig. 20, the solar power feed-in in the 50Hertz area amounted to a maximum of 22.998 MWh, and the wind power production reached a maximum of 23.065 MWh. The accumulated power feed-in from both sources reached a maximum of 35.779 MWh (PSE and 50Hertz, 2014: 14).

Fig. 20: Wind and Solar Power Production in Germany and Measured Load Flows to Poland in 2013 (PSE and 50Hertz, 2014: 14)

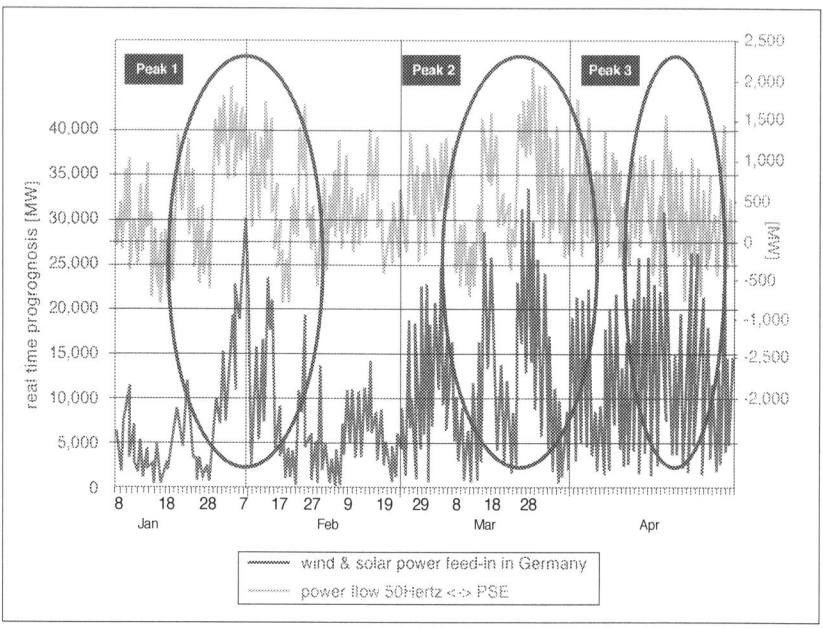

It becomes apparent from the above graphs that, in times of high wind and solar power feed-in in Germany, the observed power flows increase from the German 50Hertz area to the Polish PSE area. The amount of the physical flows varies from -729 MWh (PSE → 50Hertz) to 2.152 MWh (50Hertz→ PSE). The bulk of values range between -500 MWh and 1.600 MWh, amounting to a monthly sum of nearly 600.000 MWh in unplanned flows (50Hertz → PSE) in March 2013 (PSE and 50Hertz, 2014: 11 and 13).

The matter of unplanned flows cannot be reduced to a mere German–Polish phenomenon, but has to be construed in a European context. The above case of unplanned flows has to be understood in the context of the common German-Austrian electricity market and the surrounding CEE region (see below) (CEPS+MAVIR+PSE+SEPS, 2013). As has been stressed in a joint study by four TSOs from Central and Eastern Europe (representing the so-called CEE region), including CEPS (Czech Republic), PSE (Poland), MAVIR (Slovakia), and SEPS (Slovakia), unplanned flows significantly affect both power flows and security conditions in the abovementioned countries (CEPS+MAVIR+PSE+SEPS, 2012: 2).'Within the CEE region these unplanned power flows are due to internal exchanges between Northern and Southern Germany, but a significant share also results from exchanges within the common market area between Germany *and Austria, thus creating an unplanned transit not coordinated with the neighbouring countries'* (CEPS+MAVIR+PSE+SEPS, 2012: 2). In consequence, and owing to the geographical shape of Poland, the Czech Republic, Slovakia, and Hungary – including network topology and generation distribution – transmission systems of these countries are excessively loaded by the large scale of unplanned flows (CEPS+MAVIR+PSE+SEPS, 2012: 3). A report of the same TSOs published in 2013 comes to the conclusion that *'the volume of commercial transactions between Germany and Austria, which amount to nearly one-third of all commercial exchanges within the whole CEE region, has a direct link to the level of unplanned flows and may increase the risk of endangering the transmission systems of neighbouring countries'* (CEPS+MAVIR+PSE+SEPS, 2013: 39).[14] This is the case because flows from the north to the south of Germany may follow a path via Poland and the Czech Republic, or via the Netherlands, Belgium or France, and because necessary extra-high voltage lines between north and south Germany are (still) missing (Bundesnetzagentur, 2014a: 33; Bundesnetzagentur, 2014b, EuPD and DCTI, 2013: 70). The discrepancy between scheduled and physical flows is also referred to as loop flows (THEMA, 2013: 4).

In order to manage or mitigate the unplanned flows or loop flows, measures applied include the installation of phase-shifting transformers, or virtual phase-shifting transformers (Bundesnetzagentur, 2014a: 33; THEMA, 2013: 2). The coordinated installation and operation of phase shifting transformers on the interconnection lines between Krajnik (PL) – Vierraden (PL) and Mikulowa (PL) – Hagenwerder (DE) are already envisioned in the above-mentioned PCI 3.15.2, including German TSO 50Hertz and Polish TSO PSE (EC, 2014d: 2). As early as 18th December 2012, 50Hertz and PSE agreed on a pilot phase of a virtual phase-shifting transformer to limit unplanned cross-border power flows between Germany and Poland in the period from 8th January to 30th April 2013 (PSE and 50Hertz, 2014: 3).

The virtual phase-shifting transformer agreement defines three trigger criteria, which have to appear in order to activate measures envisioned under the agreement (PSE and 50Hertz, 2014: 10):

- Transgression of a set power flow limit on the German–Polish border between the existing interconnectors (Germany → Poland) at 1.500 MW between 8th January and 9th April 2013, and 1.600 MW between 10th April and 30th April

- Declaration of insecure operation by PSE, when (n-1)-security violation occurs

- Net scheduled exchanges in the Polish synchronous profile in import direction are lower than a threshold set at 200 MW in January, 350 MW in February and 500 MW in March and April.

---

[14] For more details on this subject, see the reports referred to here.

During the pilot phase, virtual phase-shifting transformer measures kept the physical flow at an average of 1.458 MW. Nevertheless, the maximum recorded power flow reached 2.152 MW. During the pilot phase, the integrity of the Polish power system was violated for two days, where the measures applied were insufficient, and re-dispatching measures in Poland, Germany, Austria and Czech Republic were exhausted (PSE and 50Hertz, 2014: 4). Overall, owing to foreseen further growth or renewable feed-in in Germany, 50Hertz and PSE consider virtual phase-shifting transformers to be insufficient in order to ensure a secure operation of the interconnected power systems at all times (PSE and 50Hertz, 2014: 5). They hence recommend the installation of physical phase-shifting transformers as envisioned in European PCI 3.15.2, described above (EC, 2014d: 2).

Whilst the main concern of the TSOs in the context of unplanned flows is to maintain system security with respect to (n-1)-security violations, another equally important aspect is the unresolved distribution of costs of unplanned flows. The subject is vaguely addressed in the joint 2012 study by four TSOs from Central and Eastern Europe mentioned above (CEPS+MAVIR+PSE+SEPS, 2012). The report identifies the unplanned flows to be a result of a bad market design. Accordingly, the 'absence' of market mechanisms for unplanned flows (see scheduled flows vs. unplanned flows) does not result in both social welfare maximization and secure system operation (CEPS+MAVIR+PSE+SEPS, 2012: 2-3). With respect to electricity as a commodity, some *'problems'* of unplanned flows are identified in a report prepared for the European Commission by THEMA Consulting Group (THEMA, 2013: 7):

• The market architecture anticipates the distribution of generation and consump-tion in principal, without taking into account the grid configuration within a market area. There is no direct buyer–seller connection.

• Since electricity takes the way of least resistance, physical flows are not in line with market schedules.

• The cost distribution in the grid is organised according to national borders and defined TSO control areas, to which physical electricity flows cannot be limited.

In order to illustrate the effects of high RES-E feed-in in Germany on the power price on the spot market (day-ahead market indices), indicators both in Germany and Poland, from the European Energy Exchange (EEX) and the Polish Power Exchange (PPE), will be used. The EPEX Phelix Day Base traded, *inter alia,* at the EEX represents the average price of hours 1 to 24 in €/MWh for electricity traded on the spot market in the market area Germany/Austria, disregarding power transmission bottlenecks. The EPEX Phelix Day Peak is the average price of hours 9 to 20 on the spot market in €/MWh traded at the EEX (EPEXSPOT, 2012: 5). The POLPX Spot Base traded at the PPE quoted in €/MWh represents the average price of hours 1 to 24 for the Polish day-ahead market.

The POLPX Spot Peak represents the hours 8 to 20 in €/MWh (PPE, 2014a). The Polish indices are created for direct comparison of the price movement on the Polish and European markets, in particular with respect to EPEXSPOT indices. Market participants hence *'have the opportunity to refer directly to the markets of neighbouring countries taking decision on energy exchange [...] with standard export/import transactions based on cross-border transmission capacity auctions organized by network operators for synchronous connections Poland to Germany'* (PPE, 2014a). In Fig. 20, three peaks have been identified at which the correlation between high wind and solar power feed-in in Germany, and high measured load flows from Germany to Poland in the first four months of 2013, becomes particularly evident. To exemplify a possible correlation of high wind and solar power feed-in in Germany, high measured load flows from Germany to Poland, and low spot market prices in Germany and Poland: Fig. 21 illustrates the development of the referred to EEX spot market indices within the same period of time, and highlights the aforementioned peaks, and Fig. 22 illustrates the development of the referred to PPE spot market indices for the same time period, with respect to the identified peak situations. Whilst maintaining this general logic, Fig. 23 provides a comparison of Phelix and POLPX Day Peak, and Fig. 24 allows a direct comparison of Phelix and POLPX Day Base developments.

Fig. 21: EEX SPOT Phelix Day Peak and Base in €/MWh (EEX, 2014; Energate, 2014; Zepeda, 2014)

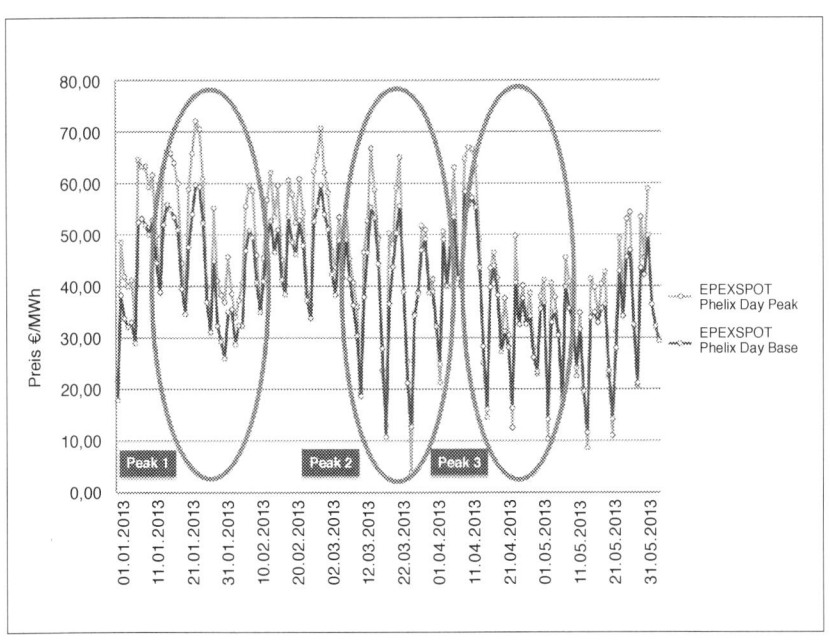

Fig. 22: PPE SPOT POLPX Day Peak and Base in €/MWh (PPE, 2014b; Zepeda, 2014)

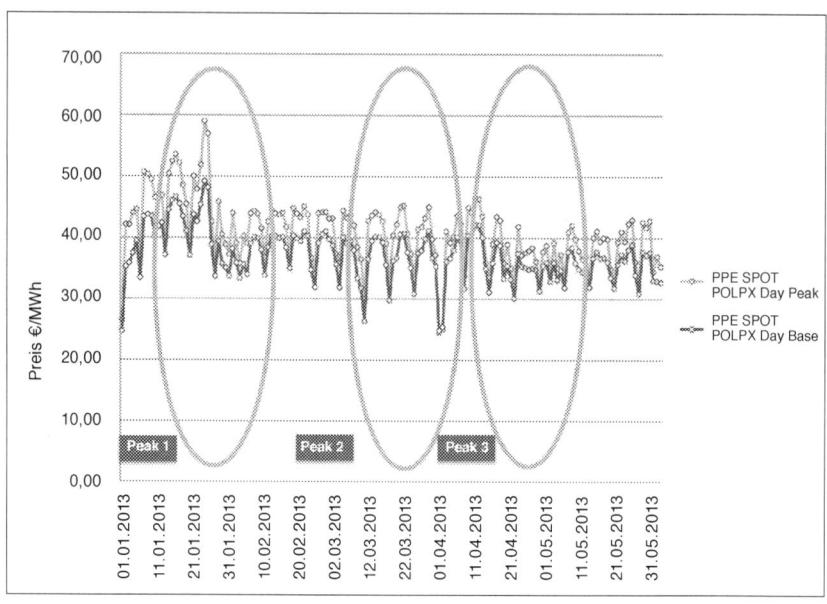

Fig. 23: SPOT Phelix and POLPX Day Peak in €/MWh (EEX, 2014; PPE, 2014b; Energate, 2014; Zepeda, 2014)

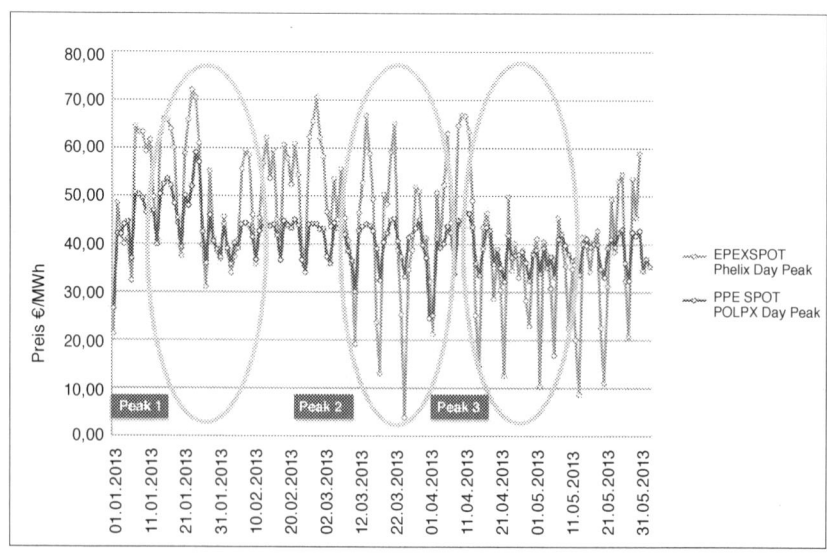

Fig. 24: SPOT Phelix and POLPX Day Base in €/MWh (EEX, 2014; PPE, 2014b; Energate, 2014; Zepeda, 2014)

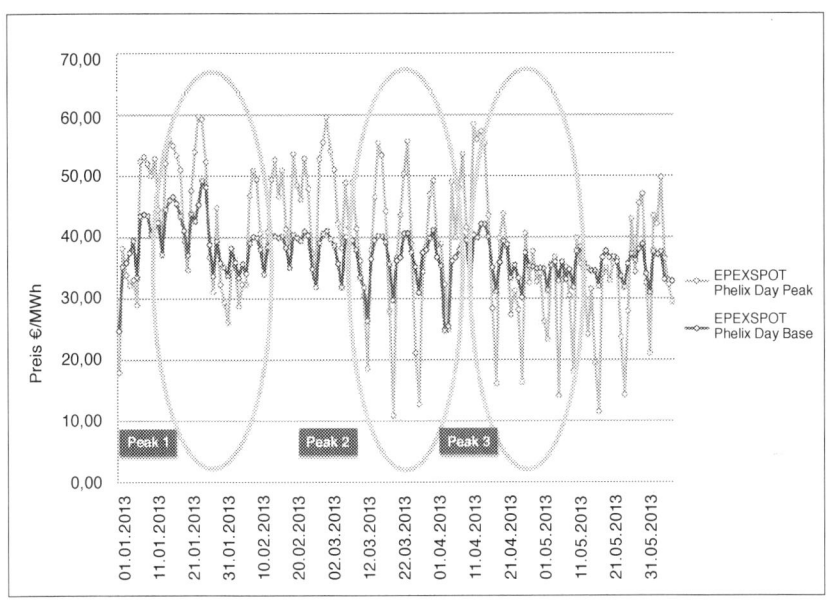

Peak 1:

As becomes apparent from Fig. 20, wind and solar feed-in in Germany, as well as measured load flows to Poland, increased sharply between 24th and 31st January 2013. EPEX Phelix Day Peak fell from 70,47 €/MWh (24th January) to 36,81 €/MWh (31st January), and Phelix Day Base from 59,25 €/MWh (24th January) to 26,06 €/MWh (31st January) in the same time period (see Fig. 21). PPE POLPX Day Peak fell from 59,00 €/MWh (24th January) to 37,32 €/MWh (31st January), and POLPX Day Base from 49,22 €/MWh (24th January) to 33,75 €/MWh (31st January) in the same time period (see Fig. 22). The correlation of EEX and PPE Day Peak and Base price developments can be observed in Fig. 23 and Fig. 24.

Peak 2:

Based on Fig. 20, another peak (high RES-E feed-in, high measured load flows) can be identified between 13th and 24th March 2013. During this time period, EPEX Phelix Day Peak fell from 66,73 €/MWh (13th March) to 3,88 €/MWh (24th March), and Phelix Day Base from 53,34 €/MWh (13th March) to 12,70 €/MWh (24th March) (see Fig. 21). PPE POLPX Day Peak fell from 44,18 €/MWh (13th March) to 33,29 €/MWh (24th March), and POLPX Day Base from 40,24 €/MWh (13th March) to 30,90 €/MWh (24th March) (see Fig. 22). The correlation of EEX and PPE Day Peak and Base price developments can be observed in Fig. 23 and Fig. 24.

Peak 3:
The third peak in this comparative synopsis has been identified between 10th and 21st April 2013. During this time period EPEX Phelix Day Peak fell from 66,68 €/MWh (10th April) to 12,50 €/MWh (21 April), and Phelix Day Base from 55,39 €/MWh (10th April) to 16,39 €/MWh (21st April) (see Fig. 21). PPE POLPX Day Peak fell from 46,48 €/MWh (10th April) to 40,69 €/MWh (21st April), and POLPX Day Base from 42,17 €/MWh (10th April) to 30,13 €/MWh (21st April) during this period (see Fig. 22). The correlation of EEX and PPE Day Peak and Base price developments can be observed in Fig. 23 and Fig. 24.

Overall, three correlations become apparent from this comparative synopsis:

a) High RES-E feed-in in Germany leads to significantly rising (unplanned) load flows from Germany to Poland, causing loop flows describing the discrepancy between scheduled and physical flows.

b) High RES-E feed-in in Germany leads to an, at times, drastic fall in the power price on the German spot market.

c) Whilst spot market price developments in response to German RES-E feed-in on the Polish Electricity Exchange are not as volatile as on the German counterpart, there is a very evident correlation in price movements.

Whilst the reason for (a) has been described above (realised schedules, measured load flows, unplanned flows → loop flows), attention shall now be given to (b) and (c).

6.3.1 Summary

German RES deployment heavily affect the Polish power system. Why do solar and wind power feed-in in Germany affect the Polish power system? In interconnected grids – as is the case between Germany and Poland (Krajnik (PL) – Vierraden (PL) and Mikulowa (PL) – Hagenwerder (DE) (50Hertz, 2014: 1) – electricity flows independently of national borders from the generator to the consumer on paths of lowest electrical resistance (PSE and 50Hertz, 2014: 6). In a zonal market approach, as used in Europe, the difference between a physical flow and a schedule (commercial transactions scheduled on a given border) is called an unplanned flow (PSE and 50Hertz, 2014: 6). Three terms were defined (CEPS+MAVIR+PSE+SEPS, 2013: 9):

- realised schedules: commercial flows through a given cross-border profile comprising long-term nominations, day-ahead nominations and intraday nominations;

- measured load flows/physical flows: measured physical cross-border flows;

- unplanned power flows: the difference of measured load flows/physical flows and realised schedules

In the case of Germany and Poland, unplanned flows can reach significant volumes, with schedules going from Poland to Germany, whilst physical flows go in the opposite direction (PSE and 50Hertz, 2014: 7). Between January 2010 and December 2012, measured physical flows on the border between the two countries in question had the opposite direction to the realised commercial schedules. During this period there was a permanent and high level of unplanned flows from Germany to Poland in between 500 and 1.500 MW (CEPS+MAVIR+PSE+SEPS, 2013: 13).Fig. 19 illustrates the development of monthly average values of realised schedules, measured load flows and unplanned flows between Germany and Poland.

Whilst the main concern of the TSOs in the context of unplanned flows is to maintain system security with respect to (n-1)-security violations, another equally important aspect is the unresolved distribution of costs of unplanned flows. In order to illustrate the effects of high RES-E feed-in in Germany on the power price on the spot market (day-ahead market indices), indicators both in Germany and Poland from the European Energy Exchange (EEX) and the Polish Power Exchange (PPE) have been compared. Overall, three correlations become apparent from the comparative synopsis illustrated in Fig. 21, Fig. 22, Fig. 23, and Fig. 24:

a) High RES-E feed-in in Germany leads to significantly rising (unplanned) load flows from Germany to Poland, causing loop flows describing the discrepancy between scheduled and physical flows.

b) High RES-E feed-in in Germany leads to an, at times, drastic fall in the power price on the German spot market.

c) Whilst spot market price developments in response to German RES-E feed-in on the Polish Electricity Exchange are not as volatile as on the German counterpart, there is a very evident correlation in price movements.

Whilst the reason for (a) has been described in the section above (realised schedules, measured load flows, unplanned flows → loop flows), attention will be given to (b) and (c) in the next section.

6.4 Merit Order Effect

Electricity spot market prices reflect the distinct nature of the power markets, which are characterised by unique price patterns as a result of the seasonality of underlying demand, the non-storability of electricity, the need of a balanced grid at all times, the challenge of intermittent renewable energy sources, and the need for base load capacities (Erni, 2012: 13). Focus shall at this stage be given to the effects the production of electricity from renewable energy sources has on the spot market price developments. When renewable energy is produced it takes precedence in the market owing to regulatory designs, if in accord with envisioned supranational implications, and has near-zero marginal costs, if support schemes such as feed-in tariffs are established (Rop, 2013: 8). Of particular relevance for the promotion of RES-E are the transparent and non-discriminatory criteria of Directive 2009/28/EC (Article 16(2) a, b and c) leading to a guarantee of transmission and distribution of electricity produced from RES, priority access or guaranteed access to the grid-system for electricity produced from RES, and priority for generating installations using RES, as long as the secure operation of the national electricity system permits it (OJoEU, 2009b: 35). Whilst the Directive does not prescribe a definite design for measures such as supporting schemes for RES-E promotion, Member States' implementation, as pointed out in Article 3(2), must lead to designs effective to fulfil or exceed the indicative trajectories (OJoEU, 2009b: 28). In Germany, both aspects mentioned above are regulated in the Law for the Readjustment of the Renewable Energy Act and corresponding legal provisions (BGBL, 2008: 2074). Transmission and non-discriminatory criteria as envisioned in Directive 2009/28/EC are transposed by respective regulations of Part II, Article 2, § 5-8, whereas support schemes for RES (feed-in tariffs) are regulated in Part III, Article 1, § 16-22 (BGBL, 2008: 2076-2080). Electricity produced in Germany thus takes precedence in the market and has near-zero marginal costs – that is, no variable costs owing to fixed feed-in tariffs pursuant to the implemented support schemes (Haas and Loew, 2012: 11). In principal, renewable energy sources such as wind energy are not subject to market pricing because, by establishing feed-in tariffs, there is a guaranteed price for every supplied renewable energy production unit to be purchased and paid by the power market participants (Traber and Kemfert, 2011: 249).

'When discussing electricity markets it is imperative to have an understanding of the merit order concept which plays a central role in the formation process of electricity spot prices. The merit order curve can be defined as the sorted marginal cost curve of electricity production which defines the price of electricity given a certain quantity' (Erni, 2012: 38).

Since the level of demand at any given time in the electricity system is usually less than the total available capacity, not all power plants are needed to meet the demand. Those plants with the lowest short-run marginal cost – that is, fuel and operating costs – will be the first in the priority or merit order to satisfy the demand (Haas and Low, 2012: 9; IEA, 2007: 47). Because RES, such as wind energy, have low marginal costs (see Fig. 25), they enter first in the merit order (Pöyry and EWEA, 2010: 11). Nuclear power plants and base-load plants, such as coal or gas-fired plants, typically follow renewable sources facilities (IEA, 2007: 47). *'The type of production technology in whose area the intersection of the demand and supply (merit order) curve is located constitutes the price setting technology'* (Erni, 2012: 38).

Fig. 25: Merit-Order and Effect of Wind Power at Different Times of the Day (Pöyry and EWEA, 2010: 11; EC, 2014e: 236)

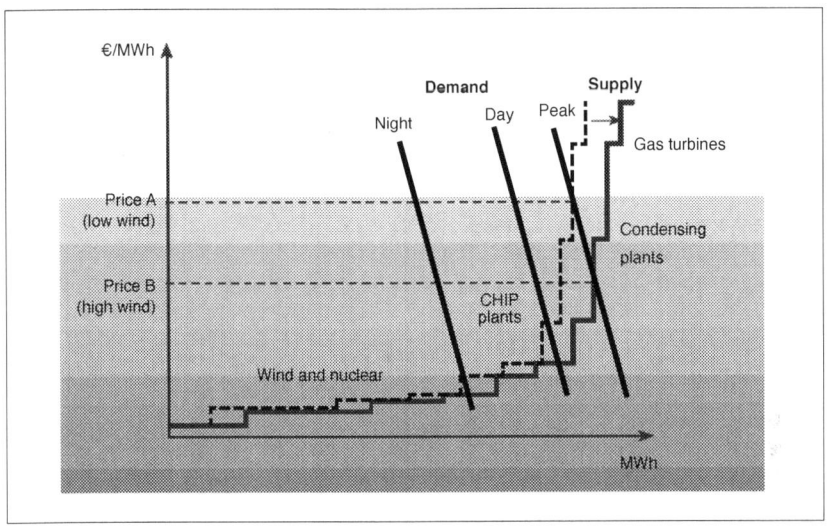

In other words (illustrated in Fig. 20, Fig. 21, Fig. 22, Fig. 23, and Fig. 24), the spot market prices for electricity falls if transmission and non-discriminatory criteria for RES are met in the underlying market design, if a period of high RES feed-in occurs (owing to, for example, windy days), and because the marginal costs of, for example, RES wind power are lower than those of conventional plants. Based on day-ahead spot market prices and short marginal costs of power generation (excluding investment costs), this is called merit-order effect. According to the merit-order, plants with the lowest costs are used first, with more costly plants being brought on-line if needed (EWEA, 2010: 135). Ultimately, this means that running a conventional plant, including base-load plans such as gas or coal fired power plants, can be noneconomic if they cannot compete with the technologies at the lower end in the merit order. In theory, only those plants will be dispatched that are profitable (IEA, 2007: 48). This is illustrated in Fig. 25.

The steep price drop in electricity prices observed in Germany, reflected in the respective graphs (see above) indicating, in principle, an over-supply of electricity, cannot be explained by the merit order alone. Another vital aspect needs to be taken into account. In 2012 and 2013, negative spot market prices for electricity could be observed in Germany during 97 h. This means that, although, according to the merit order, more expensive facilities such as gas fired power plans should not have produced electricity (for it would not have been profitable), they kept on feeding electricity into the grid. In addition, from 2012 to 2013, periods can be identified where positive electricity prices between 0 – 10 €/MWh, lower than the short-term marginal costs of nuclear and conventional thermal power plants, did not lead to a production shut-down of those facilities (Agora, 2014: 11). This phenomenon can be explained by taking into account the nature of non-intermittent non-renewable energy facilities. If producers of non-flexible base load facilities *'decide to keep their production up, they have calculated that this is the best, most cost-efficient way for them considering the costs of shutting down and restarting their plants'*, even if this de facto means paying the buyer to receive electricity (EPEXSPOT, 2014). This is the case, because to reduce output of non-flexible generators, which a so-called must-run character defines, is hardly possible from a technical perspective, or involves high shutdown costs (Sewalt and De Jong, 2003: 76). The phenomenon of operating electricity-generating units at varying load levels, including on/off load following, and minimum load operation in response to changes in system load requirements, is called cycling (Intertek and Aptech, 2012: ii). *'Every time a power plant is turned off and on, the boiler, steam lines, turbine, and auxiliary components go through unavoidably large thermal and pressure stresses, which cause damage'* (Intertek and Aptech, 2012: ii). In the context of the on-going discussion with regard to the Polish electricity market, it is interesting to note that large coal units, especially, are designed for base-load operation with on average higher cycling costs for generating units (Intertek and Aptech, 2012: 45). Generally, a higher penetration of electricity produced by intermittent renewable energy facilities increases power plant cycling of conventional production units (Intertek and Aptech, 2012: 45). In consequence, negative electricity prices on the spot markets are not just the result of an over supply of electricity owing to periods of, for example, high solar and wind feed-in, but also of the inflexible nature of conventional electricity production units and their cycling and dispatching costs (Agora, 2014: 11).

Whilst the impact of an increasing integration of renewable energies with respect to the merit order effect and negative electricity prices in the German electricity market has been implicitly touched upon in this section, deepening insights into the subject can be gained in the cited studies by, *inter alia,* Haas and Loew (2012) or Agora (2014). The consequences of the phenomena described here for the Polish electricity market will be discussed in the next section.

6.4.1 Summary

'When discussing electricity markets it is imperative to have an understanding of the merit order concept which plays a central role in the formation process of electricity spot prices. The merit order curve can be defined as the sorted marginal cost curve of electricity production which defines the price of electricity given a certain quantity' (Erni, 2012: 38).

Since the level of demand at any given time in the electricity system is usually less than the total available capacity, not all power plants are needed to meet the demand. Those plants with the lowest short-run marginal cost – that is, fuel and operating costs – will be the first in the priority or merit order to satisfy the demand (Haas and Low, 2012: 9; IEA, 2007: 47). Because RES, such as wind energy, have low marginal costs (see Fig. 25), they enter first in the merit order (Pöyry and EWEA, 2010: 11).

Since the marginal costs of, for example. RES wind power are lower than the price of the more expensive conventional plans, and if transmission and non-discriminatory criteria for RES are met in the underlying market design, in combination with high RES feed-in (owing to, for example, windy days), the average price on the spot market, as becomes apparent from the analysis above as illustrated in Fig. 20, Fig. 21, Fig. 22, Fig. 23, and Fig. 24, drops.

The steep price drop in electricity prices observed in Germany, reflected in the respective graphs (see above) indicating, in principle, an over-supply of electricity, cannot be explained by the merit order alone. Another vital aspect needs to be taken into account. In 2012 and 2013, negative spot market prices for electricity could be observed in Germany during 97 h. This phenomenon can be explained by taking into account the nature of non-intermittent non-renewable energy facilities. If producers of non-flexible base load facilities *'decide to keep their production up, they have calculated that this is the best, most cost-efficient way for them considering the costs of shutting down and restarting their plants',* even if this de facto means paying the buyer to receive electricity (EPEXSPOT, 2014).

Generally, a higher penetration of electricity produced by intermittent renewable energy facilities increases power plant cycling of conventional production units (Intertek and Aptech, 2012: 45). In consequence, negative electricity prices on the spot markets are not just the result of an over supply of electricity owing to periods of, for example, high solar and wind feed-in, but also of the inflexible nature of conventional electricity production units and their cycling and dispatching costs (Agora, 2014: 11).

6.5 Short-Term and Long-Term Impact of German RES Deployment on the Polish Electricity Market

As a result of the disputed argument discussed above, in can be stated that, whilst spot market price developments in response to German RES-E feed-in on the Polish Electricity Exchange are not as volatile as on the German counterpart, there is an evident correlation in price movements. Of course, the feed-in of installed RES-E facilities in Poland is subject to the same conditions described above. But the comparison of overall renewable energy installed capacity and gross electricity generation in Germany and Poland (see Fig.14 and Fig.15) displays the more dominant impact of German RES-E feed-in on the Polish electricity market.

Another consequence of the discussion above is the evident occurrence of unplanned physical electricity flows from Germany to Poland and, in consequence, loop flows. Because of the distinct nature of power markets, including the need for a balanced grid at all times, unplanned physical flows need to be counteracted by re-dispatching measures ultimately leading to the operation of base-load generating units at varying load levels, also referred to as cycling (Erni, 2012 : 13; Intertek and Aptech, 2012: ii). Bearing witness to the fact that re-dispatching measures, as a result of German RES-E feed-in, do not only occur in the German electricity market, but also in the Polish market, is a joint report by Polish TSO PSE and German TSO 50Hertz, already cited in this chapter. *'There were two days (i.e. 25 and 26 March 2013) where the (N-1)-secure operation of the Polish power system was violated for several straight hours [...]. On one of these days, namely on 25 March, redispatching measures in Poland, Germany, Austria and Czech Republic were exhausted (no generating capacity for up regulation available anymore after 1,600 MW curative redispatch with the help of Polish and Austrian power plants), putting the interconnected power systems at significant risk of an cascading tripping had an unforeseen outage taken place at that time'* (PSE and 50Hertz, 2014: 4). In addition to the technical implications of the period addressed in the report, the dates indicated here fall within the identified second peak of high unplanned flows (see Fig. 20), where spot market prices for electricity were significantly volatile both in the German and Polish market (see Fig. 21 and Fig. 22). It is highly likely, however not evidently verifiable, that day-ahead spot market prices during this time were lower than the short-term marginal costs of conventional thermal power plants, both for German and Polish electricity producers.

This latter assumption is confirmed by a response of the Polish Electricity Association (PKEE), and Towarzystwo Gospodarcze Polskie Elektrownie (TGPE, association of main electricity generators in Poland) to a consultation paper of the European Commission on overall generation adequacy, capacity mechanisms and the internal market in electricity (PKEE, 2013; TGPE, 2013; EC, 2012c). According to PKEE, *'[...] energy prices in Poland have reached a level below the entry price of new capacity in all energy technologies and significantly decreases profitability rates of existing units'* (PKEE, 2013: 2). TGPE declares, that *'[w]ith decreasing volume of energy sold, low energy prices and the lack of other revenues, e.g. from capacity mechanisms, maintaining many generation units becomes unprofitable'* (TGPE, 2013: 15). TGPE adds that *'[p]roblems connected with decreasing production in Polish condensation units are amplified by non-planned power flows from RES located in Germany (including so-called loop flows)'* (TGPE, 2013: 16; see also PKEE, 2013: 3).

In the short term, three effects of German RES-E deployment on the Polish electricity market can be identified:

a) The impact of periods of high RES-E feed-in (in particular wind and solar) in Germany on the Polish spot market price makes the production of electricity from conventional thermal power plants at times unprofitable.

b) The production of RES-E in Germany challenges the secure operation of the Polish power system (balanced grid at all times).

c) Re-dispatching measures as a response to German RES-E feed-in, and, in consequence, cycling of base-load generating conventional thermal power plants can lead to additional costs and potential damage in the respective production units for the Polish electricity producer.

These findings become all the more significant when taking into account that the bulk of electric power produced in Poland is derived from combined heat and power plants using traditional fuels (82% in 2012, of which hard coal is 54,5%, and lignite 25,2%; see Fig. 13) (Paiz.gov.pl, 2013: 5). This means that the Polish energy economy, representing conventional power plant facilities mainly based on relatively inflexible coal-fired units, is heavily exposed to the impact of an increasing RES deployment in economic and technical terms.

These findings also have long-term implications for the Polish power market. In the Second Appendix to its 2009 'Energy Policy of Poland until 2030', the Polish Ministry of Economy expects decommissioning of gross generation capacity in base-load power plants to reach a total 4.125 MW by 2020(Ministry of Economy, 2009b: 9). TGPE, in its 2013 response to the European Commission, however, anticipates decommissioning of base-load power plants to amount to a greater number than expected by the Ministry of Economy in 2009, namely 7.593 MW. Indeed, in the statement of TGPE, an additional possible decommissioning of 3.000 MW is considered, altogether causing a *'significant generation inadequacy threat'* (TGPE, 2013: 6).

Overall, many operators in Poland plan advanced decommissioning of generation units. On average, intended decommissioning activities for the year 2020 are pushed forward by three to five years (TGPE, 2013: 15). Respective decisions are encouraged because the market-regulatory environment (merit-order curve and implications for operating hours, and revenues of conventional generation) makes the maintenance of unprofitable generation units economically impossible in the long term (TGPE, 2013: 3 and 6; PKEE, 2013: 2).

The same is true for planned new commissioning and installation of conventional power plant capacities (TGPE, 2013: 6). Investment in new conventional generation units is stopped owing to continuing market conditions that do not guarantee return on investments made, and because futures and spot market prices do not guarantee sufficient entry prices for new generation units (TGPE, 2013: 3 and 13). Reasons for the observed discontinuation of investments in new generation units – such as. the development of 1.000 MW generation units in Ostroleka, Rybnik, Opalenie and Leczna – hence come down to investor risks being too high (TGPE, 2013: 6). *'As a result of high risk levels, more entities take decisions to suspend or discontinue development of thermal gas- or coal-fired power stations'* (TGPE, 2013: 149). Overall, it can be stated that the implications of German RES-E deployment lead to energy prices in the Polish market that are considerably lower than the short and long-term marginal costs of conventional coal-fired power plants. This makes the operation of the generation units in question unprofitable in the short-term, and causes – owing to inappropriate investment signals – strategic investments in base-load capacities to be withdrawn (TGPE, 2013:5; PKEE, 2013: 2). The long-term consequences of large RES-E generation for Poland can be summarised as follows: electricity generated by RES reduces the operating hours and profitability of Polish base-load and back-up generation technologies. This, in consequence, raises concerns about future investment decisions with respect to the technologies in question, and – in light of the overall subject matter - with respect to Polish policy objectives (Eurelectric, 2011: 4).

6.5.1 Summary

In the short term, three effects of German RES-E deployment on the Polish electricity market can be identified:

a) The impact of periods of high RES-E feed-in (in particular wind and solar) in Germany on the Polish spot market price makes the production of electricity from conventional thermal power plants at times unprofitable.

b) The production of RES-E in Germany challenges the secure operation of the Polish power system (balanced grid at all times).

c) Re-dispatching measures as a response to German RES-E feed-in, and, in consequence, cycling of base-load generating conventional thermal power plants can lead to additional costs and potential damage in the respective production units for the Polish electricity producer.

Overall, it can be stated that the implications of German RES-E deployment lead to energy prices in the Polish market that are considerably lower than the short and long-term marginal costs of conventional coal-fired power plants. This makes the operation of the generation units in question unprofitable in the short-term, and causes – owing to inappropriate investment signals – strategic investments in base-load capacities to be withdrawn (TGPE, 2013:5; PKEE, 2013: 2). The long-term consequences of large RES-E generation for Poland can be summarised as follows: electricity generated by RES reduces the operating hours and profitability of Polish base-load and back-up generation technologies. This, in consequence, raises concerns about future investment decisions with respect to the technologies in question, and – in light of the overall subject matter - with respect to Polish policy objectives (Eurelectric, 2011: 4).

6.6 Analysis

Based on the overall hypothesis in light of the multi-level policy dependence theory employed here, three subordinate hypotheses were identified in the theory section. To identify the constraints and dependencies of the European Integration of RES-(E) promotion, the explanatory factors implicitly describe three different influencing factors that follow a top-down, that is, vertical logic, in turn leading to a horizontal logic. The first subordinate hypothesis (i) claims supranational renewable energy measures and policies produce compliance at the level of Member States (Radaelli, 2003: 19). Subordinate hypothesis (ii) counterpoints the first by claiming the European Union lacks the executive authority to expect compliance at the Member States level. Instead, it is national preferences that dominate the Member States' policy outcomes. Finally, subordinate hypothesis (iii) integrates the third aspect of inter-state interdependencies into the theory framework. The underlying assumption is that successful promotion of renewable energy electricity in country A will produce adaptational economic and regulatory pressure in country B, and ultimately lead to the alignment of both systems. This chapter deals with subordinate hypothesis (iii).

As has been stressed, the overall framework includes theories derived from Liberal Intergovernmentalism and Europeanization. They are reflected in subordinate hypothesis (i) (Europeanization) and subordinate hypothesis (ii) (Liberal Intergovernmentalism). Subordinate hypothesis (iii) incorporates aspects from both theories with respect to inter-state connectedness. All three subordinate hypotheses read as follows:

(i) The supranational RES-(E) measures and policies of the European Union, when expressed in legislation, excite change in Member States' RES-(E) policies only when there is a degree of misfit between EU and domestic RES-(E) policies that causes adaptational pressure, leading to the national promotion of RES-(E) in accordance with EU level implications.

(ii) The RES-E policy of a Member State is an exclusive result of the domestic policy process that reflects energy-specific economic interests of the dominant domestic group, and is determined by geopolitical and economic interest.

(iii) European inter-state connectedness, competition and economic pressure will foster the promotion of RES-E in Europe beyond pioneering Member States if a pioneering State with sufficient power in the international system holds a dominant market position, reflected in its RES-E market design.

Since any progress in the promotion of RES-(E) can only be measured at the Member States level, a model has been derived where the EU Member States are the focal point. In the applied model, the dependent variable are the Member State RES-E policies of Poland and Germany, with and ultimate focus being on the vertical and horizontal impacts on the RES-E policy of Poland. It has been stressed that a mono-causal approach cannot explain the impact on the dependent variable 'Member State RES-E'. Therefore there cannot be a single independent variable (e.g. defined at EU level) (Radaelli, 2003: 27-28). Instead, there is a set of three decisive variables, occurring at the EU level, the national level and the inter-state level. Fig. 2 illustrates their interrelations. Since all three influencing factors for the promotion of RES-E in Member States (supranational extrinsic, national intrinsic, inter-state interdependencies) will be considered, the analysis aims at identifying general insights about constraints and dependencies of European RES-E promotion mechanisms.

Fig. 26: Applied model of inter-national adaptational pressure for Polish RES-E policy

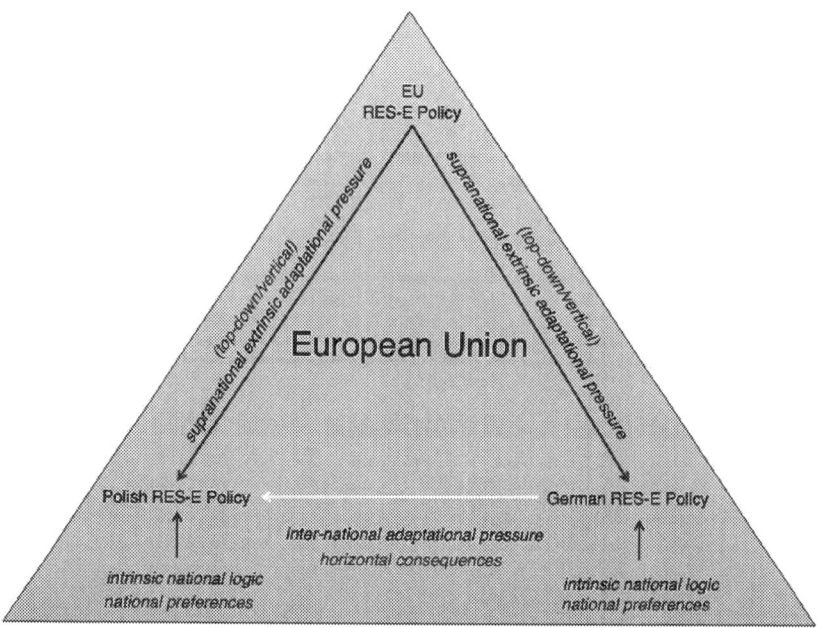

The Polish reaction to 'supranational extrinsic adaptational pressure' and – equally so – its 'intrinsic national logic' were the analytical focus of the previous chapter. The conclusion was drawn that the initial adaptational pressure arising out of the supranational requirements stipulated in Directive 2009/28/2009 did lead to a formal transposition, that is, national adaptation of European implications. However, the RES-E policy of Poland must be accepted as being – if not exclusively – primarily result of national preferences of the domestic policy process.

This chapter deals with the evaluation of subordinate hypothesis (iii). With respect to the applied model, the underlying assumption leads to the postulate of an inter-national adaptational pressure from Germany upon Poland. The analytical focus with respect to the overall model can be deduced from Fig. 26. If subordinate hypothesis (iii) serves as a valid explanation, the questions are: Can intra-European connectedness between Germany and Poland be confirmed with respect to RES-E competition and economic pressure? Can a dominant market position and market design Germany → Poland be identified? If the latter is applicable, does it lead the Polish market to adapt to the German market?

To take into account mechanisms *'starting with vertical prerequisites'* leading to *'horizontal consequences'*, a brief re-introduction to the regulatory framework conditions for the European electricity market was given. Whilst, based on these facts, no conclusions can be drawn with respect to the evaluation of subordinate hypothesis (iii), they nevertheless reflect present and planned European inter-state interconnectedness with respect to physical and economic market coupling. In particular, the *'Cluster Germany – Poland between Vierraden and Krajnik'*, and the *'Cluster Germany – Poland between Eisenhüttenstadt and Plewsika [currently known as the GerPol Power Bridge project]'*, including the respective PCIs (Projects of Common Interest) are evidence of this (OJoEU, 2013c: 35).

Whilst the trajectories for the share of energy from renewable sources in the electricity sector in the period 2015–2020 have been described independently for Poland and Germany (see Tab. 14, Tab. 15, Tab. 25, and Tab. 26), a direct comparison with regard to absolute numbers for the indicative trajectories in MW and GWh highlights the stark differences between the two countries (see Fig. 15 and Fig. 16). Overall, according to the indicative trajectories, by 2020 Germany will have installed ten times the capacity of Poland, whilst producing seven times as much electricity from renewable energies (see Tab. 29).

High German RES-E deployment to date, in particular wind and solar feed-in, has measurable and verifiable effects for the Polish electricity market. Overall three correlations become apparent from the comparative synopsis illustrated in Fig. 21, Fig. 22, Fig. 23, and Fig. 24:

a) High RES-E feed-in in Germany leads to significantly rising (unplanned) load flows from Germany to Poland, causing loop flows describing the discrepancy between scheduled and physical flows.

b) High RES-E feed-in in Germany leads to an at times drastic fall in the power price on the German spot market.

c) Whilst spot market price developments in response to German RES-E feed-in on the Polish Electricity Exchange are not as volatile as on the German counterpart, there is a evident correlation in price movements.

> As has been stressed in a joint study by four TSOs from Central and Eastern Europe (representing the so-called CEE region), including CEPS (Czech Republic), PSE (Poland), MAVIR (Slovakia), and SEPS (Slovakia), unplanned flows significantly affect both power flows and security conditions in the abovementioned countries (CEPS+MAVIR+PSE+SEPS, 2012: 2). *'Within the CEE region these unplanned power flows are due to internal exchanges between Northern and Southern Germany, but a significant share also results from exchanges within the common market area between Germany and Austria, thus creating an unplanned transit not coordinated with the neighbouring countries'* (CEPS+MAVIR+PSE+SEPS, 2012: 2). In consequence, and owing to the geographical shape of Poland, the Czech Republic, Slovakia, and Hungary – including network topology and generation distribution – transmission systems of these countries are excessively loaded

by the large scale of unplanned flows (CEPS+MAVIR+PSE+SEPS, 2012: 3). A report of the same TSOs published in 2013 comes to the conclusion that *'the volume of commercial transactions between Germany and Austria, which amount to nearly one-third of all commercial exchanges within the whole CEE region, has a direct link to the level of unplanned flows and may increase the risk of endangering the transmission systems of neighbouring countries'* (CEPS+MAVIR+PSE+SEPS, 2013: 39). Besides the impact of German RES-E deployment on the system security of the electricity grid with respect to (n-1)-security violations, another aspect could be identified in the analysis. Based on spot market (day-ahead market) indices of the European Energy Exchange (EEX) and the Polish Power Exchange (PPE) a correlation between high wind an solar power feed-in in Germany, high measured load flows from Germany to Poland, and low spot market prices in Germany and Poland could be identified (Fig. 20, Fig. 21, Fig. 22, Fig. 23, and Fig. 24). Based on these facts, the second aspect of subordinate hypothesis (iii) can be validated under usual reserve, namely, that Germany holds a dominant market position as a result of its market design with respect to the promotion of RES-E over Poland. This is considered to be a justified result not only based on the NREAP indications, but more so on the evident occurrence of loop flows and spot market developments caused by German RES-E deployment in the Polish electricity market. It is very important to understand, however, that this evaluation simply states the dominant impact of the effects of RES-E promotion in Germany, and does not give any indications as to whether these impacts are to be judged positive or negative.

As became apparent in the analysis above (illustrated in Fig. 20, Fig. 21, Fig. 22, Fig. 23, and Fig. 24), the spot market prices for electricity fall if transmission and non-discriminatory criteria for RES are met in the underlying market design, if a period of high RES feed-in occurs (owing to, for example, windy days), and because the marginal costs of, for example, RES wind power are lower than those of conventional plants. The steep drop in electricity prices observed in Germany and (to a lesser extent) in Poland, reflected in the respective graphs (see above) indicating, in principle, an over-supply of electricity, causes, inter alia, the so-called merit-order effect. According to the merit-order, plants with the lowest costs are used first, with more costly plants being brought online if needed (EWEA, 2010: 135).

Another aspect, however, needs to be taken into account. Because of the distinct nature of power markets, including the need for a balanced grid at all times, unplanned physical flows need to be counteracted by re-dispatching measures ultimately leading to the operation of base-load generating units at varying load levels, also referred to as cycling (Erni, 2012: 13; Intertek and Aptech, 2012: ii). *'Every time a power plant is turned off and on, the boiler, steam lines, turbine, and auxiliary components go through unavoidably large thermal and pressure stresses, which cause damage'* (Intertek and Aptech, 2012: ii).

> Overall, three short-term effects of German RES-E deployment on the Polish electricity market have been identified. The impact of periods of high RES-E feed-in (in particular, wind and solar) in Germany on the Polish spot market price makes the production of electricity from conventional thermal power plants at times unprofitable. The production of RES-E in Germany challenges the secure operation of the Polish power system (balanced grid at all times). Re-dispatching measures as a response to German RES-E feed-in, and, in consequence, cycling of base-load generating conventional thermal power plants lead to additional costs and potential damage in the respective production units for the Polish electricity producer. These findings become all the more significant when taking into account that the bulk of electric power produced in Poland is derived from combined heat and power plants using traditional fuels (82% in 2012, of which hard coal is 54,5%, and lignite 25,2%; see Fig. 13) (Paiz.gov.pl, 2013: 5). Regarding medium and long-term implications, many conventional power plant operators in Poland plan advanced decommissioning of generation units (TGPE, 2013: 15). The relevant decisions are encouraged because the market-regulatory environment (merit-order curve and implications for operating hours, and revenues of conventional generation) makes the maintenance of unprofitable generation units economically impossible (TGPE, 2013: 3 and 6; PKEE, 2013: 2). The same is true for planned new commissioning and installation of conventional power plant capacities (TGPE, 2013: 6). Investment in new conventional generation units is stopped owing to continuing market conditions that do not guarantee return on investments made, and because futures and spot market prices do not guarantee sufficient entry prices for new generation units (TGPE, 2013: 3 and 13). It can be concluded that the Polish energy economy, representing conventional power plant facilities mainly based on relatively inflexible coal-fired units, is heavily exposed to the impact of increasing German RES-E deployment in both economic, and physical (e.g. technical), terms. Based on these facts, the second aspect of subordinate hypothesis (iii) can be validated under usual reserve – that is, Germany holds a dominant market position as a result to its market design with respect to the promotion of RES-E over Poland.

As has been mentioned before, it is important to understand that the analysis carried out so far simply states the dominant impact of the effects of RES-E promotion in Germany, and does not give any indications as to whether these impacts are to be judged positive or negative. Furthermore, these results do not give rise to an absolute path-dependence of factual future RES-(E) development in Poland, let alone perspective RES policy developments as a result to the respective Polish policy process. The underlying subordinate hypothesis (iii), however, determines the direction of the impact in question by stating that European inter-state connectedness, competition and economic pressure will foster the promotion of RES-E in Europe beyond pioneering Member States, if a pioneering State with sufficient power in the international system holds a dominant market position, reflected in its RES-E market design. Consequently, if the hypothesis were to be verified, one would expect the share of RES-E in Poland to increase in the future, and the RES-E policy to adapt to German and European implications. Whilst the first aspect remains a matter of future developments, and therefore cannot be evaluated at present, the second aspect has been the subject of previous chapters. Under these circumstances and based on the observed, validated and reasoned impact assessment of German RES-E deployment pursuant to this chapter, only the first aspect of subordinate hypothesis (iii) can be validated under usual reserve – that is, there is inter-state connectedness, competition and economic pressure with respect to RES-(E) promotion between the European Member States Germany and Poland, where Germany holds a dominant market position in its RES-(E) policy and market design that has both economic and physical impacts on the Polish power market. Indeed, taking into account physics, the impact of German RES-E feed-in as a result of German policies in accord with supranational implications does lead to a higher penetration of RES-E in the Polish market. However, this higher penetration is originally produced in Germany and consequently enters in the balance sheet of Germany.

Tab. 30: Evaluation Matrix for Chapter 3,4,5, and 6

	European Union	GER	PL	GER and PL
Subordinate Hypothesis (I)	☺	☺	☺	X
Subordinate Hypotehsis (II)	☺	☺	☺	X
Subordinate Hypothesis (III)	X	X	X	☺

Bearing in mind the scientific constraints of the underlying theoretical design of this project, some factual speculation with respect to the second aspect of subordinate hypothesis (iii) is needed. Whilst the presented results do not give rise to an absolute path-dependence of factual future RES-(E) development in Poland, the empirically verified results allow us to define a possible development path whose constraints are defined by the validated facts. De facto, Poland faces the decommissioning of conventional power plant facilities. Investments in new conventional generation units are stopped on account of unfavourable market conditions (see above). Indeed, as Puka and Szulecki (2014: 7) argue, Poland might become a prospective electricity importer. *'Germany's generation, despite the nuclear phase-out, has been increasing, whilst Poland faces a generation deficit, and may eventually become a market for German power'* (Puka and Szulecki, 2014: 7). Based on these indications, it does not seem unlikely that RES-E technology investment in Poland should be considered a potential business opportunity for investors. A more comprehensive argument with respect to the latter deduction will be presented in the following final analysis, taking into account insights from the analysis of European, German and Polish policy implications, as well as the economic and physical impact of German–Polish electricity market interconnectedness. The conclusions of this chapter, and the results derived in this thesis so far, are visualised in Tab. 30.

7. Conclusion

The introduction posed the question of whether one can state the successful achievement of a common European energy policy, based on the lessons of the implementation of Directive 2009/28/EC. Did the European Union? Whilst analysing the progress of the promotion of renewable energies in the context of the 2020 policy framework, the European Commission released a 2013 Green Paper to reflect upon a new framework for climate and energy policies up to 2030 (EC, 2013a: 3; EC 2013c: 3). The paper raises 22 overall questions to evaluate the existing 20-20-20 policy framework, in order to draw lessons for the future design of the Union's energy and climate change policy (EC, 2013a: 13). The questions identify – it has been argued – two major concerns:

How do Member States and the European Institutions interact on the matter of the promotion of renewable energy electricity?

and

How will the Member States be encouraged to implement measures and policies for the promotion of renewable energy electricity?

In light of this thesis, the overall questions stated above were translated to meet the guiding research question:

What are the constraints and dependencies of the European Integration of RES-(E) promotion?

Whilst the latter question was considered a first necessary step to narrow down the overall subject, the following theory hypothesis served as a starting point for the scientific analysis.:

The implementation of polices for the promotion of renewable energy electricity (RES-E) in the Member States of the European Union is (i) fundamentally instigated by supranational renewable energy measures and policies leading to a general but not sufficient vertical Europeanization; (ii) predominantly influenced by national preferences determined by geopolitical and energy-specific economic interests; and (iii) primarily exposed to incremental European inter-state competition leading to horizontal adaptational pressure.

For the empirical analysis of this study, the working hypothesis reads as follows:

If (i) the European Union maintains an overall adaptational pressure by deciding on RES-E measures and policies for its Member States; (ii) pioneering Member States with sufficient power in the international system with regard to RES-E promotion (such as Germany) succeed in implementing an energy policy with a high share in electricity from renewable energy sources; and (iii) neighbouring Member States (such as Poland) are exposed to a pioneering Member State's RES-E policy and/or RES-E market design – **then,** non-pioneering Member States' RES-E measures and policies will adapt to the vertical (EU level) and horizontal (inter-state level) pressures. (These multi-level policy dependencies will encourage the harmonization of a European promotion of renewable energies and describe a reasoned path to a common promotion of renewable energies in all 28 Member States of the European Union.)

Whilst detailed information about the applied framework, theory, and model can be found in the respective chapters, an overview of the analysis scheme is reiterated in Fig. 27. Based on the analyses conducted so far, guided by the subordinate hypotheses, it is now suitable to focus on the final evaluation of the underlying working hypothesis. Following the logic of the analysis scheme assigned to each chapter, focus shall first be placed on polity dependencies and constraints.

By analysing the European polity, it has been concluded that, in principal – and notwithstanding the specific nature of the promotion of renewable energies in the EU, pursuant to Art. 5(2) and Art. 4 TEU (OJoEU, 2010: 18) – Member States' policies cannot a priori be considered an exclusive result of the domestic policy process, except for national security policies (Art. 4. (2); OJoEU, 2010: 18). In order to facilitate the attainments of the primary goals of the Treaties (Art. 288 TFEU) – that is, to exercise the Union's competencies – the Union can adopt secondary measures identical to regulations, directives, decisions, recommendations and opinions (Tobler and Beglinger, 2012: 91; OJoEU, 2008: 171). In a case law judgement, the ECJ defines the interrelations of the Union and Member States' legal order by stating, that *'the relationship between provisions of the treaty and directly applicable measures of the institutions on the on hand and the national law of the Member States on the other is such that those provisions and measures not only by their entry into force render automatically inapplicable any conflicting provisions of current national law but [...] also preclude the valid adoption of new national legislative measures to the extent to which they would be incompatible with the community provisions'* (ECJ, 1978: Summary 3.; Franzius, 2010: 43). The judgement of the ECJ was confirmed by the representatives of the governments of the Member States in the 17th Declaration of the Final Act of the Treaty of Lisbon (OJoEU, 2007a: 256): *'It results from the case-law of the Court of Justice that primacy of EC law is a cornerstone principle of Community law. [...] The fact that the principle of primacy will not be included in the future treaty shall not in any way change the existence of the principle and the existing case-law of the Court of Justice'* (OJoEU, 2007: 256). The interplay of these aspects of the Treaty on European Union, the case law of the ECJ, and the 17th Declaration of the Final Act on the Treaty of Lisbon define clear and explicit principle dependencies arising from the European polity with respect to the interaction of the supranational EU-level and the Member States.

Fig. 27: Analysis Scheme

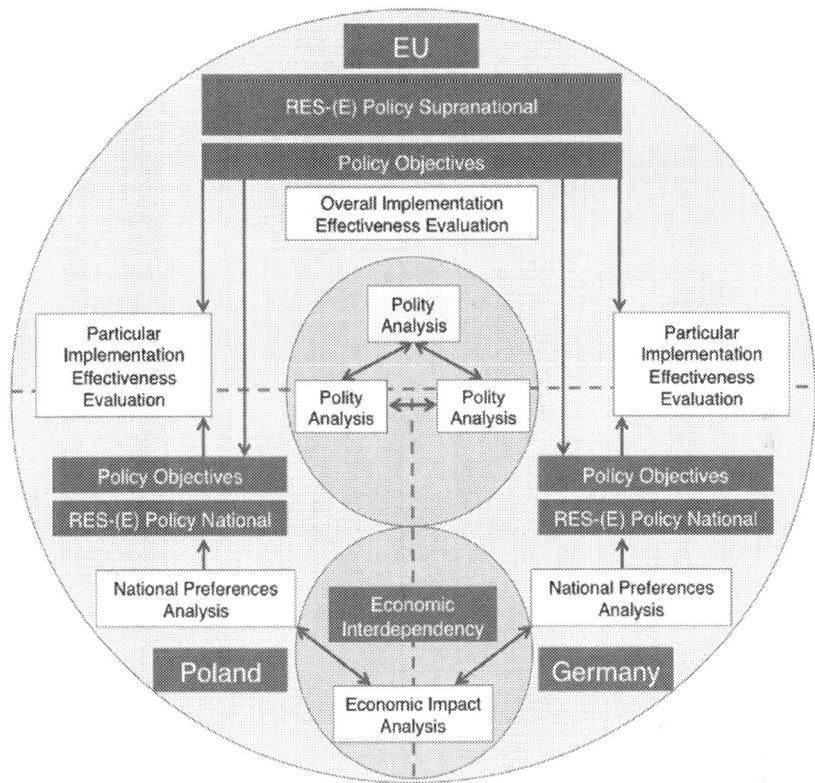

In light of the executive rights of the European bodies pursuant to the pre–Lisbon Treaty EU, the legislative provisions of the Climate Change Package – and Directive 2009/28/EC within it – reflect the non-exclusive rights of the Member States in the given policy context. With respect to Article 175 TEC (now 192(2) TFEU), it is important to emphasise that the implementation of Directive 2009/28/EC is not considered to fall within *'measures significantly affecting a Member State's choice between different energy sources and the general structure of its energy supply'* (OJoEU, 2008: 133; German Advisory Council on the Environment, 2011: 175). Directive 2009/28/EC was adopted through the ordinary legislative procedure and therefore under non-exclusive participation of the Member States, in order to excite change pursuant to the provisions content. However, <u>under Lisbon Treaty – and in particular Article 194 TFEU, which has the potential to withdraw all executive authorities from the</u>

European bodies that affect a *'Member State's right to determine the conditions for exploiting its energy resources, its choice between different energy sources and the general structure of its energy supply'* – a serious European polity constraint could be identified (OJoEU, 2008: 134-135). Whilst it is hard to imagine that Member States, with the introduction of a special energy chapter in the Lisbon Treaty, established a competence limit that would ex post take away the legal basis for pre-existing measures (e.g. Directive 2009/28/EC), the clause in question could serve as an absolute limit of EU competence because it remains questionable as to whether there is any matter in the field of energy that can be regulated without effecting Member States' rights pursuant to Article 194(2) (Czeberkus, 2013: 26; Fouquet, Nysten, Held and Johnston, 2012: 16-18; Ballesteros, 2010: 12; OJoEU: 133 and 135).

Concerning the German polity in light of the transfer of the exercise of sovereignty based on the 1949 Constitution of the Federal Republic of Germany, two provisions are relevant for European Integration (Tatham, 2013: 77; Pernice, 1998: 3-4): (a) the Preamble affirms the will of the people to *'promote world peace as an equal partner in a united Europe'* (BMJ, 2012: 1); and (b) correspondingly Art. 24(1) of the Constitution states that the *'Federation may by law transfer sovereign powers to international organisations'* (BMJ, 2012: 9). Both aspects define in principle the dependencies arising out of the German polity with respect to the interaction of the supranational EU-level and the Member State Germany. However, based on its case law, the FCC most definitely denies supremacy, direct effect, and pre-emption of European law per se. It reserves itself the right to review national legal rules that transpose European provisions with regard to whether they occurred on the basis of manifest transgressions of EU competence pursuant to the Lisbon treaty. The FCC emphasizes that *'the Federal Republic of Germany remains a member of a compound of States, the authority of which is derived from the Member States and has binding effect in German sovereign territory only by virtue of the German command to apply the law [Rechtsanwendungsbefehl]'* (BVerfGE 89 (155) quoted in University of Nevada, 1994: 21; BVerfGE, 1993: 89 (155) at 112). Not only, the FCC argues, can Germany terminate its membership, but the FFC can also examine whether instruments of the European Union exceed the limits of the sovereign rights accorded to them (Tatham, 2013: 83; Franzius, 2010: 25 and 34 and 43; Pernice, 2006: 30-31; Grimm, 2001: 288; BVerfGE 89 (155) quoted in University of Nevada, 1994: 2; BVerfGE, 1993: 89 (155) at 112). The case law judgments of the FCC highlight in principle the implicit German polity constraints against EU level aspirations reflected in the relevant aspects of the Treaty on European Union, the case law of the ECJ, and the 17th Declaration of the Final Act on the Treaty of Lisbon.

Unlike the German constitution, the Polish constitution (here Art. 90(1)) only makes mention of the delegation of *'the competence of organs of State authority in relation to certain matters'* (Sejm, 2014). Ultimately, the Polish constitution does not permit the transfer of sovereignty, but the exercise of specific competencies, as opposed to the power itself (Tatham, 2013: 223). In addition, with respect to the *'possibility of a collision between regulations of Community law and the Constitution'* the CT concludes, that *'[i]n such an event the Nation – as the sovereign, or a State authority organ authorised by the Constitution to represent the Nation, would need to decide on: amending the Constitution; or causing modifications within the Community provisions; or, ultimately, on Poland's withdrawal from the European Union'* (CT, 2005a: 9). With respect to the Polish polity, it is the rather vague nature of the Polish constitution regarding the interaction of the supranational and national legal order, and the case law judgments of the CT, that highlight principle polity constraints against EU level aspirations reflected in relevant aspects of the Treaty of European Union, the case law of the ECJ, and the 17th Declaration of the Final Act on the Treaty of Lisbon. However, the CT accepts dependencies arising out of European Union membership as long as it reflects the will of the constitution-maker who, in consequence – following a CT judgment – is responsible for solving emerging polity conflicts. The CT highlights that *'[t]he constitution-maker has decided that the system of law which is binding in the territory of the Republic of Poland will have a multi-faceted character, and will encompass, apart from legal acts constituted by Polish legislative bodies, acts of international and Community law. [...] It is the task of the Polish constitution-maker and legislator to resolve the problem of democratic legitimacy of the measures provided for in the Treaty, applied by the competent bodies of the Union'* (CT, 2010: 30-31 and 36).

There are no explicit polity constraints and dependencies enshrined in either the German or the Polish constitutions with respect to energy matters that would be comparable to the principle constraints defined in Article 194 TFEU (BMJ, 2012: 1-58; Sejm, 2014; OJoEU, 2008: 134-135).

Empirically observed dependencies and constraints in the context of the overall and country specific implementation evaluation of Directive 2009/28/EC will now be presented.

All Member States fulfilled their obligation pursuant to Article 4 of Directive 2009/28/EC to adopt a National Renewable Energy Action Plan (NREAP) in order to communicate Member States' national targets for transport, electricity, heating and cooling (OJoEU, 2009b: 28; NREAP_EU27, 2010). Generally, with the entry into force of Directive 2009/28/EC, the overall RES share in gross final consumption of energy, and the RES-E share, evidently rose in the bulk of the EU Member States (see Tab. 4, and Tab. 5). However, overall the change excited in the national arenas was and is not wholly in accordance with the change intended in Directive 2009/28/EC. Whilst Germany is among the sixteen Member States that achieved a growth rate in the RES-E sector in between 2009 and 2010, which was higher than the average growth rate needed to move from the NREAP 2010 RES-E target to the 2020 RES-E target, Poland would need *'higher than average growth rates'* to move from its 2010 to its 2020 RES-E target (Ecofys, 2012: 20 and 23). (Ecofys, 2012: 23). These empirical indications are also reflected in the transposition status of the respective Directive. With the rather timely adoption of a 'Law for the Transposition of Directive 2009/28/ EC on the Promotion of the Use of Energy from Renewable Sources' (BGBL, 2011a: 635) and the release of a NREAP (NREAP Germany, 2010: 1-182) in accordance with the template for National Renewable Energy Action Plans of the European Commission (EC, 2009: 1-40), it seems that Germany has solidly transposed Directive 2009/28/EC. With respect to Poland, it seems that the country is very reluctant to transpose supranational policy implications of the Directive. In any event, and until there is a judgment by the ECJ, the Republic of Poland has failed to notify the European Commission of provisions transposing Directive 2009/28/EC in time (OJoEU, 2009b: 44). Indeed, whilst the (non-)transposition of supranational implications does necessitate empirical constraints or dependencies in and of itself, it strongly indicates the existence of such. The analysis of national preferences shone a light on the roots of the matter and will be presented below.

Starting with the first successful initiative for a Feed-In Act for electricity from renewable energy sources (EEG-Clearingstelle, 2014a: 1) whose informal origins were based in an inter-fractional cooperation between Conservative and Green Members of Parliament (Hirschl, 2008: 131), all political parties but Die Linke (left-wing party) have been involved in fostering incrementally rising trajectories for RES-E in Germany between 1990 and 2011. Overall German RES-E policies leading to incrementally rising long-term trajectories for RES-E reflect the successful translation of issue-specific policy intentions of the dominant domestic group (Leuffen, Rittberger and Schimmelfennig, 2013: 42; Moravcsik, 1998: 24-35). Furthermore, Germany influenced the supranational agenda in accord with its national preferences, *inter alia,* during its Presidency of the Council of the European Union in the first half of 2007 (EU2007.de, 2007a: 1). By taking into account both the indicated EU-level developments in light of RES-(E), and the legislative history of German renewable energy policies (see Tab. 17), it seems that, whilst both levels are inextricably bound up to each other, the analysis of the German legislative RES-E history justifies the conclusion that the domestic national preferences of Germany reflect an independent intrinsic approach to RES-E promotion, whose principle outcome is in accord with EU implications not as a result of, but as an influential originator of, these implications.

The analysis of Polish national energy policy preferences gives rise to the conclusion that the issue-specific policy intentions of the dominant domestic group do not include the promotion of renewable energies as priority in the overlying energy policy (see also Tab. 27). Ever since the introduction of the Act on an Atomic Law, nuclear energy as an option for *'new energy sources'* has been consequently promoted in the domestic policy arena, independent of the acting governments' desire to shape the country's overall energy policy in accordance with their own agendas. The 2003 climate policy strategy, the 2005 energy policy strategy, the 2003 creation of a government representative for Alternative Sources of Energy, the 2009 energy policy, the 2011 Nuclear Power Programme, and – ultimately – the 2011 Amendment Act to the Act on Atomic Law all indicate an overall issue-specific consent for Polish national energy policy preferences for the introduction of nuclear energy (Ministry of Environment, 2003; Ministry of Economy, 2005; Ministry of Economy, 2009a; Ministry of Economy, 2011; Ancygier, 2013: 247; National Atomic Energy Agency, 2014).

It can be claimed that the domestic preferences with regard to RES-E in both Germany and Poland reflect an, if not exclusive, dominant independent intrinsic approach, which constitutes a constraint in principle. In the Polish case it reflects an empirical constraint.

With respect to German–Polish interdependencies, three short-term effects of German RES-E deployment on the Polish electricity market were identified. The impact of periods of high RES-E feed-in (in particular wind and solar) in Germany on the Polish spot market price makes the production of electricity for conventional thermal power plants at times unprofitable. The production of RES-E in Germany challenges the secure operation of the Polish power system (balanced grid at all times). Re-dispatching measures as a response to German RES-E feed-in, and, in consequence, cycling of base-load generating conventional thermal power plants lead to additional costs and potential damage in the respective production units for the Polish electricity producer.. With respect to medium and long-term implications, many conventional power plant operators in Poland plan advanced decommissioning of generation units (TGPE, 2013: 15). Investments in new conventional generation units are halted owing to uncertain market conditions that do not guarantee return on investments. It can be concluded that the Polish energy economy, representing conventional power plant facilities mainly based on relatively inflexible coal-fired units, is heavily exposed to the impact of an increasing German RES-E deployment in both economic, and physical (e.g. technical) terms. This German–Polish interdependency constitutes a constraint on the national policy preferences of Poland.

The introduction to this thesis stressed that new input to the canon of studies with regard to the subject matter is expected to arise theoretically and empirically. Based on the results presented above, a final evaluation shall now be conducted. The overall working hypothesis has been split into three subordinate hypotheses that incorporate approaches from Liberal Intergovernmentalism and Europeanization. Reflecting a top-down Europeanization framework, the following simple theory hypothesis was expressed in the theory section:

> The EU causes a change of national policies in accordance with EU level implications if there is some degree of misfit between EU and domestic policies, for EU law constitutes the law of the land (Ladrech, 2010: 35; Börzel and Risse, 2000: 5).

In a 2000 research paper, Börzel and Risse write: *'Ultimately, adapational pressures are generated by the fact that the emerging European polity encompasses structures of authoritative decision-making which might clash with national structures of policy-making and that the EU member states have no exit option given that EU law constitutes the law of the land'* (Börzel and Risse, 2000: 5). Whilst it is true that the emerging European polity *'encompasses structures of authoritative decision-making'*, as was evidently verified and becomes apparent, inter alia, in the relevant aspects of the Treaty on European Union, the case law of the ECJ, and the 17th Declaration of the Final Act on the Treaty of Lisbon, it is certainly wrong to state that *'the EU member states have no exit option'.* On the contrary, both the German and the Polish polity cases, and the respective case law judgments of the Constitutional Courts, most definitely constitute an *'exit option'* vis-à-vis the demand for supremacy, direct effect, and pre-emption of European law per se.

Nevertheless, a converse argument applies as well. Reflecting a Liberal-Intergovernmentalist approach, the following simple hypothesis was expressed in the theory section:

> The policy of a nation state is the exclusive result of the domestic policy process that reflects the issue-specific interests of the dominant domestic group (national preferences) and is determined by geopolitical and economic interests (Leuffen, Rittberger and Schimmelfennig, 2013: 42; Moravcsik, 1998: 24-35).

The conferral of powers from the European Member States to the European Institutions was a conscious act by the representatives of the governments of the Member States, ultimately confirmed by common accord during the adoption of the Lisbon Treaty in 2007 (OJoEU, 2007a: 232-248), for which *'the states have limited their sovereign rights, albeit within limited fields [...]'* (ECJ, 1963: II B, Franzius, 2010: 39). In order to attain the primary goals of the Treaties (Art. 288 TFEU) – that is, to exercise the Union's competencies – the Union can adopt secondary measures identical to regulations, directives, decisions, recommendations and opinions (Tobler and Beglinger, 2012: 91; OJoEU, 2008: 171). Furthermore, as established in the 17th Declaration of the Final Act of the Treaty of Lisbon, *'in accordance with well settled case law of the Court of Justice of the European Union, the Treaties and the law adopted by the Union on the basis of the Treaties have primacy over the law of Member States, under the conditions laid down by the said case law'* (OJoEU, 2007a: 256). Under these circumstances, the theoretical assumption that the policy of a Member State of the European Union is an exclusive result of the domestic sphere must be falsified.

If both approaches – reflecting counterfactual assumptions – must be falsified, a more balanced approach is needed. Indeed, an understanding of this fundamental polity conflict is key to conduct any reflections including the sui generis nature of the European Union. It is also the, admittedly paradoxical, fundament of the subject matter of this thesis. In the process of this analysis, the normative polity constraints identified could be reflected and further observed in contrast to the empirical example of the implementation effectiveness of Directive 2009/28/EC. It is not controversial that, with the entry into force of Directive 2009/28/EC, the overall RES share in gross final consumption of energy, and the RES-E share, rose in the bulk of the EU Member States (see Tab. 4, and Tab. 5). But overall the change excited in the national arenas was and is not wholly in accordance with the change intended in Directive 2009/28/EC. With respect to the case studies of Germany and Poland, it could be claimed that the domestic preferences with regard to RES-E in both countries reflect an, if not exclusive, clearly independent intrinsic logic. The combination of these results leads to the conclusion that the implementation of polices for the promotion of renewable energy electricity (RES-E) in the Member States of the European Union is fundamentally instigated by supranational renewable energy measures and policies, but predominantly influenced by national preferences. The theory design of this thesis involved a triadic logic, since the smallest possible reduction to represent European Integration can be achieved by taking into account the vertical and horizontal realities of the European Union, incorporating at least two neighbouring Member States. Horizontal realties were reflected in the subordinate hypothesis (iii). The following simple hypothesis reflecting both Liberal Intergovernmentalism and Europeanization in terms of horizontal integration was stated in the theory section:

> The relative power of states in the European Union, reflected in asymmetrical interdependence (Moravcsik, 1998:3), and national preferences triggered by the market and the dominant domestic group (Radaelli, 203: 17), will lead to the establishment of a dominant market position, reflected in a particular market design (as a result of policy implications).

In the applied case of a German → Polish interdependence, an asymmetrical interdependence and a dominant market position could be identified that had both economic and physical impacts on the Polish power market. However, the results presented do not give rise to an absolute path-dependence of factual future Polish RES-(E) development shaped by a dominant German market position. Nonetheless, speculations about a possible development path whose constraints are defined by the validated facts could lead to the results anticipated in subordinate hypothesis (iii).

The evaluation matrix presented in each chapter-specific analysis shows, that the proposed subordinate hypotheses could neither be verified nor falsified in its entirety (see Tab. 31). Consequently, all explanatory factors apply to a certain extinct. Since, following suggestions by Ladrech (2010: 40-42), the explanatory factors proposed consider alternative reasons or factors for empirical evidence in order to allow for counterfactual reasoning – a practice heavily emphasised in each chapter – the empirical analysis provides insights to help define the extent to which the different explanatory factors apply.

Tab. 31: Evaluation Matrix

	European Union	GER	PL	GER and PL
Subordinate Hypothesis (I)	☺	☺	☺	X
Subordinate Hypotehsis (II)	☺	☺	☺	X
Subordinate Hypothesis (III)	X	X	X	☺

Based on the analysis insights compiled above, three things become apparent, and shall be reiterated once again:

1. The change intended by supranational RES-E policies as enshrined in Directive 2009/28/EC did excite change in the EU Member States. But the change excited is not wholly in accordance with the change intended.

2. In the cases of Germany and Poland, the domestic preferences with regard to RES-E promotion in light of Directive 2009/28/EC reflect an, if not exclusive, clearly independent intrinsic logic.

3. In the applied case of a German → Polish interconnectedness, an asymmetrical interdependence and a dominant market position could be identified that have both economic and physical impacts on the Polish power market.

These deductions match the conditions implied in the overall theory-hypothesis. Thus, under the very narrow conditions of the qualitative empirical examples (Germany and Poland), with particular relevance to the subject matter of RES-E promotion, it can be claimed, under usual reserve, that:

The implementation of polices for the promotion of renewable energy electricity (RES-E) in the Member States of the European Union is fundamentally instigated by supranational renewable energy measures and policies, leading to a general, but not sufficient, vertical Europeanization, predominantly influenced by national preferences determined by geopolitical and energy-specific economic interests, and dominantly exposed to incremental European inter-state competition leading to horizontal adaptational pressure.

Based on this theoretical hypothesis, the working hypothesis relevant to the empirical examples of this thesis can be upheld:

Based on the presented and evaluated indicators, it is considered highly likely that, if the European Union maintains an overall adaptational pressure on its RES-E measures and policies for its Member States, and the Member State Germany with sufficient power in the inter-national system with regard to RES-E promotion succeeds in implementing an energy policy leading to a high share in electricity from renewable energy sources, and neighbouring Member State Poland is exposed to Germany's RES-E policy and RES-E market design effects then, Poland's RES-E measures and policies will adapt to the vertical (EU-level) and horizontal (inter-state level) pressures.

It is considered likely, and therefore proposed as subject for further analysis, that the identified multi-level policy constraints and dependencies will encourage the harmonisation of a European promotion of renewable energies, and describe a reasoned path to a common promotion of renewable energies in all 28 Member States of the European Union.

The first guiding question stressed in the introduction was whether we could state the successful achievement of a common European energy policy based on the lessons of the implementation of Directive 2009/28/EC.

In light of the results presented in this chapter, the answer must be no. This, however, applies only in so far as one considers a *common* European renewable energy policy, to be realized in comparison with nation state realities. This logic, by the very nature of European Integration, can be precluded a priori. As long as the European Union remains a compound of sovereign states, whose democratic integrity secures each Member State the right of legitimised executive authority, one cannot expect any adopted European policy to be implemented unconditionally. Such policies can be expected, however, to trigger change. It is remarkable to see that, with the entry into force of Directive 2009/28/EC the overall RES share in gross final consumption of energy, and the RES-E share, evidently rose in the bulk of the European Member States. The European policy therefore caused a simultaneous and measurable change in at least 23 European countries from 2009 to 2011. Whilst it is highly unlikely to achieve a common European RES-E policy in accordance with supranational intentions, a common European RES-E promotion will in reality be achieved by the horizontal and vertical mechanisms described here .

8. Bibliography

ACER, Agency for the Cooperation of Energy Regulators (2013), *What is Acer about in all EU official languages,* [Online], Available: http://www.acer.europa.eu/Official_documents/Publications/Pages/Publication.aspx 24 Dec. 13].

Adam, S., Kriesi, H. (2007), The Network Approach, in: Sabatier, P.A. (ed.), Theories of the Policy Process, 2nd Edition, Boulder: Westview Press, 129-154.

Addendum NREAP Poland, Addendum National Renewable Energy Action Plan of Poland (2010), *Addendum to the National Renewable Energy Action Plan of 2 December 2011,* [Online], Available: http://ec.europa.eu/energy/renewables/action_plan_en.htm [19 May 14].

Advoc (2013), *Poland: Energy Update – July 2013,* [Online], Available: http://www.advoc.com/view-news/Poland'3A+Energy+Update+-+July+2013/[19 May 14].

Agora (2014), *Negative Strompreise: Ursache und Wirkungen. Eine Analyse der aktuellen Entwicklungen – und ein Vorschlag für ein Flexibilitätsgesetz,* [Online, Available: http://www.agora-energiewende.de/fileadmin/downloads/publikationen/Studien/Negative_Strompreise/Agora_NegativeStrompreise_Web.pdf [01 Jul. 14].

Ancygier, A. (2013), *Misfit of Interests instead of the "Goodness of Fit"? Implementation of European Directives 2001/77/EC and 2009/28/EC in Poland,* Hamburg: Verlag Dr. Kovac.

Ancygier, A., Szulecki, K. (2013a), Without strong German leadership, the adoption of an ambitious renewable energy policy across Europe will be impossible, [Online], Available: http://blogs.lse.ac.uk/europpblog/2013/08/21/without-strong-german-leadership-the-adoption-of-an-ambitious-renewable-energy-policy-across-europe-will-be-impossible/ [18 Oct. 13].

Ancygier, A., Szulecki, K. (2013b), *German-Polish Cooperation In Renewables: Towards Policy Convergence? Dahrendorf Symposium Paper Series Summary,* [Online,Available:http://papers.ssrn.com/sol3/papers.cfm?abstractid=2377981 [01 Jul. 14].

Banaszkiewicz, B. (2005), *Die Rechtsprechung des polnischen Verfassungsgerichtshofes angesichts der Mitgliedschaft des Landes in der Europäischen Union: Erste Erfahrungen und neue Probleme,* in: EIF Working Paper Series, Working Paper Nr: 15, 4-18, [Online], Available: http://eif.univie.ac.at/downloads/projekte/wp15.pdf [19 May 14].

Barkin, J. S. (2010), *Realist Constructivism. Rethinking International Relations Theory,* Cambridge: Cambridge University Press.

BEMIP, Baltic Energy Market Interconnection Plan (2012), 4th progress report, [Online],Available:http://ec.europa.eu/energy/infrastructure/doc/20121016_4rd_ bemip_progress_report_final.pdf [18 Oct. 13].

BGBL, Bundesgesetzblatt (1990), *Gesetz über die Einspeisung von Strom aus erneuerbaren Energien (Stromeinspeisungsgesetz) vom 7. Dezember 1990,* [Online], Available: http://www.bgbl.de/Xaver/text.xav?startbk=Bundesanzeiger_ BGBl&jumpTo=bgbl190s2633b.pdf [10 Mar. 14].

BGBL, Bundesgesetzblatt (2000), *Gesetz für den Vorrang Erneuerbarer Energien (Erneuerbare-Energien-Gesetz – EEG) sowie zur Änderung des Energiewirt-schaftsgesetzes und des Mineralölsteuergesetzes vom 29. März 2000,* [Online, Available:http://www.bgbl.de/Xaver/text.xav?startbk=Bundesanzeiger_ BGBl&jump To=bgbl100s0305.pdf [10 Mar. 14].

BGBL, Bundesgesetzblatt (2002), *Gesetz zur geordneten Beendigung der Kernenergienutzung zur gewerblichen Erzeugung von Elektrizität vom 22. April 2002,* [Online], Available: http://www.bgbl.de/Xaver/start.xav?startbk=Bundes-anzeiger_BGBl&jumpTo=bgbl102s1351.pdf [10 Mar. 14].

BGBL, Bundesgesetzblatt (2004), *Gesetz zur Neuregelung des Rechts der Erneuerbaren Energien im Strombereich vom 21. Juli 2004,* [Online], Available:http://www.bgbl.de/Xaver/start.xav?startbk=Bundesanzeiger_BGBl& jumpTo=bgbl104s1918.pdf [10 Mar. 14].

BGBL, Bundesgesetzblatt (2008), *Gesetz zur Neuregelung des Rechts der Erneuerbaren Energien im Strombereich und zur Änderung damit zu-sammenhängender Vorschriften vom 25. Oktober 2008,* [Online],Available:http:// www.bgbl.de/Xaver/start.xav?startbk=BundesanzeigerBGBl&jumpTo=bgbl10 8s2074.pdf [10 Mar. 14].

BGBL,Bundesgesetzblatt(2010), *Elftes Gesetz zur Änderung des Atomgesetzes vom 8. Dezember 2010,* Online], Available: http://www.bgbl.de/Xaver/start. xav?startbk=Bundesanzeiger_BGBl&jumpTo=bgbl110s1814.pdf [10 Mar. 14].

BGBL, Bundesgesetzblatt (2011a), *Gesetz zur Umsetzung der Richtlinie 2009/28/EG zur Förderung der Nutzung von Energie aus erneuerbaren Quellen (Europarechtsanpassungsgesetz Erneuerbare Energien EAGEE) vom 12 April 2011,* [Online], Available: http://www.bgbl.de/Xaver/start.xav?startbk=Bundes-anzeiger_BGBl&jumpTo=bgbl111s0619.pdf [10 Mar. 14].

BGBL, Bundesgesetzblatt (2011b), *Dreizehntes Gesetz zur Änderung des Atom-gesetzes vom 31. Juli 2011,* [Online], Available: http://www.bgbl.de/Xaver/start. xav?startbk=Bundesanzeiger_BGBl&jumpTo=bgbl111s1704.pdf [10 Mar. 14].

BGBL, Bundesgesetzblatt (2011c), *Gesetz zur Neuregelung des Rechtsrahmens für die Förderung der Stromerzeugung aus erneuerbaren Energien,* [Online], Available: http://www.bgbl.de/xaver/bgbl/start.xav?startbk=Bundesanzeiger_ BGBl&jumpTo=bgbl111s1633.pdf [10 Mar. 14].

Biernat, S. (1998), *Constitutional Aspects of Poland's Future Membership in the European Union,* in: Archiv des Völkerrechts, Band 36, Heft 4, Tübingen: Mohr Siebeck, 398-424, [Online], Available: http://www.jstor.org/stable/40799098 19 May 14].

Biernat, S. (undated), *Impact of EU Accession on the National Legal Orders of New Member States: Poland, [Online], Available: http://www.pspe.org.pl/ dokumenty/110_asser1.pdf [19 May 14].*

BMJ, Bundesministerium der Justiz und für Verbraucherschutz (2012), Basic Law for the Federal Republic of Germany, [Online], Available: http://www. gesetze-im-internet.de/englisch_gg/basic_law_for_the_federal_republic_of_ germany.pdf [11 Mar. 14].

BMU, Bundesministerium für Umwelt, Naturschutz, Bau und Reaktorsicherheit (2001), *Vereinbarung zwischen der Bundesregierung und den Energieversorgungsunternehmen vom 14. Juni 2000,* [Online], Available: http://www. bmub.bund.de/fileadmin/bmu-import/files/pdfs/allgemein/application/pdf/ atomkonsens.pdf [11 Mar. 14].

BMU, Bundesministerium für Umwelt, Naturschutz, Bau und Reaktorsicherheit (2002a), *Pressemitteilung: Neues Atomgesetz tritt in Kraft,* [Online], Available: www.bmub.bund.de/N1472/ [11 Mar. 14].

BMU, Bundesministerium für Umwelt, Naturschutz, Bau und Reaktorsicherheit (2002b), *Bericht über den Stand der Markteinführung und der Kostenentwicklung von Anlagen zur Erzeugung von Strom aus erneuerbaren Energien (Erfahrungsbericht zum EEG),* [Online], Available: http://www.erneuerbare-energien.de/fileadmin/ee-import/files/pdfs/allgemein/application/pdf/eeg_ erfahrungsbericht.pdf [11 Mar. 14].

BMU, Bundesministerium für Umwelt, Naturschutz, Bau und Reaktorsicherheit (2007), *Key Elements of an Integrated Energy and Climate Programme. Decision of German Cabinet on August 23rd/24th 2007 at Meseberg,* [Online], Available: http://www.bmub.bund.de/fileadmin/bmu-import/files/english/pdf/application/ pdf/klimapaket_aug2007_en.pdf [11 Mar. 14].

BMU, Bundesministerium für Umwelt, Naturschutz, Bau und Reaktorsicherheit (2009), *Vorausschätzung der Bundesrepublik Deutschland zur Nutzung der flexiblen Kooperationsmechanismen zur Zielerreichung gemäß Art. 4 Abs. 3* der Richtlinie 2009/28/EG (veröffentlicht am 21. Dezember 2009), [Online], Available: http://www.erneuerbare-energien.de/fileadmin/ee-import/files/pdfs/ allgemein/application/pdf/vorausschaetzung_kooperationsmechanismen_ bf.pdf 11 Mar. 14].

BMU, Bundesministerium für Umwelt, Naturschutz, Bau und Reaktorsicherheit (2010a), Bundesrepublik Deutschland. Nationaler Aktionsplan für erneuerbare Energien gemäß der Richtlinie 2009/28/EG zur Förderung der Nutzung von Energie aus erneuerbaren Quellen, [Online], Available: http://www.erneuer- bare-energien.de/fileadmin/ee-import/files/pdfs/allgemein/application/pdf/ nationaler_aktionsplan_ee.pdf [11 Mar. 14].

BMU, Bundesministerium für Umwelt, Naturschutz, Bau und Reaktorsicherheit (2010b), Vorblatt. Entwurf eines Gesetzes zur Umsetzung der Richtlinie 2009/ 28/EG zur Förderung der Nutzung der Energie aus erneuerbaren Quellen (Europa- rechtsanpassungsgesetz Erneuerbare Energien – EAG EE), [Online], Available: http://www.erneuerbare-energien.de/fileadmin/ee-import/files/pdfs/allgemein/ application/pdf/entwurf_eag_ee_bf.pdf [11 Mar. 14].

BMU, Bundesministerium für Umwelt, Naturschutz, Bau und Reaktorsicherheit (2010c), The Federal Government's energy concept of 2010 and the transfor- mation of the energy system of 2011, [Online], Available: http://www.bmub.bund. de/fileadmin/bmu-import/files/english/pdf/application/pdf/energiekonzept_ bundesregierung_en.pdf [11 Mar. 14].

BMU, Bundesministerium für Umwelt, Naturschutz, Bau und Reaktorsicherheit (2011), Dreizehntes Gesetz zur Änderung des Atomgesetzes, [Online], Avail- able: www.bmub.bund.de/N47463/ [11 Mar. 14].

BMWi, Bundesministerium für Wirtschaft und Energie (2012), Germany's new energy policy. Heading towards 2050 with secure, affordable and environmentally sound energy, [Online], Available: http://www.bmwi.de/English/Redaktion/Pdf/ germanys-new-energy-policy,property=pdf,bereich=bmwi2012,sprache=en,r wb=true.pdf [18 Oct. 13].

BMWi, Bundesministerium für Wirtschaft und Energie (2014a) [based on AGEB, Arbeitsgemeinschaft Energiebilanzen, 2013], Primärenergieverbrauch nach Energieträgern, [Online], Available: http://www.bmwi.de/DE/ThemenEnergie/ Energiedaten-und-analysen/Energiedaten/energiegewinnung-energiever- brauch.html [11 Mar. 14].

BMWi, Bundesministerium für Wirtschaft und Energie (2014b) [based on AGEB, Arbeitsgemeinschaft Energiebilanzen, 2013], Heimische Energiegewinnung und Importabhängigkeit, [Online], Available: http://www.bmwi.de/DE/Themen/ Energie/Energiedaten-und-analysen/Energiedaten/energiegewinnung-energieverbrauch.html [11 Mar. 14].

BMWi, Bundesministerium für Wirtschaft und Energie (2014c) [based on AGEB, Arbeitsgemeinschaft Energiebilanzen, 2013 and BDEW, Bundesverband der Energie und Wasserwirtschaft, undated], *Erneuerbare Energien auf einen Blick,* [Online], Available: http://www.bmwi.de/DE/Themen/Energie/Erneuerbare-Energien/erneuerbare-energien-auf-einen-blick.html 11 Mar. 14].

BPB, Bundeszentrale für politische Bildung (2013), Artikel 23 und 146 im Grundgesetz vor 1990, [Online], Available: http://www.bpb.de/geschichte/ deutsche-geschichte/grundgesetz-und-parlamentarischer-rat/39014/warum-keine-verfassung?type=galerie&show=image&i=38988 [11 Mar. 14].

Börzel, T. A. (2002), *Pace-Setting, Foot-Dragging, and Fence-Sitting:* Member State Responses to Europeanization, [Online], Available: http://web.grinnell. edu/courses/pol/f02/pol295-01/Borzel%20-%20Member%20State%20 Responses%20to%20Europeanization.pdf [04 Nov. 13].

Börzel, T and Risse, T (2000), *When Europe Hits Home: Europeanization and Domestic Change,* in: European Integration Online Papers (EIoP), vol. 4, no. 15, [Online], Available: http://eiop.or.at/eiop/pdf/2000-015.pdf [11 Mar. 14].

Börzel, T and Risse, T (2009), *Conceptualizing the Domestic Impact of Europe,* [Online], Available: http://www.gs.uni.wroc.pl/files/BOERZEL%20-%2RISSE%20 Conceptualizing%20the%20domestic%20impact%20of%20Europe.pdf [11 Mar. 14].

Brunner, S. (2008), *Understand Policy Change: Multiple Streams and Emission Trading in Germany,* Global Environmental Change, vol. 18, pp. 501-507.

Buchanan J. M. (1981), *Politics without Romance: A Sketch of Positive Public Choice Theory and its Normative Implications,* in: Buchanan, J. M. and Tollison, R. D. (ed.), The Theory of Public Choice II, Michigan: University of Michigan Press, 11-22.

Bundesnetzagentur (2014a), *Monitoringreport 2013 in accordance with § 63 Abs. 3 i.V. m. § 35 EnWG and § 48 Abs. 3 i.V. m § 53 Abs. 3 GWB As of January 2014,* [Online], Available: http://www.bundesnetzagentur.de/SharedDocs/Downloads/ EN/BNetzA/PressSection/ReportsPublications/2013/MonitoringReport2013. pdf?__blob=publicationFile&v=10 [01 Jul. 14].

Bundesnetzagentur (2014b), *Expansion of the electricity transmission networks,* [Online], Available: http://www.bundesnetzagentur.de/cln_1422/EN/Areas/Energy/Companies/ElectricityGridExpansion/ElectricityGridExpansion-node.html [01 Jul. 14].

Bundesregierung (2002), *Perspectives for Germany. Our Strategy for Sustainable Development,* [Online], Available: http://www.bundesregierung.de/Content/DE/_Anlagen/Nachhaltigkeit-wiederhergestellt/perspektives-for-germany-langfassung.pdf?__blob=publicationFile&v=2 [11 Mar. 14].

Bundesrat (2014), *Mediation Committee,* [Online, Available: http://www.bundesrat.de/static/Web/EN/national-en/va-en/va-en-node_nnn-true.html [11 Mar. 14].

Bundesregierung (2011a), Bulletin der Bundesregierung. Nr. 27-1 vom 17. März 2011. Regierungserklärung von Bundeskanzlerin Dr. Angela Merkel, [Online], Available: http://www.bundesregierung.de/Content/DE/Bulletin/2011/03/Anlagen/27-1a-bk.pdf?__blob=publicationFile [11 Mar. 14].

Bundesregierung (2011b), *Bulletin der Bundesregierung. Nr. 59-1 vom 9. Juni 2011. Regierungserklärung von Bundeskanzlerin Dr. Angela Merkel,* [Online], Available: https://www.bundesregierung.de/Content/DE/Bulletin/2011/06/Anlagen/59-1-bk.pdf;jsessionid=DADAF187D3055C663DE85904468EA322.s3t2?__blob=publicationFile&v=2 [11 Mar. 14].

BVerfGE, Bundesverfassungsgericht (1971), *BVerfGE 31, 145 - Milchpulver,* [Online], Available: http://www.servat.unibe.ch/dfr/bv031145.html [11 Mar. 14].

BVerfGE, Bundesverfassungsgericht (1974), *BVerfGE 37, 271 – Solange I,* [Online], Available: http://www.servat.unibe.ch/dfr/bv037271.html [11 Mar. 14].

BVerfGE, Bundesverfassungsgericht (1986), *BVerfGE 73, 339 – Solange II,* [Online], Available: http://www.servat.unibe.ch/dfr/bv073339.html [11 Mar. 14].

BVerfGE, Bundesverfassungsgericht (1993), *BVerfGE 89, 155 – Maastricht,* [Online], Available: http://www.servat.unibe.ch/dfr/bv089155.html [11 Mar. 14].

BVerfGE, Bundesverfassungsgericht (2009), *Headnotes to the judgment of the Second Senate of 30 June 2009,* [Online], Available: http://www.bundesverfassungsgericht.de/entscheidungen/es20090630_2bve000208en.html [11 Mar. 14].

CEPS+MAVIR+PSE+SEPS (2012), *Position of CEPS, MAVIR, PSE Operator and SEPS regarding the issue of Bidding Zones Definition. Response to the study commissioned by Bundesnetzagentur and authored by Frontier Economics and Consentec: "Relevance of established national bidding areas for European power market integration – an approach to welfare oriented evaluation"* , [Online], Available: http://www.pse.pl/uploads/pliki/17225Position_of_CEPS_ MAVIR_PSEO_SEPS-Bidding_Zones_Definition.pdf [01 Jul. 14].

CEPS+MAVIR+PSE+SEPS (2013), *Joint study by CEPS, MAVIR, PSE and SEPS regarding the issue of Unplanned flows in the CEE region In relation to the common market area Germany – Austria,* [Online], Available: http://www.pse. pl/uploads/pliki/Unplanned_flows_in_the_CEE_region.pdf [01 Jul. 14].

CIA, Central Intelligence Agency (2014a), *The World Factbook. Europe. Poland,* [Online], https://www.cia.gov/library/publications/the-world-factbook/geos/pl.html [01 Jul. 14].

CIA, Central Intelligence Agency (2014b), *The World Factbook. Europe. Germany,* [Online], https://www.cia.gov/library/publications/the-world-factbook/ geos/gm.html [01 Jul. 14].

Cohen, M. D., March J. G. and Olsen, J. P. (1972), *A Garbage Can Model of Organizational Choice,* Administrative Science Quarterly, vol. 17, 1-25.

Council of the European Union (2000), *Presidency Conclusions. Lisbon European Council 23 and 24 March 2000,* [Online], Available: http://www. consilium.europa.eu/uedocs/cms_data/docs/pressdata/en/ec/00100-r1.en0. htm [24 Dec. 13].

Council of the European Union (2007), *Presidency Conclusions. Brussels European Council 8 and 9 March 2007,* [Online], Available: http://register. consilium.europa.eu/pdf/en/07/st07/st07224-re01.en07.pdf
[24 Dec. 13].

Council of the European Union (2008), *Presidency Conclusions. Brussels European Council 11 and 12 December 2008,* [Online], Available: http://europa. eu/rapid/press-release_DOC-08-5_en.htm [24 Dec. 13].

CT, Constitutional Tribunal (2003), *Judgement of 27th May 2003, REFERENDUM ON POLAND'S ACCESSION TO THE EUROPEAN UNION,* K 11/03, [Online, Available: http://trybunal.gov.pl/fileadmin/content/omowienia/K_11_03_GB.pdf [19 May 14].

CT, Constitutional Tribunal (2005a), *Judgment of 11th May 2005, POLAND'S MEMBERSHIP IN THE EUROPEAN UNION (THE ACCESSION TREATY),* K 18/05, [Online], Available: http://trybunal.gov.pl/fileadmin/content/omowienia/K_18 _04_GB.pdf [19 May 14].

CT, Constitutional Tribunal (2005b), *JUDGMENT dated 27 April 2005*, P 11/05, [Online], Available: http://trybunal.gov.pl/fileadmin/content/omowienia/P_1_05_full_GB.pdf [19 May 14].

CT, Constitutional Tribunal (2010), *JUDGMENT of 24 November 2010*, K 32/09, [Online], Available: http://trybunal.gov.pl/fileadmin/content/omowienia/K_32_09_EN.pdf [19 May. 14].

CT, Constitutional Tribunal (2011), *JUDGMENT IN THE NAME OF THE REPUBLIC OF POLAND*, SK 45/09, [Online], Available: http://trybunal.gov.pl/fileadmin/content/omowienia/SK_45_09_EN.pdf [19 May 14].

Cwiek-Karpowicz J., Gawlikowska-Fyk A. and Westphal K. (2013), *German and Polish Energy Policies: Is Cooperation Possible?*, [Online], Available: http://www.pism.pl/files/?id_plik=12692 [18 Oct. 13].

Dabrowska, P. (2009), *The Polish Legal System Facing the Treaty of Lisbon – Remarks on Necessary Modifications and Possible Problems*, [Online], Available: http://www.ce.uw.edu.pl/pliki/pw/y12_dabrowska.pdf [19 May 14].

Dena, German Energy Agency (2013), *dena Grid Study II. Integration of Renewable Energy Sources in the German Power Supply System from 2015 – 2020 with an Outlook to 2025.*, [Online], Available: http://www.dena.de/fileadmin/user_upload/Projekte/Erneuerbare/Dokumente/dena_Grid_Study_II_-_final_report.pdf [01 Jul. 14].

DG Climate Action (2012), *Climate Change*, [Online], Available: http://ec.europa.eu/clima/publications/docs/factsheet_climate_change_2012_en.pdf [14 Jun. 13].

DG Energy (2006), *Study On The Technical Security Rules Of The European Electricity Network. Final Report 62236A/001 REV 2*, [Online], Available: http://ec.europa.eu/energy/gas_electricity/studies/doc/electricity/2006_02_security_rules_pb_power.pdf [01 Jul. 14].

DG Energy (2012a), List of projects submitted to be considered as potential Projects of Common Interest in energy infrastructure – Electricity, [Online], Available: http://ec.europa.eu/energy/infrastructure/consultations/doc/pci_list_electricity.pdf [18 Oct. 13].

DIP, Dokumentations- und Informationssystem des Deutschen Bundestages (2002), *Gesetzbeschluss des Deutschen Bundestages. Gesetz zur geordneten Beendigung der Kernenergienutzung zur gewerblichen Erzeugung von Elektrizität*, [Online], Available: http://dipbt.bundestag.de/doc/brd/2002/D7+02.pdf [10 Mar. 14].

DIP, Dokumentations- und Informationssystem des Deutschen Bundestages (2004a), *Gesetzentwurf der Bundesregierung. Entwirf eines Gesetzes zur Neuregelung des Rechts der Erneuerbaren-Energien im Strombereich,* [Online], Available: http://dipbt.bundestag.de/doc/brd/2004/0015-04.pdf [10 Mar. 14].

DIP, Dokumentations- und Informationssystem des Deutschen Bundestages (2004b), *Gesetzentwurf der Fraktionen SPD und BÜNDNIS 90/Die GRÜNEN. Entwirf eines Gesetzes zur Neuregelung des Rechts der Erneuerbaren-Energien im Strombereich,* [Online], Available: http://dipbt.bundestag.de/doc/btd/15/023/1502327.pdf [10 Mar. 14].

DIP, Dokumentations- und Informationssystem des Deutschen Bundestages (2004c), *Gesetzbeschluss des Deutschen Bundestages. Gesetz zur Neuregelung des Rechts der Erneuerbaren Energien,* [Online], Available: http://dipbt.bundestag.de/doc/brd/2004/0290-04.pdf [10 Mar. 14].

DIP, Dokumentations- und Informationssystem des Deutschen Bundestages (2004d), *Empfehlung der Ausschüsse zu Punkt ..... der 799. Sitzung des Bundesrates am 14. Mai 2004. Gesetz zur Neuregelung des Rechts der Erneuerbaren Energien im Strombereich,* [Online], Available: http://dipbt.bundestag.de/doc/brd/2004/0290-1-04.pdf [10 Mar. 14].

DIP, Dokumentations- und Informationssystem des Deutschen Bundestages (2004e), *Beschlussempfehlung und Bericht des Vermittlungsausschusses zu dem Gesetz zur Neuregelung des Rechts der Erneuerbaren Energien im Strombereich,* [Online], Available: http://dipbt.bundestag.de/doc/btd/15/033/1503385.pdf [10 Mar. 14].

DIP, Dokumentations- und Informationssystem des Deutschen Bundestages (2004f), *Beschluss des Deutschen Bundestages. Gesetz zur Neuregelung des Rechts der Erneuerbaren Energien im Strombereich,* [Online], Available: http://dipbt.bundestag.de/doc/brd/2004/0512-04.pdf [10 Mar. 14].

DIP, Dokumentations- und Informationssystem des Deutschen Bundestages (2004g), *Beschluss des Bundesrates. Gesetz zur Neuregelung des Rechts der Erneuerbaren Energien im Strombereich,* [Online], Available: http://dipbt.bundestag.de/doc/brd/2004/0512-04B.pdf [10 Mar. 14].

DIP, Dokumentations- und Informationssystem des Deutschen Bundestages (2008a), *Gesetzentwurf der Bundesregierung. Entwurf eines Gesetzes zur Neuregelung des Rechts der Erneuerbaren Energien im Strombereich und zur Äderung damit zusammenhängender Vorschriften,* [Online], Available: http://dipbt.bundestag.de/dip21/brd/2008/0010-08.pdf [10 Mar. 14].

DIP, Dokumentations- und Informationssystem des Deutschen Bundestages (2008b), *Gesetzbeschluss des Deutschen Bundestages. Gesetz zur Neuregelung des Rechts der Erneuerbaren Energien im Strombereich und zur Äderung damit zusammenhängender Vorschriften,* [Online], http://dipbt.bundestag.de/dip21/brd/2008/0418-08.pdf [10 Mar. 14].

DIP, Dokumentations- und Informationssystem des Deutschen Bundestages (2010a), *Deutscher Bundestag. Stenografischer Bericht. 71. Sitzung. Berlin, Donnerstag, den 11. November 2010,* [Online], http://dipbt.bundestag.de/dip21/btp/17/17071.pdf#P.7744 [10 Mar. 14].

DIP, Dokumentations- und Informationssystem des Deutschen Bundestages (2010b), *Stellungnahme des Bundesrates. Entwurf eines Gesetzes zur Umsetzung der Richtlinie 2009/28/EC zur Förderung der Nutzung von Energie aus erneuerbaren Quellen (Europarechtsanpassungsgesetz Erneuerbare Energien – EAG EE),* [Online], http://dipbt.bundestag.de/dip21/brd/2010/0647-10B.pdf [10 Mar. 14].

DIP, Dokumentations- und Informationssystem des Deutschen Bundestages (2010c), *Unterrichtung durch die Bundesregierung. Entwurf eines Gesetzes zur Umsetzung der Richtlinie 2009/28/EC zur Förderung der Nutzung von Energie aus erneuerbaren Quellen (Europarechtsanpassungsgesetz Erneu-erbare Energien – EAG EE),* [Online], http://dipbt.bundestag.de/dip21/btd/17/042/1704 233.pdf [10 Mar. 14].

DIP, Dokumentations- und Informationssystem des Deutschen Bundestages (2010d), *Gesetzentwurf der Fraktionen der CDU/CSU und der FDP,* [Online], http://dipbt.bundestag.de/dip21/btd/17/030/1703051.pdf [10 Mar. 14].

DIP, Dokumentations- und Informationssystem des Deutschen Bundestages (2010e), *Basisinforation über den Vorgang. 17. Wahlperiode. Vorgangstyp: Gesetzgebung. Elftes Gesetz zur Änderung des Atomgesetzes,* [Online], http://dipbt.bundestag.de/extrakt/ba/WP17/294/29435.html [10 Mar. 14].

DIP, Dokumentations- und Informationssystem des Deutschen Bundestages (2010f), *Deutscher Bundestag. Stenografischer Bericht. 68. Sitzung. Berlin, Donnerstag, den 28. Oktober 2010,* [Online], http://dipbt.bundestag.de/dip21/btp/17/17068.pdf#P.7159 [10 Mar. 14].

DIP, Dokumentations- und Informationssystem des Deutschen Bundestages (2010g), *Gesetzbeschluss des Deutschen Bundestages. Elftes Gesetz zur Änderung des Atomgesetzes,* [Online], http://dipbt.bundestag.de/dip21/brd/20 10/0683-10.pdf [10 Mar. 14].

DIP, Dokumentations- und Informationssystem des Deutschen Bundestages (2010h), *Beschluss des Bundesrates. Elftes Gesetz zur Änderung des Atomgesetzes,* [Online], http://dipbt.bundestag.de/dip21/brd/2010/0683-10B.pdf [10 Mar. 14].

DIP, Dokumentations- und Informationssystem des Deutschen Bundestages (2011a), *Deutscher Bundestag. Stenografischer Bericht. 93. Sitzung. Berlin, Donnerstag, den 24. Februar 2011,* [Online], http://dipbt.bundestag.de/dip21/btp/17/17093.pdf#P.10566 [10 Mar. 14].

DIP, Dokumentations- und Informationssystem des Deutschen Bundestages (2011b), *Beschlussempfehlung und Bericht des Ausschusses für Umwelt, Naturschutz und Reaktorsicherheit (16. Ausschuss)zu dem Gesetzentwurf der Bundesregierung – Drucksachen 17/3629, 17/4233 – Entwurf eines Gesetzes zur Umsetzung der Richtlinie 2009/28/EG zur Förderung der Nutzung von Energie aus erneuerbaren Quellen (Europarechtsanpassungsgesetz Erneuerbare Energien – EAG EE),* [Online, http://dipbt.bundestag.de/dip21/btd/17/048/1704895.pdf [10 Mar. 14].

DIP, Dokumentations- und Informationssystem des Deutschen Bundestages (2011c), *Bundesrat. Stenografischer Bericht. 881. Sitzung. Berlin, Freitag, den 18. März 2011,* [Online], http://dipbt.bundestag.de/dip21/brp/881.pdf#P.99 [10 Mar. 14].

DIP, Dokumentations- und Informationssystem des Deutschen Bundestages (2011d), *Empfehlung der Ausschüsse zu Punkt ... der 881. Sitzung des Bundesrates am 18. März 2011. Gesetz zur Umsetzung der Richtlinie 2009/28/EG zur Förderung der Nutzung von Energie aus erneuerbaren Quellen (Europarechtsanpassungsgesetz Erneuerbare Energien – EAG EE),* [Online], http://dipbt.bundestag.de/dip21/brd/2011/0105-1-11.pdf [10 Mar. 14].

DIP, Dokumentations- und Informationssystem des Deutschen Bundestages (2011e), *Deutscher Bundestag. Stenografischer Bericht. 117. Sitzung. Berlin, Donnerstag, den 30. Juni 2011,* [Online], http://dipbt.bundestag.de/dip21/btp/17/17117.pdf#P.13404 [10 Mar. 14].

DIP, Dokumentations- und Informationssystem des Deutschen Bundestages (2011f), *Gesetzentwurf der Fraktionen der CDU/CSU und FDP. Entwurf eines Dreizehnten Gesetzes zur Änderung des Atomgesetzes,* [Online], http://dipbt.bundestag.de/dip21/btd/17/060/1706070.pdf [10 Mar. 14].

DIP, Dokumentations- und Informationssystem des Deutschen Bundestages (2011g), *Gesetzentwurf der Fraktionen der CDU/CSU und FDP. Entwurf eines Gesetzes zur Neuregelung des Rechtsrahmens für die Förderung der Stromerzeugung aus erneuerbaren Energien,* [Online], http://dipbt.bundestag.de/dip21/btd/17/060/1706071.pdf [10 Mar. 14].

DIP, Dokumentations- und Informationssystem des Deutschen Bundestages (2011h), *Gesetzbeschuss des Deutschen Bundestages. Gesetz zur Neuregelung des Rechtsrahmens für die Förderung der Stromerzeugung aus erneuerbaren Energien,* [Online], http://dipbt.bundestag.de/dip21/brd/2011/0392-11.pdf [10 Mar. 14].

DIP, Dokumentations- und Informationssystem des Deutschen Bundestages (2011i), *Beschluss des Bundesrates. Gesetz zur Neuregelung des Rechtsrahmens für die Förderung der Stromerzeugung aus erneuerbaren Energien*, [Online], http://dipbt.bundestag.de/dip21/brd/2011/0392-11B.pdf [10 Mar. 14].

DIP, Dokumentations- und Informationssystem des Deutschen Bundestages (2014), *Basisinforation über den Vorgang. 17. Wahlperiode. Vorgangstyp: Gesetzgebung. Gesetz zur Umsetzung der Richtlinie 2009/28/EG zur Förderung der Nutzung von Energie aus erneuerbaren Quellen (Europarechtsanpassungsgesetz Erneuerbare Energien – EAG EE)*, [Online], http://dipbt.bundestag.de/extrakt/ba/WP17/301/30143.html [10 Mar. 14].

Duffield, J. S. and Westphal, K. (2011), *Germany and EU Energy Policy in the 2000s: Implications of German Energy Policy*, in: Birchfeld, V. L. and Duffield, J. S. (ed.), Toward A Common European Union Energy Policy, New York: Palgrave MacMillan, 175-186.

EC, European Commission (2001), *Green Paper: Towards a European strategy for the security of energy supply*, [Online], Available: http://ec.europa.eu/energy/green-paper-energy-supply/doc/green_paper_energy_supply_en.pdf [13 Mar. 13].

EC, European Commission (2006), *Green Paper: A European Strategy for Sustainable*, Competitive and Secure Energy, [Online], Available: http://europa.eu/documents/comm/green_papers/pdf/com2006_105_en.pdf [24 Dec. 13].

EC, European Commission (2007a), *Communication from the Commission to the European Council and the European Parliament: An Energy Policy for Europe*, [Online], Available: http://ec.europa.eu/energy/energy_policy/doc/01_energy_policy_for_europe_en.pdf [14 Jun. 13].

EC, European Commission (2007b), *Communication from the Commission to the Council, the European Parliament, the European Economic and Social Committee and the Committee of the Regions: Limiting Global Climate Change to 2 degrees Celsius. The way ahead for 2020 and beyond*, [Online], Available: http://eur-lex.europa.eu/LexUriServ/LexUriServ.do?uri=COM:2007:0002:FIN:EN:PDF [24 Dec. 13].

EC, European Commission (2007c), POLAND – *Energy Mix Fact Sheet* [Online], Available: http://ec.europa.eu/energy/energy_policy/doc/factsheets/mix/mix_pl_en.pdf [20 May 14].

EC, European Commission (2008), *Communication from the Commission to the European Parliament, the Council, the European Economic and Social Committee and the Committee of the Regions: 20 20 by 2020 Europe's climate change opportunity*, [Online], Available: http://eur-lex.europa.eu/LexUriServ/LexUriServ.do?uri=COM:2008:0030:FIN:EN:PDF [05 Jul. 13].

EC, European Commission (2009), *Commission Decision of 30.6.2009 establishing a template for National Renewable Energy Action Plans under Directive 2009/28/ EC,* [Online], Available: http://ec.europa.eu/energy/renewables/doc/nreap__ adoptedversion__30_june_en.pdf [24 Dec. 13].

EC, European Commission (2011a), *Communication from the Commission to the European Parliament, the Council, the European Economic and Social Committee and the Committee of the Regions: Energy Roadmap 2050,* [Online], Available: http:// eur-lex.europa.eu/LexUriServ/LexUriServ.do?uri=COM:2011:0885:FIN:EN:PDF [14 Jun. 13].

EC, European Commission (2011b), *Commission Staff Working Document: 2009-2010 Report on progress creating the internal gas an electricity market,* [Online, Available: http://ec.europa.eu/energy/gas_electricity/legislation/doc/20100609_ internal_market_report_2009_2010.pdf [05 Jul. 13].

EC, European Commission (2011c), *Commission Staff Working Document: Recent progress in developing renewable energy sources and technical evaluation of the use of biofuels and other renewable fuels in transport in accordance with Article 3 of Directive 2001/77/EV and Article 4(2) of Directive 2003/30/EC. Accompanying document to the Communication from the Commission to the European Parliament and the Council,* [Online, Available: http://ec.europa.eu/ energy/renewables/reports/doc/sec_2011_0130.pdf [24 Oct. 13].

EC, European Commission (2011d), *Energy Markets in the European Union in 2011,* [Online, Available: http://ec.europa.eu/energy/gas_electricity/doc/20121217_ energy_market_2011_lr_en.pdf [20 May 14].

EC, European Commission (2012a), *Communication from the Commission to the European Parliament, the Council, the European Economic and Social Committee and the Committee of the Regions: Making the internal energy market work,* Online], Available: http://eur-lex.europa.eu/LexUriServ/LexUriServ. do?uri=COM:2012:0663:FIN:EN:PDF [05 Jul. 13].

EC, European Commission (2012b), *European Commission – Press Release, Renewable energy: Finish, Greek and Polish legislation still not in line with EU rules,* [Online], Available: http://europa.eu/rapid/press-release_IP-12-278_en.htm?loca le=fr [20 May 14].

EC, European Commission (2012c), *Consultation Paper on generation adequacy, capacity mechanisms and the internal market in electricity,* [Online], Available: http://ec.europa.eu/energy/gas_electricity/consultations/doc/20130207_ generation_adequacy_consultation_document.pdf [01 Jul. 14].

EC, European Commission (2013a), Green Paper. A 2030 *framework for climate and energy policies,* [Online], Available: http://eur-lex.europa.eu/LexUriServ/LexUriServ.do?uri=COM:2013:0169:FIN:EN:PDF [14 Jun. 13].

EC, European Commission (2013b), *Report from the Commission to the European Parliament, the Council, the European Economic and Social Committee and the Committee of the Regions: Renewable energy progress report,* [Online, Available: http://eur-lex.europa.eu/LexUriServ/LexUriServ.do?uri=COM:2013:0175:FIN:EN:PDF [14 Jun. 13].

EC, European Commission (2013c), *Commission Staff Working Document. Accompanying the document: Report from the Commission to the European Parliament, the Council: Renewable energy progress report,* [Online], Available: http://eur-lex.europa.eu/LexUriServ/LexUriServ.do?uri=SWD:2013:0102:FIN:EN:PDF [14 Jun. 13.

EC, European Commission (2013d), *Table of infringement procedures open for non-communication of national transposition measures (situation 21/02/2013),* [Online], Available: http://ec.europa.eu/energy/infringements/doc/infringements_energy.pdf [05 Jul. 13].

EC, European Commission (2013e), *Renewable Energy: Commission refers Poland and Cyprus to Court for failing to transpose EU rules,* Press Releases RAPID, [Online], Available: http://europa.eu/rapid/press-release_IP-13-259_en.htm [05 Jul. 13].

EC, European Commission (2013f), *May infringements package: main decisions,* Press Releases RAPID, [Online], Available: http://europa.eu/rapid/press-release_MEMO-13-470_en.htm [05 Jul. 13].

EC, European Commission (2014a), *Energy. Projects of Common Interest – Interactive map,* [Online], Available: http://ec.europa.eu/energy/infrastructure/transparency_platform/map-viewer/ [01 Jul. 14].

EC, European Commission (2014b), *01 Implementation Plan – for publication. For Project of Common Interest (number and name): 3.14.1 Interconnection between Eisenhüttenstadt (DE) and Plewiska (PL),* [Online], Available: http://ec.europa.eu/energy/infrastructure/pci/doc/Annexes/pci_Annex_3_14_1_en.pdf [01 Jul. 14].

EC, European Commission (2014c), *01 Implementation Plan – for publication. For Project of Common Interest (number and name): 3.15.1 Interconnection between Vierradent (DE) and Krajnik (PL),* [Online], Available: http://ec.europa.eu/energy/infrastructure/pci/doc/Annexes/pci_Annex_3_15_1_en.pdf [01 Jul. 14].

EC, European Commission (2014d), *01 Implementation Plan – for publication. For Project of Common Interest (number and name): 3.15.2 Coordinated installation and operation of phase shifting transformers on the interconnection lines between Krajnik (PL) – Vierraden (DE) and Mikulowa (PL) – Hagenwerder (DE)*, [Online], Available: http://ec.europa.eu/energy/infrastructure/pci/doc/Annexes/pci_Annex_3_15_2_en.pdf [01 Jul. 14].

EC, European Commission (2014e), *Commisison Staff Working Document Energy prices and costs report Accompanying the document Communication From The Commission To The European Parliament, The Council, And the European Economic And Social Committee And The Committee Of The Regions Energy prices and costs in Europe,* [Online], Available: http://ec.europa.eu/energy/doc/2030/20140122_swd_prices.pdf [01 Jul. 14].

Ecofys (2012), *Renewable Energy Progress and Biofuels Sustainability,* [Online], Available: http://ec.europa.eu/energy/renewables/reports/doc/2013_renewable_energy_progress.pdf [18 Oct. 13].

ECJ, European Court of Justice (1963), *Judgment of the Court of 5 February 1963. - NV Algemene Transport- en Expeditie Onderneming van Gend & Loos v Netherlands Inland Revenue Administration. - Reference for a preliminary ruling: Tariefcommissie - Pays-Bas. - Case 26-62.,* [Online], Available: http://eur-lex.europa.eu/LexUriServ/LexUriServ.do?uri=CELEX:61962CJ0026:EN:HTML [24 Dec. 13].

ECJ, European Court of Justice (1964), *Judgment of the Court of 15 July 1964. Flaminio Costa v E.N.E.L. Reference for a preliminary ruling: Giudice conciliatore di Milano – Italy. Case 6-64..,* [Online, Available: http://eur-lex.europa.eu/LexUriServ/LexUriServ.do?uri–CELEX:61962CJ0026:EN:HTML [24 Dec. 13].

ECJ, European Court of Justice (1978), *Judgment of the Court of 9 March 1978. - Amministrazione delle Finanze dello Stato v Simmenthal SpA. - Reference for a preliminary ruling: Pretura di Susa - Italy. - Discarding by the national court of a law contrary to Community law. - Case 106/77.,* [Online], Available: http://eur-lex.europa.eu/LexUriServ/LexUriServ.do?uri=CELEX:61964J0006:EN:NOT [11 Mar. 14].

EEG, Energy Economics Group at Vienna University of Technology, Institute of Power Systems and Energy Economics (2007), *Short characterisation of the model Green-X,* [Online], Available: http://www.green-x.at/downloads/Short%20characterisation%20of%20the%20Green-X%20model%20(March%202007).pdf [24 Dec. 13].

EEG-Clearingstelle (2014a), Gesetzentwurf der Fraktion der CDU/CSU und FDP. Entwurf eines Gesetzes über die Einspeisung von Strom aus erneuerbaren Energien in das öffentliche Netz (Stromeinspeisungsgesetz), Online], Available: https://www.clearingstelle-eeg.de/files/private/active/0/2-GesEntw_BT-Ds_11-7816.pdf [10 Mar. 14].

EEG-Clearingstelle (2014b), *Gesetzentwurf der Fraktion der SPD und BÜNDIS 90/DIE GRÜNEN. Entwurf eines Gesetzes zur Förderung der Stromerzeugung aus erneuerbaren Energien (Erneuerbare-Energien-Gesetz – EEG),* [Online], Available: https://www.clearingstelle-eeg.de/files/private/active/0/2-GesEntw_BT-Ds_14-2341.pdf [10 Mar. 14].

EEG-Clearingstelle (2014c), *Parlamentarische Gesetzesdokumentation: Signatur XIV/99,* [Online], Available: https://www.clearingstelle-eeg.de/files/1-Gesetzgebungsverfahren-Uebersicht.pdf [10 Mar. 14].

EEX, European Energy Exchange (2014), Auction/EPEX SPOT, [Online], Available: http://www.eex.com/en/market-data/power/spot-market/auction#!/2014/06/07 [01 Jul. 14].

EG Science, EU Climate Change Expert Group (2008), *The 2°C target. Information Reference Document. Background on Impacts, Emission Pathways, Mitigation Options and Costs,* Online], Available: http://ec.europa.eu/clima/policies/international/negotiations/future/docs/brochure_2c_en.pdf [14 Jun. 13].

EIA, U.S. Energy Information Administration (2013), Poland, [Online], Available: http://www.eia.gov/countries/country-data.cfm?fips=pl [20 May 14].

Energate (2014), *EPEX SPOT Indizes,* [Online], Available: http://www.energate-messenger.de/markt/preise/index.php?id=39362 [01 Jul. 14].

ENTSO-E, European Network of Transmission System Operators for Electricity (2013a), *Scenario Outlook & Adequacy Forecast 2013-2030,* [Online], Available: https://www.entsoe.eu/about-entso-e/system-development/system-adequacy-and-market-modeling/soaf-2013-2030/ [18 Oct. 13].

ENTSO-E, European Network of Transmission System Operators for Electricity (2013b), *Winter Outlook Report 2013/2014 and Summer Review,* [Online], Available:https://www.entsoe.eu/publications/system-development-reports/outlook-reports/Pages/default.aspx [01 Jul. 14].

ENTSO-E, European Network of Transmission System Operators for Electricity (2014), *Summer Outlook 2014 and Winter Review 2013/14,* [Online], Available: https://www.entsoe.eu/publications/system-development-reports/outlook-reports/Pages/default.aspx [01 Jul. 14].

EP, European Parliament (2007a), *P6_TA(2007)0038. Climate Change. European Parliament resolution on climate change*, [Online], Available: http://www.europarl. europa.eu/sides/getDoc.do?pubRef=-//EP//NONSGML+TA+P6-TA-2007-0038+0+DOC+PDF+V0//EN 24 Dec. 13].

EP, European Parliament (2007b), P6 TA(2007)0326. *Internal gas and electricity market. European Parliament resolution of 10 July 2007 on prospects for the internal gas and electricity market (2007/2089(INI))*, [Online], Available: http:// www.europarl.europa.eu/sides/getDoc.do?pubRef=-//EP//NONSGML+TA+P6-TA-2007-0326+0+DOC+PDF+V0//EN [24 Dec. 13].

EP, European Parliament (2008), *EP seals Climate Change Package,* [Online], Available:http://www.europarl.europa.eu/sides/getDoc.do?pubRef=-// EP//NONSGML+IM-PRESS+20081208BKG44004+0+DOC+PDF+V0// EN&language=EN [24 Dec. 13].

EP, European Parliament (2009), *Third Energy Package gets finale approval from MEPs,* [Online], Available: http://www.europarl.europa.eu/sides/getDoc. do?pubRef=-//EP//NONSGML+IM-PRESS+20080616FCS31737+0+DOC+PDF +V0//EN&language=EN [24 Dec. 13].

EP, European Parliament (2013), ****I Report on the proposal for a regulation of the European Parliament and the Council on guidelines for trans-European energy infrastructure and repealing Decision No 1364/2006/EC (COM(2011)0658 – C7-0471/2011–2011/0300(COD)),*[Online],Available:http://www.europarl. europa.eu/sides/getDoc.do?pubRef=-//EP//NONSGML+REPORT+A7-2013-0036+0+DOC+PDF+V0//EN [24 Dec. 13].

EP, European Parliament (2014), *Legislative Observatory,* [Online], Available: http://www.europarl.europa.eu/oeil/home/home.do [03 Dec. 14].

EPEX, European Power Exchange (2012), *Descrption Of Indices Derived From EPEX SPOT Markets,* [Online], Available: http://static.epexspot.com/ document/12852/EPEXSpot_Indices.pdf [01 Jul. 14].

EPEX, European Power Exchange (2014), *Negative Prices Q&A,* [Online], Available: http://www.epexspot.com/en/company-info/basics_of_the_power_ market/negative_prices 01 Jul. 14].

Erni, D. (2012), *Day-Ahead Electricity Spot Prices – Fundamental Modelling and the Role of Expected Wind Electricity Infeed at the European Energy Exchange. Dissertation of the University of St. Gallen, Scholl of Management, Economics, Law, Social Sciences and International Affairs to obtain the title of Doctor of Philosophy in Management,* [Online], Available: http://verdi.unisg.ch/ www/edis.nsf/SysLkpByIdentifier/4076/$FILE/dis4076.pdf [01 Jul. 14].

EU Energy Security Unit (2012), *Unconventional Gas: Potential Energy Market Impacts in the European Union. A Report by the Energy Security Unit of the European Commission's Joint Research Centre,* [Online], Available: http://ec.europa.eu/dgs/jrc/downloads/jrc_report_2012_09_unconventional_gas.pdf [17 Jun. 13].

EuPD and CTI, EuPD Research and Deutsches CleanTech Institut (2013), *Auswirkungen der Energiewende auf Ostdeutschland. Endbericht,* [Online], Available:http://www.bmi.bund.de/SharedDocs/Downloads/DE/Broschueren/2013/auswirkungen-energiewende-ostdeutschland.pdf?__blob=publicationFile [01 Jul. 14].

Eurelectric (2011), *RES Integration And Market Design: Are Capacity Remuneration Mechanisms Needed To Ensure Generation Adequacy?,* [Online], Available: http://www.eurelectric.org/media/26300/res_integration_lr-2011-030-0464-01-e.pdf [01 Jul. 14].

Eurostat (2012), *Energy production and imports,* Online], Available: http://epp.eurostat.ec.europa.eu/statistics_explained/index.php/Energy_production_and_imports [17 Jun. 13].

Eurostat (2013a), *Renewable energy: Share of renewable energy up to 13% of energy consumption in the EU27 in 2011,* [Online], Available: http://europa.eu/rapid/press-release_STAT-13-65_en.htm [24 Dec. 13].

Eurostat (2014), *Energieabhängigkeit%* , [Online], Available: http://epp.eurostat.ec.europa.eu/tgm/table.do?tab=table&plugin=1&language=de&pcode=tsdcc310 [20 May 14].

Eurostat (2013b), *Electricity generated from renewable sources% of gross electricity consumption,* [Online], Available: http://epp.eurostat.ec.europa.eu/tgm/table.do?tab=table&init=1&plugin=1&language=en&pcode=tsdcc330 [24 Dec. 13].

Europa.eu (2001), *Renewable energy: White Paper laying down a Community strategy and action plan,* [Online], Available: http://europa.eu/legislation_summaries/other/l27023_en.htm [11 Mar. 14].

EU2007.de (2007), *What is the Presidency?,* [Online], Available: http://www.eu2007.de/en/The_Council_Presidency/What_is_the_Presidency/index.html [11 Mar. 14].

EWEA, The European Wind Energy Association (2010), *Powering Europe: wind energy and the electricity grid,* [Online], Available: http://www.ewea. org/fileadmin/ewea_documents/documents/publications/reports/Grids_ Report_2010.pdf [01 Jul. 14].

Franzius, C. (2010), *Europäisches Verfassungsrechtsdenken,* Tübingen: Mohr Siebeck.

Franzius, C. (2013), *Recht und Politik in der transnationalen Konstellation. Skizze eines Forschungsgebiets,* in: Archiv des öffentlichen Rechts, Band 128, Heft 2, Tübingen: Mohr Siebeck, 205-288.

Franzius, C., Preuß, U. K. (2012), *Die Zukunft der Europäischen Demokratie,* in: Heinrich Böll Stiftung: Schriften zu Europa, Band 7, Berlin: Heinrich Böll Stiftung e.V..

Gaus, D. (2011), *The State's Existence Between Facts and Norms. A reflection on some problems tot he analysis oft he state,* ARENA Working Paper (online), [Online], Available: http://www.sv.uio.no/arena/english/research/publications/ arena-publications/workingpapers/working-papers2011/wp-11-11.pdf [4 Nov. 13].

German Advisory Council on the Environment (2011), *Pathways towards a 100% renewable electricity system,* [Online], Available: http://www.umweltrat.de/ SharedDocs/Downloads/EN/02_Special_Reports/2011_10_Special_Report_ Pathways_renewables.pdf?__blob=publicationFile [26 Dec. 13].

Graziano, P. R. and Vink, P. M. (2013), *Europeanization: Concept, Theory, and Methods,* in: Bulmer, S. and Lequesne, C. (ed.), The Member States of the European Union, Oxford: Oxford University Press, 31-54.

Grimm, D. (2001), Die Verfassung und die Politik. Einsprüche in Störfällen, München: C.H. Beck.

Haas, R. and Loew, T. (2012), *Diskussionspapier. Die Auswirkung Der Energie-wende Auf Die Strommärkte Und Die Rentabilität Von Konventionellen Kraftwer-ken,*[Online],Available:http://www.nachhaltigkeit.wienerstadtwerke.at/fileadmin/ user_upload/Downloadbereich/Haas-Loew-Auswirkungen Energiewende-auf-Energiemaerkte2012.pdf [01 Jul. 14].

Hajer, M. (2003), *Policy without polity? Policy analysis and the institutional void,* Policy Science, vol. 36, pp. 175-195.

Helm, D. (2013), *The European framework for energy and climate policies,* Energy Policy (Article in Press), [Online], Available: http://dx.doi.org/ 10.1016/j. enpol.2013.05.063 [5 Jul. 13].

Henningsen, J. (2013), *Commission in do-nothing mode on climate change policy,* Euobserver.com, [Online], Available: http://euobserver.com/opinion/11 9671 [5 Jul. 13].

Heritier, A. (2001), *Differential Europe: The European Union Impact on National Policymaking,* in: Heritier, A., Kerwer, D., Knill, C., Lehmkuhl, D., Teutsch, M. and Douillet, A.C. (ed.), *Differential Europe: The European Union Impact on National Policymaking,* Lanham, Boulder, New York and Oxford: Rowman & Littefield Publishers, INC., 1-21.

Heritier, A. and Knill, C. (2001), *Differential Responses to European Policies: A Comparison,* in: Heritier, A., Kerwer, D., Knill, C., Lehmkuhl, D., Teutsch, M. and Douillet, A.C. (ed.), Differential Europe: The European Union Impact on National Policymaking, Lanham, Boulder, New York and Oxford: Rowman & Littefield Publishers, INC., 257-294.

Hirschl, B. (2007), *Erneuerbare Energien-Politik. Eine Multi-Level Policy-Analyse mit Fokus auf den deutschen Strommarkt,* Wiesbaden: VS Verlag für Sozialwissenschaften.

Hobbes (1909, reprinted from the edition of 1651), *Leviathan,* University of Toronto Libraries, [Online], Available:https://archive.org/details/hobbessleviath an00hobbuoft [4 Nov. 13]

Howlett, M., Ramesh, M. and Perl, A. (2009), *Studying Public Policy, Policy Cycles & Policy Subsystems,* 3rd edition, Oxford: Oxford University Press.

Hönninge, C., Kneip, S. and Lorenz, A. (2011), *Formen, Ebenen, Interaktionen – eine erweiterte Analyse des Verfassungswandels,* in: Hönninge, C., Kneip, S. and Lorenz, A. (ed.), Verfassungswandel im Mehrebenensystem, Wiesbaden: VS Verlag für Sozialwissenschaften.

IfIL, Institute for International Law (2007a), *Case: BVerfGE 31, 145 2 BvR 225/69 Milk powder-decision,* [Online], Available: https://www.utexas.edu/ law/academics/centers/transnational/work_new/german/case.php?id=590 [11 Mar. 14].

IfIL, Institute for International Law (2007b), *Case: BVerfGE 37, 271 2 BvL 52/71 Solange I-Beschluß,* [Online], Available: http://www.utexas.edu/law/academics/ centers/transnational/work_new/german/case.php?id=588 [11 Mar. 14].

IfIL, Institute for International Law (2007c), *Case: BVerfGE 73, 339 2 BvR 197/83 Solange II-decision,* [Online], Available: http://www.utexas.edu/law/ academics/centers/transnational/work_new/german/case.php?id=572[11 Mar. 14].

Intertek and Aptech (2012), *Power Plant Cycling Costs,* [Online], Available: http://wind.nrel.gov/public/wwis/aptechfinalv2.pdf [01 Jul. 14].

IEA, International Energy Agency (2007), *Climate Policy Uncertainty and Investment Risk,* [Online], Available: http://www.iea.org/publications/freepublications/publication/Climate_Policy_Uncertainty.pdf [01 Jul. 14].

IPPNW, International Physicians for the Prevention of Nuclear War (2014), *Fukushima Disaster,* [Online], Available: http://www.fukushima-disaster.de/information-in-english/press-releases.html [12 Mar. 14].

Jänicke, M., Kunig, P. and Stitzel, M. (2003), *Umweltpolitik. Politik, Recht und Management des Umweltschutzes in Staat und Unternehmen,* Bonn: J.H.W. Dietz Nachf. GmbH.

Kingdon, J. (2004): Agendas, *Alternatives and Public Policies,* Boston: Longman Classics in Political Science.

Knill, C., Lenschow, A. (2005), *Coercion, Competition and Communication: Different Approaches of European Governance and their Impact on National Institutions,* Journal of Common Market Studies, vol. 43, no. 3, pp. 583-606.

Ladrech, R. (2010), *Europeanization and National Politics,* Hampshire: Palgrave Macmillan.

Leuffen, D., Rittberger, B. and Schimmelfennig, F. (2013), *Differentiated Integration. Explaining Variation in the European Union,* Hampshire: Palgrave Macmillan.

Lexology (2013), *Small Power Tripack becomes effective,* [Online], Available: http://www.lexology.com/library/detail.aspx?g=6c994a3d-1f4e-4419-b452-7b4ba832be01 [20 May 14].

Maier, W. (2001), *Staats- und Verfassungsgrecht,* Achim: Erisch Fleischer Verlag.

Meyer, T. (2010), Was ist Politik?, 3rd edition, Wiesbaden: VS Verlag für Sozialwissenschaften.

Mielczarski, W. (2010), *Project of European Interest. Poland-Lithuania Link and Germany – Poland Power Lines,* [Online], Available: http://ec.europa.eu/energy/infrastructure/tent_e/doc/power_link/2011_power_link_annual_report_2009_2010_en.pdf 04 Nov. 13].

Miller, V. (2011), *The EU's Acquis Communautaire,* [Online], Available: www.parliament.uk/briefing-papers/sn05944.pdf [20 May 14].

Minister of Economic Affairs (2010a), *FORECAST PURSUANT TO ARTICLE 4(3) OF DIRECTIVE 2009/28/EC OF 23 APRIL 2009 OF THE EUROPEAN PARLIAMENT AND OF THE COUNCIL ON THE PROMOTION OF THE USE OF ENERGY FROM RENEWABLE SOURCES AND AMENDING AND SUBSEQUENTLY REPEALING DIRECTIVES 2001/77/EC AND 2003/30/EC,* [Online], Available: http://ec.europa.eu/energy/renewables/action_plan_en.htm [20 May 14].

Minister of Economic Affairs (2010b), KRAJOWY PLAN DZIAŁANIA W ZAKRESIE ENERGII ZE ŹRÓDEŁ ODNAWIALNYCH, [Online], Available: http://www.imp.gda.pl/bkeeold/wydarzenia10_pliki/Krajowyplan.pdf [20 May 14].

Minister of Economic Affairs (2014), *RES Act adopted by the Council of Ministers,* [Online], Available: http://www.mg.gov.pl/node/20480# [20 May 14].

Ministry of Economy (2005), *ENERGY POLICY OF POLAND UNTIL 2025,* [Online], Available:http://energieodnawialne.pl/pliki/energetyka_do_2025_ang.doc [20 May 14].

Ministry of Economy (2009a), *Energy Policy of Poland until 2030,* [Online], Available:http://www.mg.gov.pl/files/upload/8134/Polityka%20energetyczna%20ost_en.pdf [20 May 14].

Ministry of Economy (2009b), *Projection Of Demand For Fuels And Energy Until 2030. Appendix 2 to draft "Energy Policy of Poland until 2030",* [Online], Available: http://www.mg.gov.pl/files/upload/8134/Appendix2.pdf [01 Jul. 14].

Ministry of Economy (2011), *Polish Nuclear Power Program,* [Online], Available: http://www.mugv.brandenburg.de/cms/media.php/lbm1.a.3310.de/pl_kep_en.pdf [20 May 14].

Ministry of Environment (2003), *POLAND'S CLIMATE POLICY. The strategies for greenhouse gas emission reductions in Poland until 2020,* [Online], Available:http://www.mos.gov.pl/g2/big/2009_04/cf234906b019de170218bf79f913990c.pdf [20 May 14].

Moravcsik, A. (1998), *The Choice for Europe. Social Purpose & State Power from Messina to Maastricht,* Ithaca: Cornell University Press.

Moravcsik, A. and Schimmelfenig, F. (2009), *Liberal Intergovernmentalism,* in: Wiener, A. and Diez, T. (ed.), European Integration Theory, [Online], Available: http://www.princeton.edu/~amoravcs/library/intergovernmentalism.pdf [04 Nov. 13].

Morris, C. and Pehnt, M. (2012), *Energy Transition. The German Energiewende (An initiative of the Heinrich Böll Foundation)*, [Online], Available: http://energytransition.de [12 Mar. 14].

National Atomic Energy Agency (2014), *Act of Parliament of 29 November 2000 on the Atomic Law,* [Online], Available: http://www.paa.gov.pl/en/node/254 [20 May 14].

NREAP_EU27, National Renewable Energy Action Plan of EU 27 (2010), [Online], Available: http://ec.europa.eu/energy/renewables/action plan_en.htm [26 Dec. 13].

NREAP Germany, National Renewable Energy Action Plan of Germany (2010), *National Renewable Energy Action Plan,* [Online], Available: http://ec.europa.eu/energy/renewables/action_plan_en.htm [04 Nov. 13].

NREAP Poland, National Renewable Energy Action Plan of Poland (2010), *National Renewable Energy Action Plan,* [Online], Available: http://ec.europa.eu/energy/renewables/action_plan_en.htm [04 Nov. 13].

OECD (2008), *Nuclear Legislation in OECD Countries. Regulatory and Institutional Framework for Nuclear Activities. Poland,* [Online], Available: https://www.oecd-nea.org/law/legislation/poland.pdf [20 May 14].

Ostrom, E. (2007): *Institutional Rational Choice,* in: Sabatier, P.A. (ed.), Theories of the Policy Process, 2nd Edition, Boulder: Westview Press, 21-64.

OJoEC, Official Journal of the European Communities (2001), *Directive 2001/77/ EC of the European Parliament and of the Council of 27 September 2001 on the promotion of electricity produced from renewable energy sources in the internal electricity market,* L 283, vol. 44, 33-40, [Online], Available: http://eur-lex.europa.eu/LexUriServ/LexUriServ.do?uri=OJ:L:2001:283:0033:0040:EN:PDF [26 Dec. 13].

OJoEU, Official Journal of the European Union (2003a), *Directive 2003/30/EC of the European Parliament and of the Council of 8 May 2003 on the promotion of the use of biofuels or other renewable fuels for transport,* L 123, vol. 46, 42-46, [Online], Available: http://eur-lex.europa.eu/LexUriServ/LexUriServ.do?uri= OJ:L:2003:123:0042:0046:EN:PDF [26 Dec. 13].

OJoEU, Official Journal of the European Union (2003b), *ACT concerning the conditions of accession of the Czech Republic, the Republic of Estonia, the Republic of Cyprus, the Republic of Latvia, the Republic of Lithuania, the Republic of Hungary, the Republic of Malta, the Republic of Poland, the Republic of Slovenia and the Slovak Republic and the adjustments to the Treaties on which the European Union is founded,* L 236, vol. 46, 33-49, [Online], Available: http://eur-lex.europa.eu/resource.html?uri=cellar:2578ec46-e068-4949-b290-fbd013e18e6c.0004.02/DOC_1&format=PDF [20 May 14].

OJoEU, Official Journal of the European Union (2006a), *Decision No 1364/2006/ EC of the European Parliament and of the Council of 6 September 2006 laying down guidelines for trans-European energy networks and repealing Decision*

96/391/EC and Decision No 1229/2003/EC, L 262, vol. 49, 1-23, [Online], Available: http://eur-lex.europa.eu/LexUriServ/LexUriServ.do?uri=OJ:L:2006:2 62:0001:0023:EN:PDF [26 Dec. 13].

OJoEU, Official Journal of the European Union (2006b), *EUROPEAN UNION — CONSOLIDATED VERSIONS OF THE TREATY ON EUROPEAN UNION AND OF THE TREATY ESTABLISHING THE EUROPEAN COMMUNITY,* C 321E, vol. 49, 1-331, [Online], Available: http://eur-lex.europa.eu/JOHtml.do?uri=OJ:C:20 06:321E:SOM:EN:HTML [26 Dec. 13].

OJoEU, Official Journal of the European Union (2007a), *Treaty of Lisbon amending the Treaty on European Union and the Treaty establishing the European Community, signed at Lisbon, 13 December 2007,* C 306, vol. 50, 1-231, [Online], Available: http://eur-lex.europa.eu/LexUriServ/LexUriServ.do? uri=OJ:C:2007:306:FULL:EN:PDF [26 Dec. 13].

OJoEU, Official Journal of the European Union (2007b), *Directive 96/92/EC of the European Parliament and of the Council of 19 December 1996 concerning common rules for the internal market in electricity,* [Online], Available: http://eur-lex.europa.eu/LexUriServ/LexUriServ.do?uri=CELEX:31996L0092:EN:HTML [12 Mar. 14].

OJoEU, Official Journal of the European Union (2007c), *Council Decision of 1 January 2007 determining the order in which the office of President of the Council shall be held, L 1, vol. 50, 11-12,* [Online], Available: http://eur-lex. europa.eu/JOHtml.do?uri=OJ:L:2007:001:SOM:EN:HTML [12 Mar. 14].

OJoEU, Official Journal of the European Union (2008), *Consolidated version of the Treaty on the Functioning of the European Union,* C 115, vol. 51, 47-200, [Online], Available: http://eur-lex.europa.eu/JOIndex.do?year=2008&s erie=C&textfield2=115&Submit=Search&_submit=Search&ihmlang=en [26 Dec. 13].

OJoEU, Official Journal of the European Union (2009a), *Regulation (EC) No 443/2009 of the European Parliament and of the Council of 23 April 2009 setting emission performance standards for new passenger cars as part of the Community's integrated approach to reduce CO_2 emissions from light-duty vehicles,* L 140, vol. 52, 1-15, [Online, Available: http://eur-lex.europa. eu/LexUriServ/LexUriServ.do?uri=OJ:L:2009:140:FULL:EN:PDF [26 Dec. 13].

OJoEU, Official Journal of the European Union (2009b), *Directive 2009/28/ EC of the European Parliament and of the Council of 23 April 2009 on the promotion of the use of energy from renewable sources and amending and subsequently repealing Directives 2001/77/EC and 2003/30/EC,* L 140, vol. 52, 16-62, [Online], Available: http://eur-lex.europa.eu/LexUriServ/LexUriServ.do ?uri=OJ:L:2009:140:FULL:EN:PDF [26 Dec. 13].

OJoEU, Official Journal of the European Union (2009c), *Directive 2009/29/ EC of the European Parliament and of the Council of 23 April 2009 amending Directive 2003/87/EC so as to improve and extend the greenhouse gas emission allowance trading scheme of the Community,* L 140, vol. 52, 63-87, [Online, Available: http://eur-lex.europa.eu/LexUriServ/LexUriServ.do?uri=OJ :L:2009:140:FULL:EN:PDF [26 Dec. 13].

OJoEU, Official Journal of the European Union (2009d), *Directive 2009/30/ EC of the European Parliament and of the Council of 23 April 2009 amending Directive 98/70/EC as regards the specification of petrol, diesel and gas-oil and introducing a mechanism to monitor and reduce greenhouse gas emissions and amending Council Directive 1999/32/EC as regards the specification of fuel used by inland waterway vessels and repealing Directive 93/12/EEC,* L 140, vol. 52, 88-113, [Online], Available: http://eur-lex.europa.eu/LexUriServ/ LexUriServ.do?uri=OJ:L:2009:140:FULL:EN:PDF [26 Dec. 13].

OJoEU, Official Journal of the European Union (2009e), *Directive 2009/31/ EC of the European Parliament and of the Council of 23 April 2009 on the geological storage of carbon dioxide and amending Council Directive 85/337/ EEC, European Parliament and Council Directives 2000/60/EC, 2001/80/EC, 2004/35/EC, 2006/12/EC, 2008/1/EC and Regulation (EC) No 1013/2006,* L 140, vol. 52, 114-135, [Online], Available: http://eur-lex.europa.eu/LexUriServ/ LexUriServ.do?uri=OJ:L:2009:140:FULL:EN:PDF [26 Dec. 13].

OJoEU, Official Journal of the European Union (2009f), *Decision No 406/2009/ EC of the European Parliament and of the Council of 23 April 2009 on the effort of Member States to reduce their greenhouse gas emissions to meet the Community's greenhouse gas emission reduction commitments up to 2020,* L 140, vol. 52, 136-148, [Online], Available: http://eur-lex.europa.eu/LexUriServ/ LexUriServ.do?uri=OJ:L:2009:140:FULL:EN:PDF [26 Dec. 13].

OJoEU, Official Journal of the European Union (2009g), *Regulation (EC) No 713/2009 of the European Parliament and of the Council of 13 July 2009 establishing an Agency for the Cooperation of Energy Regulators,* L 211, vol. 52, 1-14, [Online], Available: http://eur-lex.europa.eu/LexUriServ/LexUriServ. do?uri=OJ:L:2009:211:FULL:EN:PDF [26 Dec. 13].

OJoEU, Official Journal of the European Union (2009h), *Regulation (EC) No 714/2009 of the European Parliament and of the Council of 13 July 2009 on conditions for access to the network for cross-border exchanges in electricity and repealing Regulation (EC) No 1228/2003,* L 211, vol. 52, 15-35, [Online], Available: http://eur-lex.europa.eu/LexUriServ/LexUriServ.do?uri=OJ:L:2009: 211:FULL:EN:PDF [26 Dec. 13].

OJoEU, Official Journal of the European Union (2009i), *Regulation (EC) No 715/2009 of the European Parliament and of the Council of 13 July 2009 on* conditions for access to the natural gas transmission networks and repealing Regulation (EC) No 1775/2005, L 211, vol. 52, 36-54, [Online, Available: http://eur-lex.europa.eu/LexUriServ/LexUriServ.do?uri=OJ:L:2009:211:FULL:EN:P DF 26 Dec. 13].

OJoEU, Official Journal of the European Union (2009j), Directive 2009/72/EC of the European Parliament and of the Council of 13 July 2009 concerning common rules for the internal market in electricity and repealing Directive 2003/54/EC, L 211, vol. 52, 55-93, [Online], Available: http://eur-lex.europa. eu/LexUriServ/LexUriServ.do?uri=OJ:L:2009:211:FULL:EN:PDF [26 Dec. 13].

OJoEU, Official Journal of the European Union (2009k), Directive 2009/73/EC of the European Parliament and of the Council of 13 July 2009 concerning common rules for the internal market in natural gas and repealing Directive 2003/55/EC, L 211, vol. 52, 94-136, [Online], Available: http://eur-lex.europa. eu/LexUriServ/LexUriServ.do?uri=OJ:L:2009:211:FULL:EN:PDF [26 Dec. 13].

OJoEU, Official Journal of the European Union (2010), Consolidated version of the Treaty on European Union, C 83, vol. 53, 13-46, [Online], Available: http://eur-lex.europa.eu/LexUriServ/LexUriServ.do?uri=OJ:C:2010:083:FULL :EN:PDF [28 Dec. 13].

OJoEU, Official Journal of the European Union (2011), Regulation (EU) No 1227/2011 of the European Parliament and of the Council of 25 October 2011 on wholesale energy market integrity and transparency, L 326, vol. 54, 1-16, [Online], Available: http://eur-lex.europa.eu/LexUriServ/LexUriServ.do? uri=OJ:L:2011:326:FULL:EN:PDF [28 Dec. 13].

OJoEU, Official Journal of the European Union (2013a), Regulation (EU) No 347/2013 of the European Parliament and of the Council of 17 April 2013 on guidelines for trans-European energy infrastructure and repealing Decision No 1364/2006/EC and amending Regulations (EC) No 713/2009, (EC) No 714/2009 and (EC) No 715/2009, L 115, vol. 56, 39-75, [Online, Available: http://eur-lex.europa.eu/LexUriServ/LexUriServ.do?uri=OJ:L:2013 :115:FULL:EN:PDF [28 Dec. 13].

OJoEU, Official Journal of the European Union (2013b), Case C–320/13: Action brought on 12 June 2013 – European Commission v Republic of Poland, C 226, vol. 56, 8, [Online], Available: http://eur-lex.europa.eu/legal-content/EN/TXT/;jsessionid=pnyvSTtQTVPrmpJLsDRD4ZJKzLyHLD2JvKyPQvN0D6snL1 d2L7dV!-80792881?uri=uriserv:OJ.C_.2013.226.01.0008.01.ENG [28 Dec. 13].

OJoEU, Official Journal of the European Union (2013c), COMMISSION DELEGATED REGULATION (EU) No 1391/2013 of 14 October 2013 amending Regulation (EU) No 347/2013 of the European Parliament and of the Council on guidelines for trans-European energy infrastructures as regards the Union *list of projects of common interest*, L 349, vol. 56, 28-43, [Online, Available: http://eur-lex.europa.eu/legal-content/EN/TXT/PDF/?uri=CELEX:32013R1391 &from=EN [28 Dec. 13].

Paiz.gov.pl, POLISH INFORMATION AND FOREIGN INVESTMENT AGENGCY (2013), *Encrgy Sector in Poland. Sector profile* [Online], Available: http://www.paiz.gov.pl/files/?id_plik=21681[28 Dec. 13].

Pelte, D. (2010), *Die Zukunft unserer Energieversorgung. Eine Analyse aus matematisch-naturwissenschaftlicher Sicht,* Wiesbaden: GWV Fachverlage GmbH.

Pernice, I. (1998), *Constitutional law implications for a state participating in a process of regional integration. German constitution and "multilevel constitutionalism",* [Online], Available: http://www.whi-berlin.eu/documents/pernice-regional-integration.pdf [12 Mar. 14].

Pernice, I. (2001), *Multilevel Constitutionalism in the European Union,* [Online], Available: http://whi-berlin.de/documents/whi-paper0502.pdf [05 Jul. 13].

Pernice, I. (2006), *Das Verhältnis europäischer zu nationalen Gerichten im europäischen Verfassungsverbund,* [Online], Available: http://books.google.de/books?id=MySiCZR7Mn4C&printsec=frontcover&hl=de&source=gbs_ge_summary_r&cad=0#v=onepage&q&f=false [12 Mar. 14].

PKEE, Polish Electricity Association (2013), *Response of the Polish Electricity Association concerning public consultation on generation adequacy, capacity mechanism and the internal market in electricity,* [Online], Available: http://www.pkee.pl/upload/files/PKEE_capacity_markets_fin.pdf [01 Jul. 14].

Podrygala, I. (2008), *Erneuerbare Energien im polnischen Stromsektor. Analyse der Entstehung und Ausgestaltung der Instrumente zur Förderung der Stromerzeugung aus erneuerbaren Energien,* in: Danyel Reiche (ed.), Ecological Energy Policy, Band 8, Stuttgart: ibidem-Verlag.

Pöyry and EWEA, Pöyry and European Wind Energy Association (2010), *Wind Energy and Electricity Prices. Exploring the ,merit order effect',* [Online], Available:http://www.ewea.org/fileadmin/ewea_documents/documents/publications/reports/MeritOrder.pdf [04 Jul. 14].

PPE, Polish Power Exchange (2014a), *POLPX indices for the Polish Day-Ahead and Intraday Market,* [Online], Available: http://wyniki.tge.pl/en/wyniki/euroindex/description/ [01 Jul. 14].

PPE, Polish Power Exchange (2014b), *Market Data,* [Online], Available: http://wyniki.tge.pl/en/wyniki/euroindex/spot/?date=2013-06-03 [01 Jul. 14].

Premier.gov.pl (2013), *Prime minister on Climate and Energy Package,* [Online], Available:https://www.premier.gov.pl/en/news/news/premier-ws-pakietu-klima tyczno-energetycznego.html [20 May 14].

Princeton.edu (interlinked with Wikipedia) (2014), List of political parties in Poland, [Online], Available: https://www.princeton.edu/~achaney/tmve/wiki100 k/docs/List_of_political_parties_in_Poland.html [20 May 14].

Prognos and EWI, Prognos and Energiewirtschaftliches Institut an der Universität zu Köln (2007), *Endbericht. Energieszenarien für den Energiegipfel 2007 (Inklusive Anhang 2% -Variante),* [Online], Available: http://www.ewi.uni-koeln. de/fileadmin/user_upload/Publikationen/Studien/Politik_und_Gesellschaft/2007/ EWI_2007-07-17_Energieszenarien_Energiegipfel.pdf [04 Jul. 14].

PSE and 50Hertz (2014), Report on vPST *pilot phase experience,* [Online, Available: http://www.50hertz.com/de/file/vPST_pilot_phase_report.pdf [11 Jun. 14].

Puka, L. and Szulecki, K. (2014), *Beyond the „Grid-Lock" in Electricity Interconnectors. The Case of Germany and Poland,* in: DIW Berlin Discussion Papers, Discussion Paper No. 1378, [Online], Available: http://www.diw.de/ documents/publikationen/73/diw_01.c.463023.de/dp1378.pdf [04 July 14].

PWEA, Polish Wind Energy Association (2010), *Opinion of the Polish Wind Energy Association concerning the draft National Renewable Energy Action Plan (NREAP) version dated 25 May 2010,* [Online], Available: http://www. pwea.pl/files/opinia_psew_eng.pdf [20 May 14].

Radaelli, C. M. (2003), *The Europeanization of Public Policy,* in: Featherstone, K. and Radaelli, C. M. (ed.), The Politics of Europeanization, Oxford Scholarship Online, [Online], Available: DOI: 10.1093/0199252092.003.0002 [04 Nov. 13].

Reuters (2013), *Poland government backs energy law changes as EU fines loom,* [Online], Available: http://www.reuters.com/article/2013/04/10/poland-energy-idUSL5N0CX1S520130410 [20 May 14].

Riedl, R. (2008), *When environmental challenges spill over into energy policy problems – the case of the Polish (potential) veto on the EU climate-energy package during the council summit in December 2008,* Online], Available: http://www.jhubc.it/ecpr-porto/virtualpaperroom/150.pdf [20 May 14].

Rob, N. (2013), *The effect of intermittent renewables on the energy markets,* in: EDI Quarterly, Volume 4, No. 4, January 2013, [Online], Available: http://www.energydel ta.org/uploads/bestanden/6af45b2b-caf0-41fa-bcbe-b87b794dba6d [04 July 14].

Rybski, R. and Stoczkiewicz, M. (2013), *Legal Analysis. The "Small Tri-pack" and the Directive 2009/28/EC of 23 April 2009 on the promotion of the use of energy from renewable sources,* [Online], Available: http://www.clientearth.org/ reports/130816-climate-and-energy-clientearth-small-tripack-legal-analysis-eng-final.pdf [20 May 14].

Sabatier, P. A. (2007), *The Need for Better Theories,* in: Sabatier, P.A. (ed.), Theories of the Policy Process, 2nd Edition, Boulder: Westview Press, 3-17.

Sabatier, P. A. and Weible, C. M. (2007), *The Advocacy Coalition Framework: Innovations and Clarifications,* in: Sabatier, P.A. (ed.), Theories of the Policy Process, 2nd Edition, Boulder: Westview Press, 189-220.

Sejm (2014), *THE CONSTITUION OF THE REPUBLIC OF POLAND OF 2nd APRIL, 1997,* [Online], Available: http://www.sejm.gov.pl/prawo/konst/angielski/kon1. htm [20 May 14].

Sewalt, M and De Jong, C. (2003), *Negative Prices in Electricity Markets,* [Online], Available: https://www.google.de/search?client=safari&rls=en&q=Se walt,+M+and+De+Jong,+C.+(2003),+Negative+Prices+in+Electricity+Market s&ie=UTF-8&oe=UTF-8&gfe_rd=cr&ei=K6a2U9yhHJGLswbJ9YHwDQ [04 Jul. 14].

Soritrov, M./Memmler, M. (2011), *The Advocacy Coalition Framework in Natural Resource Policy Studies - Recent Experiences and Further Prospects,* [Online], Available: http://foper.sfb.rs/pdf/ACF_October_2010.pdf [04 Nov. 13].

Schaps, K. (2013), EU *could save billions with cross-border renewables coopera-tion,* Reuters, [Online], Available: http://www.reuters.com/article/2013/03/14/ europe-renewables-cooperation-idUSL5N0BDDXU20130314 [5 Jul. 13].

Schimmelfennig, F. and Sedelmeier, U. (2005), *The Europeanization of Central and Eastern Europe,* Ithaca and London: Cornell University Press.

Schlager, E. (2007), *A Comparison of Frameworks, Theories, and Models of Policy Processes,* in: Sabatier, P.A. (ed.), Theories of the Policy Process, 2nd Edition, Boulder: Westview Press, 293-319.

Schnell, R., Paul, B. H. and Esser, E. (2005), *Methoden der empirischen Sozialforschung,* München: Oldenbourg Wissenschaftsverlag GmbH.

SPIEGEL Online (2007a), *Energy Summit: Merkel Nudges for Nuclear Power Comeback,* [Online], Available: http://www.spiegel.de/international/germany/ energy-summit-merkel-nudges-for-nuclear-power-comeback-a-492202.html [12 Mar. 14].

SPIEGEL Online (2007b), Grönland-Reise: Merkel auf Eis, [Online], Available: http://www.spiegel.de/politik/deutschland/groenland-reise-merkel-auf-eis-a-500231.html [12 Mar. 14].

Spinoza, B. (1883), *Tractatus Politicus*, [Online], Available: http://www.constitution.org/bs/poltreat.txt [4 Nov. 13].

Tatham, A. F. (2013), *Central European Constitutional Courts in the Face of EU Membership. The Influence of the German Model in Hungary and Poland*, Leiden and Boston: Martinus Nijhoff Publishers.

TGPE, Towarzystwo Gospodarcze Polskie Elektrownie (2013), *European Commission Consultation Paper on generation adequacy, capacity mechanisms and internal market in electricity. TGPE response paper*, in: DG Energy, Public Consultation: Generation adequacy, capacity mechanisms and internal market in electricity; Related Documents, [Online], Available: http://ec.europa.eu/energy/gas_electricity/consultations/20130207_generation_adequacy_en.htm [04 Jul. 14].

THEMA (2013), *Loop flows – Final advice. Prepared for The European Commission* [Online], Available: http://ec.europa.eu/energy/gas_electricity/studies/doc/electricity/201310_loop-flows_study.pdf [04 Jul. 14].

Tobler, C. and Beglinger, J. (2012), *Essential EU Law in Charts*, 2nd "Lisbon" edition, Budapest: HVG-ORAC.

Traber, T and Kemfert, C. (2011), *Gone with the wind? – Electricity market prices and incentives to invest in thermal power plants under increasing wind energy supply*, in: *Energy Economics*, vol. 33, pp. 249-256 , [Online], Available: DOI: 0.1016/j.eneco.2010.07.002 [04 Jul. 14].

University of Nevada (1994), *Decision of the German Federal Constitutional Court of October 12, 1993. In Re Maastricht Treaty*, [Online], Available: DOI: [11 Mar. 14].

Uken, M. (2012), *Deutschland nervt Polen mit der Energiewende*, Die Zeit, [Online],Available:http://www.zeit.de/wirtschaft/2011-11/stromnetz-ring-fluesse/seite-3 [5 Jul. 13].

UN, United Nations Sustainable Development (1992), *United Nations Conference on Environment & Development Rio de Janerio, Brazil, 3 to 14 June 1992,* [Online, Available: http://www.un.org/esa/sustdev/documents/agenda21/english/Agenda21.pdf [20 May 14].

Venjakob, J. (2012), *Qualitativ-narrative Szenarios für die langfristige Entwicklung des polnischen Energiesektors. Eine energiegeographische Untersuchung,* in: Danyel Reiche (ed.), Ecological Energy Policy, Band 13, Stuttgart: ibidem-Verlag.

Wetzel, D. (2012), *Polen macht die Grenzen für deutschen Strom dicht,* Die Welt, [Online], Available:http://www.welt.de/wirtschaft/article112279952/Polen-macht-die-Grenze-fuer-deutschen-Strom-dicht.html [5 Jul. 13].

Weber, M. (1994), *Wissenschaft als Beruf 1917/1919, Politik als Beruf 1919,* Tübingen: J.C.B Mohr (Paul Siebeck).

World Energy Council (2013), *2013 World Energy Issues Monitor,* [Online], Available:http://www.worldenergy.org/documents/2013_world_energy_issues_monitor_report_feb2013.pdf [17 Jun. 13].

Windhoff-Heritier, A. (1987), *Policy-Analyse. Eine Einführung,* Frankfurt & New York: Campus Verlag.

Youngs, R. (2007), *Europe's External Energy Policy: Between Geopolitics and the Market: CEPS Working Document No. 278/November 2007,* [Online], Available: http://aei.pitt.edu/7579/1/Wd278.pdf [5 Jul. 13].

Zepeda, F. (2014): *Renewable Energy Development in Germany and Poland 2005-2020. Comparison, estimations and objectives to their National Renewable Energy Action Plans (NREAP),* internal analysis.

ZEIT Online (2006), *Ökostrom: Das unterschätze Gesetz,* [Online], Available: http://www.zeit.de/online/2006/39/EEG/komplettansicht [12 Mar. 14].

50Hertz (2012), *50Hertz and PSE Operator work together to improve management of cross-border energy flows through use of phase shifters,* Press release, [Online], Available: http://www.50hertz.com/en/file/20121222_PM_Phasenschieber_EN.pdf [5 Jul. 13].

9. Appendix I – Summary English

The introduction posed the question of whether one can state the successful achievement of a common European energy policy, based on the lessons of the implementation of Directive 2009/28/EC. Whilst analysing the progress of the promotion of renewable energies in the context of the 2020 policy framework, the European Commission released a 2013 Green Paper to reflect upon a new framework for climate and energy policies up to 2030 (EC, 2013a: 3; EC 2013c: 3). The paper raises 22 overall questions to evaluate the existing 20-20-20 policy framework, in order to draw lessons for the future design of the Union's energy and climate change policy (EC, 2013a: 13). In light of this thesis, these overall questions were translated to meet the guiding research question:
What are the constraints and dependencies of the European Integration of RES-(E) promotion?
For the empirical analysis of this study, the working hypothesis reads as follows:

If (i) the European Union maintains an overall adaptational pressure by deciding on RES-E measures and policies for its Member States; (ii) pioneering Member States with sufficient power in the international system with regard to RES-E promotion (such as Germany) succeed in implementing an energy policy with a high share in electricity from renewable energy sources; and (iii) neighbouring Member States (such as Poland) are exposed to a pioneering Member State's RES-E policy and/or RES-E market design – **then,** non-pioneering Member States' RES-E measures and policies will adapt to the vertical (EU level) and horizontal (inter-state level) pressures. (These multi-level policy dependencies will encourage the harmonization of a European promotion of renewable energies and describe a reasoned path to a common promotion of renewable energies in all 28 Member States of the European Union.)

Based on the analysis insights, three things become apparent:

1.The change intended by supranational RES-E policies as enshrined in Directive 2009/28/EC did excite change in the EU Member States. But the change excited is not wholly in accordance with the change intended.

2. In the cases of Germany and Poland, the domestic preferences with regard to RES-E promotion in light of Directive 2009/28/EC reflect an, if not exclusive, clearly independent intrinsic logic.

3. In the applied case of a German → Polish interconnectedness, an asymmetrical interdependence and a dominant market position could be identified that have both economic and physical impacts on the Polish power market.

These deductions match the conditions implied in the overall theory-hypothesis.

Thus, under the very narrow conditions of the qualitative empirical examples (Germany and Poland), with particular relevance to the subject matter of RES-E promotion, it can be claimed, under usual reserve, that:

The implementation of polices for the promotion of renewable energy electricity (RES-E) in the Member States of the European Union is fundamentally instigated by supranational renewable energy measures and policies, leading to a general, but not sufficient, vertical Europeanization, predominantly influenced by national preferences determined by geopolitical and energy-specific economic interests, and dominantly exposed to incremental European inter-state competition leading to horizontal adaptational pressure.

Based on this theoretical hypothesis, the working hypothesis relevant to the empirical examples of this thesis can be upheld.
It is considered likely, and therefore proposed as subject for further analysis, that the identified multi-level policy constraints and dependencies will encourage the harmonisation of a European promotion of renewable energies, and describe a reasoned path to a common promotion of renewable energies in all 28 Member States of the European Union.
The first guiding question stressed in the introduction was whether we could state the successful achievement of a common European energy policy based on the lessons of the implementation of Directive 2009/28/EC.

In light of the results presented in this chapter, the answer must be no. This, however, applies only in so far as one considers a *common* European renewable energy policy, to be realized in comparison with nation state realities. This logic, by the very nature of European Integration, can be precluded *a priori*. As long as the European Union remains a compound of sovereign states, whose democratic integrity secures each Member State the right of legitimised executive authority, one cannot expect any adopted European policy to be implemented unconditionally. Such policies can be expected, however, to trigger change. It is remarkable to see that, with the entry into force of Directive 2009/28/EC the overall RES share in gross final consumption of energy, and the RES-E share, evidently rose in the bulk of the European Member States. The European policy therefore caused a simultaneous and measurable change in at least 23 European countries from 2009 to 2011. Whilst it is highly unlikely to achieve a common European RES-E policy in accordance with supranational intentions, a common European RES-E promotion will in reality be achieved by the horizontal and vertical mechanisms described here .

10. Appendix II – Summary German

In der Einführung der zugrunde liegenden Arbeit wird die Frage gestellt, ob man auf Grundlage der Lehren aus der Umsetzung der Richtlinie 2009/28/EG die erfolgreiche Umsetzung einer gemeinsamen europäischen Energiepolitik konstatieren kann. Im Kontext der Fortschrittsanalyse der Förderung von erneuerbaren Energien als Teil der europäischen 2020-Strategie veröffentlichte die Europäische Kommission im Jahr 2013 ein Grünbuch, um die Rahmenbedingungen für eine europäische Klima- und Energiepolitik bis 2030 zu erörtern (EC, 2013a:3; EC 2013c: 3). Das Papier wirft insgesamt 22 Fragen auf, die bestehenden politischen Rahmen der 2020-Strategie zu bewerten und daraus Lehren für die künftige Gestaltung der europäischen Klima- und Energiepolitik zu ziehen (EC, 2013a: 13).

Vor dem Hintergrund dieser Arbeit wurden diese Fragen in eine übergeordnete Forschungsfrage übersetzt:
Was sind die Einschränkungen und Abhängigkeiten für eine Europäische Integration der Förderung von (Strom aus) erneuerbaren Energien? Für die empirische Analyse gilt folgende Arbeitshypothese:

Wenn (i) die Europäische Union einen Gesamtanpassungsdruck durch Entscheidungen über EE-Maßnahmen und Strategien für die Mitgliedstaaten aufrecht erhält; (ii) Pioniermitgliedstaaten mit ausreichender Durchsetzungsfähigkeit in den zwischenstaatlichen Beziehungen im Hinblick auf die Förderung von EE (Deutschland) eine erfolgreiche Politikimplementierung mit einem – in Konsequenz – hohen Anteil EE errei-chen; und (iii) benachbarte Mitgliedstaaten (Polen) den Effekten der unter (ii) beschriebenen Politik und/oder des etablierten Markt-Designs ausgesetzt sind – **dann** werden sich die EE Politik und das Markt-Design des unter (iii) beschriebenen Mitgliedstaates den Implikationen entsprechend des vertikalen (EU-Ebene) und horizontalen (zwischenstaatliche Ebene) Anpassungsdrucks anpassen.

Die Analyseerkenntnisse der zugrunde liegenden Arbeit lassen sich in drei wesentlichen Punkten zusammenfassen:

1. Die supranationale EE-Politik, wie in der Richtlinie 2009/28/EG festgeschrieben, wird in den Mitgliedstaaten in Teilen umgesetzt. Allein, die nationale Politikimplementierung ist nicht grundsätzlich vollumfassend im Sinne der supranationalen Vorgaben.

2. In den Fällen von Deutschland und Polen folgt die Ausgestaltung der nationalen EE-Politik signifikant einer eigenständigen nationalen Logik.

3. Im Kontext zwischenstaatlicher Interdependenzen zwischen Deutschland → Polen kann eine asymmetrische Beziehung, das heißt markt-

dominierende Stellung Deutschlands, konstatiert werden, die sowohl wirtschaftliche als auch physische Auswirkungen auf den polnischen Energiemarkt haben.

Die Analyseergebnisse stimmen mit den in der Theorie-Hypothese implizierten Einschränkungen und Abhängigkeiten überein. Somit kann, innerhalb des empirischen Gegenstandsbereiches (Deutschland und Polen), folgende Theorie-Hypothese als verifiziert angesehen werden:

Die Umsetzung von Maßnahmen zur Förderung von (Strom aus) erneuerbaren Energien in den Mitgliedstaaten der Europäischen Union wird grundsätzlich aber nicht vollumfänglich von supranationalen Implikationen beeinflusst, vor allem von nationalen Interessen unter Berücksichtigung geostrategischer und energiewirtschaftlicher Präferenzen bestimmt und in steigendem Maße von zwischenstaatlichen Abhängigkeiten beeinflusst.

Die für die empirische Analyse zuvor formulierte Arbeitshypothese wird vor diesem Hintergrund aufrechterhalten.

Es wird als wahrscheinlich angesehen, dass die identifizierten horizontalen und vertikalen Mechanismen der beschriebenen Mehrebenenstruktur letztendlich zu einer Harmonisierung, dass heißt abschließenden Europäischen Integration, der Förderung von (Strom aus) erneuerbaren Energien in den Mitgliedstaaten führen. Dies zu verifizieren bedarf jedoch einer weitergehenden Analyse und wird als Gegenstand für zukünftige Forschung vorgeschlagen.

Mit Bezug auf die eingangs formulierte Frage, ob man auf Grundlage der Lehren aus der Umsetzung der Richtlinie 2009/28/EG die erfolgreiche Umsetzung einer gemeinsamen europäischen Energiepolitik konstatieren kann, fällt die Antwort negativ aus. Dies gilt allerdings nur in sofern, als dass man eine gemeinsame europäische Energiepolitik an dem Maßstab einer Politikentfaltung vor dem Hintergrund nationalstaatlicher Exekutivgewalten misst. Solange die Europäische Union ein verfassungsrechtlich ungeklärter Zusammenschluss von souveränen Staaten ist, deren demokratische Integrität jedem Mitgliedsland die Möglichkeit zur Nichtverabschiedung von Rechtsanwendungsbefehlen einräumt, kann nicht erwartet werden, dass verabschiedete europäische Gesetzespakete bedingungslos umgesetzt werden. Es kann jedoch erwartet werden, dass europäische Politikimplikationen Veränderungen in der nationalstaatlichen Arena anregen. Seit Inkrafttreten der Richtlinie 2009/28/EG ist der Anteil EE am Bruttoendenergieverbrauch in einem Großteil der EU-Mitgliedsstaaten nachweislich gestiegen. Abschließend bedeutet dies: Während es unwahrscheinlich ist, die Erreichung einer gemeinsame europäische EE-Politik allein vor dem Hintergrund supranationaler Implikationen zu erwarten, wird es als wahrscheinlich angesehen, dass die Realisierung einer gemeinsamen euro-päischen EE-Politik durch die in dieser Arbeit beschriebenen vertikalen und horizontalen Mechanismen etabliert wird.